Ninth International Workshop on Health Text Mining and Information Analysis (LOUHI 2018)

Brussels, Belgium
31 October 2018

ISBN: 978-1-5108-7423-7

EMNLP 2018

Ninth International Workshop on Health Text Mining and Information Analysis (LOUHI)

Proceedings of the Workshop

31 October, 2018
Brussels, Belgium

Order copies of this and other ACL proceedings from:

Association for Computational Linguistics (ACL)
209 N. Eighth Street
Stroudsburg, PA 18360
USA
Tel: +1-570-476-8006
Fax: +1-570-476-0860
acl@aclweb.org

Introduction

The International Workshop on Health Text Mining and Information Analysis (LOUHI) provides an interdisciplinary forum for researchers interested in automated processing of health documents. Health documents encompass electronic health records, clinical guidelines, spontaneous reports for pharmacovigilance, biomedical literature, health forums/blogs or any other type of health-related documents. The LOUHI workshop series fosters interactions between the Computational Linguistics, Medical Informatics and Artificial Intelligence communities. The eight previous editions of the workshop were co-located with SMBM 2008 in Turku, Finland, with NAACL 2010 in Los Angeles, California, with Artificial Intelligence in Medicine (AIME 2011) in Bled, Slovenia, during NICTA Techfest 2013 in Sydney, Australia, co-located with EACL 2014 in Gothenburg, Sweden, with EMNLP 2015 in Lisbon, Portugal, with EMNLP 2016 in Austin, Texas; and in 2017 was held in Sydney, Australia. This year the workshop is co-located with EMNLP 2018 in Brussels, Belgium.

The aim of the LOUHI 2018 workshop is to bring together research work on topics related to health documents, particularly emphasizing multidisciplinary aspects of health documentation and the interplay between nursing and medical sciences, information systems, computational linguistics and computer science. The topics include, but are not limited to, the following Natural Language Processing techniques and related areas:

- Techniques supporting information extraction, e.g. named entity recognition, negation and uncertainty detection

- Classification and text mining applications (e.g. diagnostic classifications such as ICD-10 and nursing intensity scores) and problems (e.g. handling of unbalanced data sets)

- Text representation, including dealing with data sparsity and dimensionality issues

- Domain adaptation, e.g. adaptation of standard NLP tools (incl. tokenizers, PoS-taggers, etc) to the medical domain

- Information fusion, i.e. integrating data from various sources, e.g. structured and narrative documentation

- Unsupervised methods, including distributional semantics

- Evaluation, gold/reference standard construction and annotation

- Syntactic, semantic and pragmatic analysis of health documents

- Anonymization/de-identification of health records and ethics

- Supporting the development of medical terminologies and ontologies

- Individualization of content, consumer health vocabularies, summarization and simplification of text

- NLP for supporting documentation and decision making practices

- Predictive modeling of adverse events, e.g. adverse drug events and hospital acquired infections

The call for papers encouraged authors to submit papers describing substantial and completed work but also focus on a contribution, a negative result, a software package or work in progress. We also encouraged to report work on low-resourced languages, addressing the challenges of data sparsity and language characteristic diversity.

This year we received a high number of submissions (49), therefore the selection process was very competitive. Due to time and space limitations, we could only choose a small number of the submitted papers to appear in the program.

Each submission went through a double-blind review process which involved three program committee members. Based on comments and rankings supplied by the reviewers, we accepted 23 papers. Although the selection was entirely based on the scores provided by the reviewers, we regretfully had to set a relatively high threshold for acceptance. The overall acceptance rate is 46%. During the workshop, 13 papers will be presented orally, and 10 papers will be presented as posters.

Our special thanks go to Goran Nenadic for accepting to give an invited talk.

Finally, we would like to thank the members of the program committee for providing balanced reviews in a very short period of time, and the authors for their submissions and the quality of their work.

Organizers:

Alberto Lavelli, FBK, Trento, Italy
Anne-Lyse Minard, IRISA, CNRS, Rennes, France
Fabio Rinaldi, University of Zurich, Switzerland & FBK, Trento, Italy

Program Committee:

Sophia Ananiadou, University of Manchester, UK
Georgeta Bordea, Université de Bordeaux, France
Leonardo Campillos Llanos, LIMSI, CNRS, France
Wendy Chapman, University of Utah, USA
Vincent Claveau, IRISA, CNRS, France
Kevin B Cohen, University of Colorado/School of Medicine, USA
Francisco Couto, University of Lisbon, Portugal
Hercules Dalianis, Stockholm University, Sweden
Martin Duneld, Stockholm University, Sweden
Filip Ginter, University of Turku, Finland
Natalia Grabar, CNRS UMR 8163, STL Université de Lille3, France
Gintaré Grigonyté, Stockholm University, Sweden
Cyril Grouin, LIMSI, CNRS, Université Paris-Saclay, Orsay, France
Thierry Hamon, LIMSI, CNRS, Université Paris-Saclay, Orsay, France & Université Paris 13, Villetaneuse, France
Aron Henriksson, Stockholm University, Sweden
Rezarta Islamaj-Dogan, NIH/NLM/NCBI, USA
Antonio Jimeno Yepes, IBM Research, Australia
Yoshinobu Kano, Shizuoka University, Japan
Jin-Dong Kim, Research Organization of Information and Systems, Japan
Dimitrios Kokkinakis, University of Gothenburg, Sweden
Martin Krallinger, Spanish National Cancer Research Centre (CNIO)
Michael Krauthammer, Yale University, USA
Ivano Lauriola, University of Padova and FBK, Trento, Italy
Analia Lourenco, Universidade de Vigo, Spain
David Martinez, University of Melbourne and MedWhat.com, Australia
Sérgio Matos, University of Aveiro, Portugal
Marie-Jean Meurs, UQAM & Concordia University, QC, Canada
Timothy Miller, Harvard Medical School, USA
Hans Moen, University of Turku
Diego Molla, Maquaire University, Australia
Roser Morante, VU Amsterdam, Netherlands
Danielle L Mowery, University of Utah, USA
Henning Müller, University of Applied Sciences Western Switzerland, Switzerland
Goran Nenadic, University of Manchester, UK
Aurélie Névéol, LIMSI, CNRS, Université Paris-Saclay, Orsay, France
Mariana Lara Neves, German Federal Institute for Risk Assessment, Germany
Richard Nock, CSIRO, Australia
Øystein Nytrø, NTNU, Norway

Naoaki Okazaki, Tokyo Institute of Technology, Japan
Jong C. Park, KAIST Computer Science, Korea
Thomas Brox Røst, Norwegian University of Science and Technology, Norway
Patrick Ruch, SIB Swiss Institute of Bioinformatics, Switzerland
Tapio Salakoski, University of Turku, Finland
Sanna Salanterä, University of Turku, Finland
Stefan Schulz, Graz General Hospital and University Clinics, Austria
Isabel Segura-Bedmar, Universidad Carlos III de Madrid, Spain
Maria Skeppstedt, Linneus University, Sweden, and Potsdam University, Germany
Manfred Stede, University of Potsdam, Germany
Hanna Suominen, CSIRO, Australia
Sumithra Velupillai, KTH, Royal Institute of Technology, Sweden, and King's College London, UK
Özlem Uzuner, MIT, USA
Pierre Zweigenbaum, LIMSI, CNRS, Université Paris-Saclay, Orsay, France

Invited Speaker:

Goran Nenadic, University of Manchester, UK

Table of Contents

Workshop Program

October 31, 2018

9:00–10:30   Session 1

9:00   ***Introduction***

9:05   *Detecting Diabetes Risk from Social Media Activity*
                 Dane Bell, Egoitz Laparra, Aditya Kousik, Terron Ishihara, Mihai Surdeanu and
                 Stephen Kobourov

9:30   *Treatment Side Effect Prediction from Online User-generated Content*
                 Hoang Nguyen, Kazunari Sugiyama, Min-Yen Kan and Kishaloy Halder

9:55   ***Poster booster***

10:15   ***Poster session***

                 Revisiting neural relation classification in clinical notes with external information
                 Simon Suster, Madhumita Sushil and Walter Daelemans

                 *Supervised Machine Learning for Extractive Query Based Summarisation of
                 Biomedical Data*
                 Mandeep Kaur and Diego Molla

                 *Comparing CNN and LSTM character-level embeddings in BiLSTM-CRF models
                 for chemical and disease named entity recognition*
                 Zenan Zhai, Dat Quoc Nguyen and Karin Verspoor

                 *Deep learning for language understanding of mental health concepts derived from
                 Cognitive Behavioural Therapy*
                 Lina M. Rojas Barahona, Bo-Hsiang Tseng, Yinpei Dai, Clare Mansfield, Osman
                 Ramadan, Stefan Ultes, Michael Crawford and Milica Gasic

                 Investigating the Challenges of Temporal Relation Extraction from Clinical Text
                 Diana Galvan, Naoaki Okazaki, Koji Matsuda and Kentaro Inui

                 De-identifying Free Text of Japanese Dummy Electronic Health Records
                 Kohei Kajiyama, Hiromasa Horiguchi, Takashi Okumura, Mizuki Morita and
                 Yoshinobu Kano

Unsupervised Identification of Study Descriptors in Toxicology Research: An Experimental Study
Drahomira Herrmannova, Steven Young, Robert Patton, Christopher Stahl, Nicole Kleinstreuer and Mary Wolfe

Identification of Parallel Sentences in Comparable Monolingual Corpora from Different Registers
Rémi Cardon and Natalia Grabar

Evaluation of a Prototype System that Automatically Assigns Subject Headings to Nursing Narratives Using Recurrent Neural Network
Hans Moen, Kai Hakala, Laura-Maria Peltonen, Henry Suhonen, Petri Loukasmäki, Tapio Salakoski, Filip Ginter and Sanna Salanterä

Automatically Detecting the Position and Type of Psychiatric Evaluation Report Sections
Deya Banisakher, Naphtali Rishe and Mark A. Finlayson

10:30–11:00   Break

11:00–12:30   Session 2

11:00   *Iterative development of family history annotation guidelines using a synthetic corpus of clinical text*
Taraka Rama, Pål Brekke, Øystein Nytrø and Lilja Øvrelid

11:25   *CAS: French Corpus with Clinical Cases*
Natalia Grabar, Vincent Claveau and Clément Dalloux

11:40   *Analysis of Risk Factor Domains in Psychosis Patient Health Records*
Eben Holderness, Nicholas Miller, Kirsten Bolton, Philip Cawkwell, Marie Meteer, James Pustejovsky and Mei Hua-Hall

12:05   *Patient Risk Assessment and Warning Symptom Detection Using Deep Attention-Based Neural Networks*
Ivan Girardi, Pengfei Ji, An-phi Nguyen, Nora Hollenstein, Adam Ivankay, Lorenz Kuhn, Chiara Marchiori and Ce Zhang

12:30–14:00   Lunch

14:00–15:30   Session 3

14:00   *Invited Talk - Distributed text mining in healthcare: linking data, methods and people*
Goran Nenadic

14:50   *Syntax-based Transfer Learning for the Task of Biomedical Relation Extraction*
Joël Legrand, Yannick Toussaint, Chedy Raïssi and Adrien Coulet

15:15   *In-domain Context-aware Token Embeddings Improve Biomedical Named Entity Recognition*
Golnar Sheikhshabbafghi, Inanc Birol and Anoop Sarkar

15:30–16:00   Break

16:00–17:30   Session 4

16:00   *Self-training improves Recurrent Neural Networks performance for Temporal Relation Extraction*
Chen Lin, Timothy Miller, Dmitriy Dligach, Hadi Amiri, Steven Bethard and Guergana Savova

16:25   *Listwise temporal ordering of events in clinical notes*
Serena Jeblee and Graeme Hirst

16:40   *Time Expressions in Mental Health Records for Symptom Onset Extraction*
Natalia Viani, Lucia Yin, Joyce Kam, Ayunni Alawi, André Bittar, Rina Dutta, Rashmi Patel, Robert Stewart and Sumithra Velupillai

16:55   *Evaluation of a Sequence Tagging Tool for Biomedical Texts*
Julien Tourille, Matthieu Doutreligne, Olivier Ferret, Aurélie Névéol, Nicolas Paris and Xavier Tannier

17:10   *Learning to Summarize Radiology Findings*
Yuhao Zhang, Daisy Yi Ding, Tianpei Qian, Christopher D. Manning and Curtis P. Langlotz

Detecting Diabetes Risk from Social Media Activity

Dane Bell[1], Egoitz Laparra[2], Aditya Kousik[3], Terron Ishihara[3],
Mihai Surdeanu[3], and Stephen Kobourov[3]
[1]Department of Linguistics, University of Arizona
[2]School of Information, University of Arizona
[3]Department of Computer Science, University of Arizona
{dane, laparra, adityak, tishihara, msurdeanu, kobourov}@email.arizona.edu

Abstract

This work explores the detection of individuals' risk of type 2 diabetes mellitus (T2DM) directly from their social media (Twitter) activity. Our approach extends a deep learning architecture with several contributions: following previous observations that language use differs by gender, it captures and uses gender information through domain adaptation; it captures recency of posts under the hypothesis that more recent posts are more representative of an individual's current risk status; and, lastly, it demonstrates that in this scenario where activity factors are sparsely represented in the data, a bag-of-word neural network model using custom dictionaries of food and activity words performs better than other neural sequence models. Our best model, which incorporates all these contributions, achieves a risk-detection F_1 of 41.9, considerably higher than the baseline rate (36.9).

1 Introduction

The prevalence of diabetes is increasing in the US, mounting to 30.3 million cases in 2015, of whom 7.2 million were undiagnosed (Centers for Disease Control and Prevention, 2017). Diabetes caused over 79 thousand US deaths in 2015, in addition to \$245 billion in economic costs in 2012 (American Diabetes Association, 2013). Along with genetic factors, lifestyle factors such as diet and physical activity are one of the important drivers of risk for Type 2 Diabetes Mellitus (T2DM), the most common type of diabetes. At the same time, the widespread use of social media has produced a digital record of these factors, offering potential insight into how these factors interact to contribute to health risk over time. These publicly available data present an opportunity to detect diabetes risk and similar health risks at scale.

This work shows that the detection of *individuals'* diabetes risk solely from their public Twit-ter activity is possible, demonstrating that at-risk individuals use language differently from less at-risk individuals. Importantly, this detection is a first, crucial component in a larger battery of social media-based, public-health intervention tools that will work toward disease prevention on a large scale. Specifically, our contributions are:

(1) We introduce a process that creates a novel dataset, which pairs individuals' T2DM risk with their social media activity. We measured individuals' T2DM risk using a well-established, validated questionnaire (Bang et al., 2009), and aligned the result with the corresponding Twitter accounts. To our knowledge, this is the first dataset that directly links T2DM risk with social media activity.

(2) We introduce the first machine learning (ML) approach for classifying individuals' T2DM risk based solely on their Twitter activity. Our deep learning approach has several novel contributions: (a) following previous observations that language use differs by gender, it captures and uses gender information through domain adaptation[1] (Daumé, 2007); (b) it captures recency of posts under the hypothesis that more recent posts are more representative of an individual's current risk status; and, lastly (c) it demonstrates that in this scenario where words representing real-life risk factors are sparsely represented in the data, a bag-of-word (BOW) model that uses custom dictionaries of food and physical activity words is a better solution than recurrent neural networks (RNN). Our best model, which incorporates all these contributions, achieves a risk-detection F_1 of 41.9, considerably higher than the baseline rate (36.9). In comparison, a realistic ceiling model based on the true age, gender, and Body Mass Index (BMI, $\frac{kg}{m^2}$) of each respondent, achieves only 62.7 on this task.

---

[1]In our experiments, domain adaptation for age did not improve performance.

Proceedings of the 9th International Workshop on Health Text Mining and Information Analysis (LOUHI 2018), pages 1–11
Brussels, Belgium, October 31, 2018. ©2018 Association for Computational Linguistics

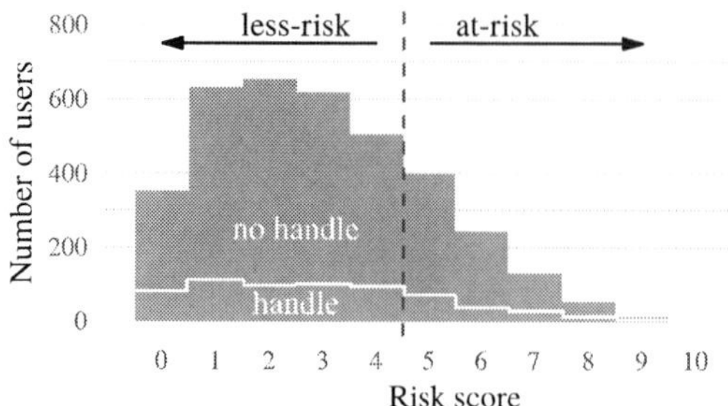

Figure 1: Histogram of respondents' risk scores, with labels as assigned based on Bang et al. (2009).

	less-risk	at-risk
accounts	467	137
tweets (mean)	893 K (1,912)	282 K (2,059)
tokens (mean)	15.2 M (32.5 K)	5.1 M (37.0 K)
# women (%)	312 (67%)	73 (53%)
mean age	36.4	51.1
mean BMI	25.6	34.8

Table 1: A summary of the size and qualities of the *less-risk* and *at-risk* accounts in the dataset collected for this work. *BMI*: Body Mass Index, $\frac{kg}{m^2}$.

(3) We provide a feature analysis based on Layerwise Relevance Propagation (Bach et al., 2015; Binder et al., 2016; Arras et al., 2016, 2017), revealing that relevance aligns, albeit inconsistently, to expected food and activity values on average.

2 Data

We collected the dataset used in this work on a voluntary basis through a Qualtrics survey.[2] Participants self-selected by following an URL in an invitation tweet, and after consenting to participate, provided their Twitter handles, demographic information, and answers to an established questionnaire that estimates T2DM risk (Bang et al., 2009). The questionnaire provides an easy-to-understand measure of diabetes risk from data such as age and physical activity level, ranging from 0 to 10, with a score of 5 or higher representing elevated risk. Each participant received a risk assessment, including a summary of the sources of their risk, an explanation of how to get diagnosed (i.e., through a blood test), and a link to further information.

Of the 3,612 respondents who completed surveys, 736 (20.4%) supplied a Twitter handle. After removing respondents who provided no handle, an obviously false handle,[3] or a handle with no public tweets, 604 (16.7%) respondents with handles remained. The relatively modest dataset size is a natural consequence of the complexity of the data and the sensitivity of its collection. The distribution of risk scores among respondents is summarized in Fig. 1.

The complex relationship between height, weight, and risk score is illustrated in Fig. 2. Although BMI is a major risk factor for diabetes, the existence of other factors means that there is considerable risk variation within BMI categories, and the discretization of BMI into categories necessarily obscures variation within categories. Many respondents would change BMI categories if an inch were added to or subtracted from their height, for example.

We used the Twitter API to collect the tweet and profile text for each handle. The tweets and profile descriptions were tokenized and part-of-speech tagged using ARK Tweet NLP (Owoputi et al., 2013). Each account was labeled *at-risk* if the owner's questionnaire risk score was 5 or greater, or *less-risk* otherwise. A summary of account statistics is shown in Table 1.

3 Approach

We predict individual-level T2DM risk from individual-level data (i.e., individual Twitter accounts), as opposed to transferring from community level statistics (e.g., county diabetes rate as dependent variable; all tweets in that region as input). Intuitively, using a community-level model should be a viable strategy: much more data is available for training; previous work has shown that exploring this data leads to good community-level estimations (Fried et al., 2014). However, our initial experiments showed that individual variation *within* communities was considerable, overshadowing the variation *across* communities and limiting the effectiveness of such methods. In our preliminary experiments the community-level model did not perform better than chance for estimating individual risk.

As a result of this initial analysis, in this work we focus on predicting T2DM risk from individual Twitter accounts. To this end, we propose a neural network (NN) architecture tailored to T2DM risk estimation, which relies on the following resources.

---

[2]The collection and analysis was approved by an institutional research board (IRB).

[3]These were inspected manually. Some examples of excluded handles are @jack (the example handle given), @realdonaldtrump, and @no.

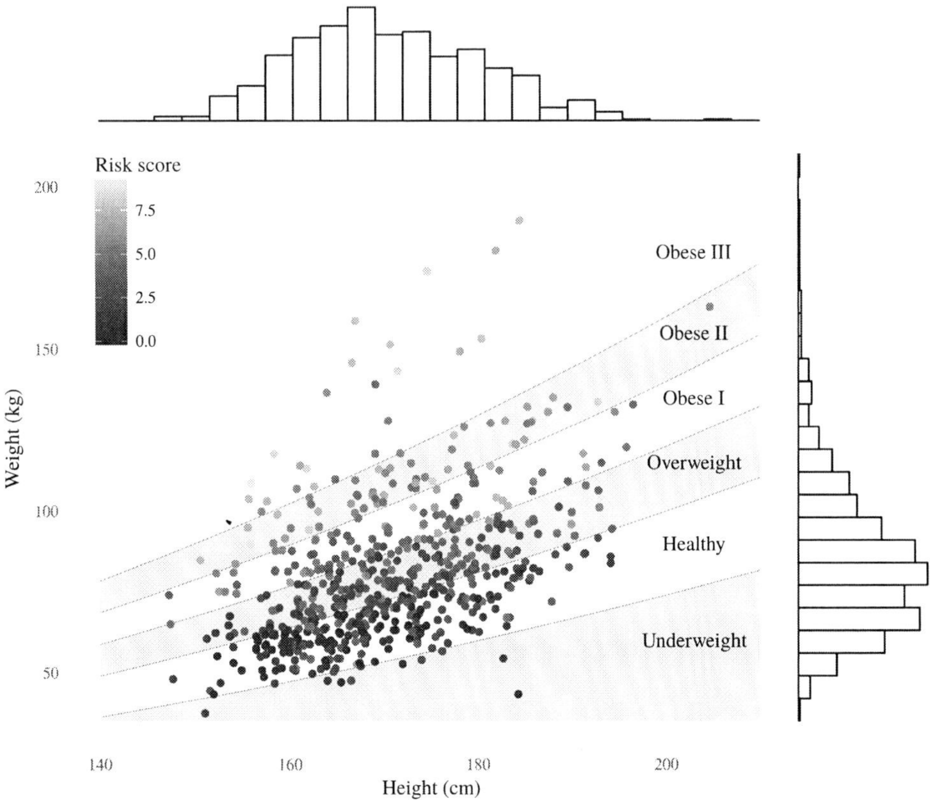

Figure 2: An illustration of the relationship between height, weight, BMI, and risk score for respondents who provided a valid Twitter handle. The BMI categories (underweight, healthy, overweight, etc.) are assigned according to boundaries set by the World Health Organization. The marginal histograms denote the distribution of height (top) and weight (right) in the sample. This figure is best viewed in color.

3.1 Resources

Custom dictionaries: In early experiments, we observed that no model that trained on the posts' entire content outperformed a simple baseline. We explain this result by the fact that indicators of risk factors (e.g., diet or activity words) are sparsely represented in this data, and the models cannot reliably identify them. To mitigate this problem, we created domain-specific dictionaries of words and hashtags indicating foods (*pizza*), exercise (*#5k*), chain restaurant names (*#mcdonalds*), and hashtags related to being overweight (*#fatguyproblems*). The food words were derived from a domain-specific Spanish-English glossary[4] and food vocabulary set[5], following Fried et al. (2014). Exercise words and restaurant names were adapted from Wikipedia lists of sports[6] and restaurants[7]. The smaller list of 13 overweight-related terms were hand-chosen based on Twitter searches.

To adapt the food dictionary to Twitter, we automatically expanded it using semantic vectors. We trained the `word2vec` algorithm (Mikolov et al., 2013) over an independent dataset of 12.3 M food-related tweets[8], creating 200-dimension vectors for each word. From each existing dictionary term, we found the 5 closest candidate words, as measured by cosine distance. Each candidate could appear in multiple lists (e.g. *#breakfastburrito* is similar to both *burrito* and *taco*), so we calculated the softmax of the distances for each candidate. We then expanded our dictionary with the top 500 candidates, which included words such as

---

[4] `www.lingolex.com/spanishfood/a-b.htm`
[5] `www.enchantedlearning.com/wordlist/food.shtml`

[6] `en.wikipedia.org/wiki/List_of_sports`
[7] `en.wikipedia.org/wiki/List_of_the_largest_fast_food_restaurant_chains`
[8] Collected automatically using a set of seven diet-related hashtags such as *#breakfast* and *#lunch*.

halloumi, *muesli*, and *sriracha*. After these additions, there was a total of 2,871 features.

Gender: It is well established that language use differs by gender (Rao et al., 2010; Burger et al., 2011; Volkova et al., 2013; Johannsen et al., 2015). On the hypothesis that conditioning classification on these secondary variables would maximize the informativeness of other features[9], we automatically annotated each account for gender. We predicted gender using a SVM model trained on a separate corpus of 1,000 Twitter accounts hand-annotated with gender information (man or woman).[10] This gender classifier used solely unigram features extracted from the account description and its tweets. The macro-averaged F_1 of this model is 75.79 on the T2DM dataset (c.f. human annotators, who averaged 71% accuracy on a similar task (Nguyen et al., 2014)).

3.2 Neural network architecture

We propose a feedforward neural network with one hidden layer, which captures both post recency (by weighing each input word by the recency of the corresponding post) and gender information (captured through domain adaptation). The proposed architecture is depicted and summarized in Fig. 3. This network uses pre-trained word embeddings of 200 dimensions generated using `word2vec` (Mikolov et al., 2013) on the above corpus of food-related tweets. The *tanh* layer has 128 neurons, and was trained under a 40% dropout. Importantly, this network uses only account words that matched entries in the above custom dictionaries.[11]

Recency weighting: Our preliminary analysis indicated that more recent tweets are more relevant for classification. We attribute the effect of recency to transitions from high to low risk or vice versa due to lifestyle changes, in which case more recent tweets are more representative. To capture recency, we introduce a simple attention mechanism where each word is weighted by its recency, defined as normalized tweet position in the corresponding account. More formally, the recency weight (r_i) of a word w_i is defined as:

$$r_i = \frac{position\_of\_tweet\_containing\_w_i}{\#tweets\_in\_account}$$

where the newest tweet in an account has the highest position. The average embedding ($\overline{x}$) is calculated as:

$$x_i = w_i r_i, \qquad \overline{x} = \frac{\sum_{i=1}^{n} x_i}{\sum_{i=1}^{n} r_i}$$

Domain adaptation: We capture gender information using the domain adaptation method of Daumé (2007), adapted to neural networks. As shown in Figure 3, we replicate the output of the *tanh* layer $\langle t \rangle$ to have a domain-independent version, and one version specific to each domain modeled. For example, the concatenated vector for a female account is $\langle t, t, 0 \rangle$, where 0 is the zero vector corresponding to the male-account domain. This routing process is automatically implemented using the gender classifier described in the previous subsection. All in all, this allows the top sigmoid layer to detect information that generalizes across all domains, in which case the domain-independent vector (t_g) receives a larger update during backpropagation, or is specific to a domain, in which case the corresponding domain-specific vector (t_{d1} or t_{d2}) is updated more.

3.3 Baselines

We implemented three baselines:

(1) All at risk: This baseline assumes all individuals are at risk, i.e., they have a score 5 or higher.

(2) Support vector machines (SVM): This baseline model uses a linear SVM with unigram features from words and hashtags that match our custom dictionaries.[12] Similarly, following the domain adaption method of Daumé (2007), we incorporate gender information by prepending each feature name with the account's gender annotation (in addition to keeping the original feature). For example, an account annotated as a woman who used the word *coffee* 16 times would yield an unigram feature `coffee` in all models, and additionally a feature `gender:woman_coffee`, both with a feature value of 16. (The accounts feature `gender:man_coffee` would have a value of 0.) This allows these models to discover the best generalization for this task, e.g., if *coffee* is an important classification word for women only, the models will put the greatest weight on the *gender:woman_coffee* feature; conversely, if coffee is

---

[9]We additionally tested this hypothesis, classifying participants' ages into 5 classes (0-20, 21-30, 31-40, 51-60, 61+). Although the age classifier itself performed better than chance, initial experiments showed that age provided no benefit in classifying diabetes risk, and so the influence of age was left to future work.

[10]Non-binary individuals represented $< 1\%$ of our dataset.

[11]Implemented in PyTorch: http://pytorch.org/.

[12]Other kernels, larger n-grams, and using all words did not improve performance.

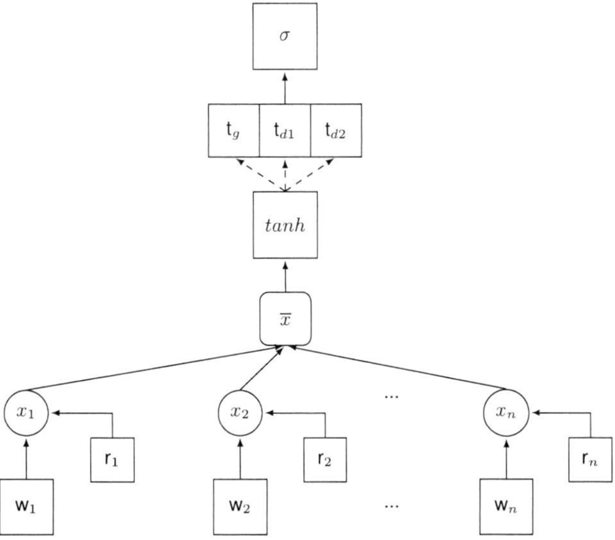

Figure 3: An illustration of the proprosed NN architecture. The bag of words from a user's account matching our custom dictionaries is translated into a set of word embeddings $w_1, w_2, ..., w_n$. The embeddings are multiplied by recency weights $r_1, r_2, ..., r_n$. The resulting vectors are averaged ($\overline{x}$) and passed to the $tanh$ hidden layer. The output of this layer is replicated, producing copies for the general domain, t_g, and for each of the domains, t_{d1} and t_{d2}, e.g., $d1$ = female, and $d2$ = male. If an account belongs to domain $d1$, the copy t_{d2} is set to the zero vector, and vice versa. The copies are then concatenated and fed to the top sigmoid (σ) layer.

always important, the generic unigram feature *coffee* will be assigned greater weight.

(3) Convolutional neural networks (CNN): For this baseline, we apply a CNN layer to the sequence of embeddings of dictionary words that occur in the corresponding account, followed by a rectified linear operator (ReLU). We implement domain adaptation for gender by augmenting the output of the ReLU layer, similarly to the $tanh$ layer in Figure 3. The resulting vector feeds a top $sigmoid$ layer that makes the prediction.[13]

3.4 Ceiling models

We also developed two ceiling models against which to compare our text-based approaches. The first model (Ceiling) is an SVM trained with all the risk assessment variables collected in the survey mentioned in Section 2. This dataset is maximally informative, because these are precisely the variables that determine the risk score (Bang et al., 2009). However, it is not realistic, because most of these features are not available in social media, neither directly nor through machine learning techniques. For this reason, we also implemented an alternative and more realistic version of the ceiling system (Realistic Ceiling) that incorpo-

Feature	Type	Ceiling	Realistic Ceiling
age	Integer	✓	✓
gender	Boolean	✓	✓
BMI	Float	✓	✓
diabetic relatives	Boolean	✓	
high blood pressure	Boolean	✓	
little physical activity	Boolean	✓	
gestational diabetes	Boolean	✓	

Table 2: Features available to each ceiling system.

rates only those features that have previously been predicted by automatic systems through social media text or images (see Section 5). The features are summarized in Table 2.

4 Results

We used 10-fold cross-validation to train and evaluate each model on the binary classes *at-risk* and *less-risk* (see Section 2), using the same folds across all models. For each of the 10 runs, we reserve one fold for development, to tune hyperparameters such as classifier confidence cutoff, one fold for testing, and the rest for training. Table 3 summarizes the results of the proposed models, compared against the baselines described in Section 3. In the table, -R marks models that have *r*ecency information (models without recency used uniform r_i weights), -GG marks

---

models that used the *g*old *g*ender information collected during the questionnaire, and -PG marks models that used *p*redicted *g*ender information. The SVM-U is an SVM model using all available words except a stoplist of closed-class words.

The table underlines several observations:

(1) The proposed NN models outperform all baselines, demonstrating that our NNs generalize better on this task dominated by sparse signals. Importantly, most of the strong baselines we include are below the performance of the simple "all at risk" baseline, highlighting again the difficulty of the task. The only baseline that outperformed "all at risk" is CNN-GG, which uses gold gender information, which would not be available in real-world deployments. Interestingly, our approach, which essentially relies on a (recency-weighted) bag-of-word model outperforms all the baselines that rely on sequence models. Similar observations about bag-of-word models outperforming sequence models on complex NLP tasks have been made in the past (Iyyer et al., 2015; Wang and Manning, 2012, inter alia).

(2) Both recency and gender information help. Our best model includes both, validating our original hypotheses. Surprisingly, models using predicted gender performed slightly better than models using gold gender information, but this difference was not statistically significant.

(3) This bag-of-word NN that uses only words/hashtags from relevant dictionaries outperforms considerably other complex NN sequence models that had access to the entire account texts (CNN-all). This highlights the importance of task-specific information (food and activity dictionaries in our case), which, in turn, emphasizes the need of collaboration between NLP researchers and domain (i.e., nutritional science and health care) experts.

(4) Even the Ceiling and Realistic Ceiling classifiers have considerably less than perfect performance at 68.1 and 62.7, respectively. Better performance would be likely with a larger dataset, which would likely also improve the performance of the proposed classifiers.

4.1 Feature analysis

To understand the influence of individual features (tokens) to the classification of an account by the best-performing neural net (using predicted gender and recency-weighted averaging), we adapted

Model	P	R	F_1
All at-risk baseline	22.59	100.00	36.86
Ceiling	67.14	69.12	68.12**
Realistic Ceiling	62.96	62.50	62.73**
SVM-U	**32.82**	31.62	32.21
SVM	28.90	36.77	32.36
SVM-GG	27.95	33.09	30.30
SVM-PG	30.91	37.50	33.89
CNN-GG	26.67	**76.47**	39.54
CNN-PG	24.17	75.00	36.56
NN	28.10	63.24	38.91
NN-R	30.94	60.29	40.90
NN-GG-R	29.39	71.32	41.63*
NN-PG-R	29.38	72.79	**41.86****

Table 3: The precision, recall, and F_1 score of each model in predicting the *at-risk* label. See Section 3 for a description of the models. The *s indicate that the difference in F_1 score between the corresponding model and the best baseline is statistically significant (* indicates $p < 0.05$, and ** indicates $p < 0.01$). All significance values were determined through a one-tailed bootstrap resampling test with 100,000 iterations.

the Layerwise Relevance Propagation (Bach et al., 2015; Binder et al., 2016; Arras et al., 2016, 2017) technique. LRP has the advantage of maintaining both positive and negative relevances, representing in this case contribution to the at-risk and less-risk class scores, respectively. In contrast, the commonly used Sensitivity Analysis (Dimopoulos et al., 1995; Gevrey et al., 2003; Simonyan et al., 2013; Li et al., 2015) measures relevance to the decision, rather than to a given class's score, and is therefore always non-negative. LRP assigns relevance to each neuron (including input values) as a function of how much they contribute to the final layer's values, as a share of its layer's contribution. To accomplish this, the neuron's activation must be divided by the sum of whole layer's activation, which can lead to unbounded values when a layer's activations sum to near zero. For this reason, we employ Bach et al. (2015)'s equation 58, which applies a small smoothing constant to the layer's summed activation to the avoid this value explosion.

Examples of accounts' most recent words marked with their relevances according to the NN-PG-R model are shown in Table 4. As the table shows, the health value of words broadly aligns to relevance scores. However, because of the recency weighting of this model, making older tweets' words progressively less relevant, and because of variance in the training of different cross-validation folds, these relevance scores are highly variable. The result is that sometimes a given

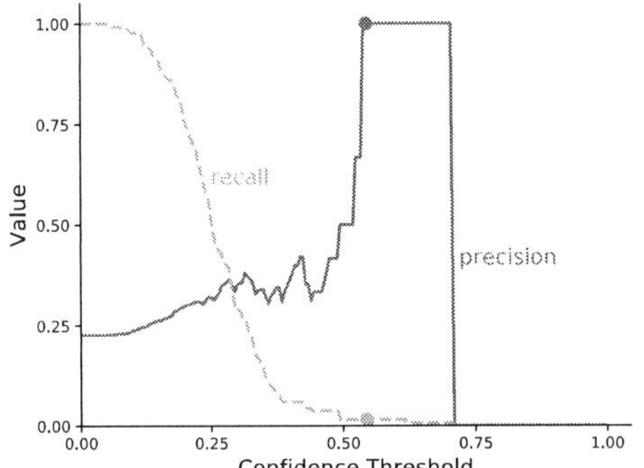

Figure 4: Precision and recall in the NN-PG-R ensemble model, as a function of classifier confidence. The dots mark a threshold of 0.55, at which precision is 100%, and recall is 1.47%. Note that no accounts meet the highest confidence thresholds, leading to a precision and recall of 0, which explains the steep drop of precision for high confidence thresholds.

token is counted as relevant to one classification (e.g., *at-risk*), and other times another (e.g., *less-risk*). This is likely due to both the modest dataset size and to the indirectness of the connection between language use and health.

4.2 Real-world deployment using a high-precision model

The practical application of this risk detection system would involve pointing high-risk individuals toward the Bang et al. (2009) survey, and, if at-risk, to further medical diabetes screening (Rains et al., 2018). To mitigate the drawbacks of false positives (i.e., unnecessary and stressful medical testing), it is likely that in real-world deployments of this technology a high-precision variant of the learned model would be used.

In Figure 4, we show the classification performance at different thresholds for the classifier confidence. In this experiment, in order to increase stability, we have built an ensemble of models through bagging (Breiman, 1996): we generated 50 different versions of the training set by resampling it with replacement, and we trained a different model of the NN-PG-R architecture on each sampled training set. The final predictions are obtained averaging the outputs of the resulting models. As shown in Fig. 4, a threshold of 0.55, for example, yields a precision of 100% and a recall of 1.47%.

Despite the modest recall of such a high-

precision model, this classifier would detect a large number of individuals at risk, if applied to the all of Twitter. Assuming the 28.2% prediabetes rate of (Rowley et al., 2017), and the 11% prediabetes diagnosis rate of Li et al. (2013), there are approximately 62 million undiagnosed prediabetic individuals in the US. If we further assume a lower-than-average Twitter adoption of 15%—compared to (Pew Research Center, 2018)'s estimate of approximately 25%—there are roughly 9.2 million Americans who use Twitter and have undiagnosed prediabetes. A similar application to the estimated 7.2 million Americans with undiagnosed diabetes (National Center for Health Statistics, 2017) produces an estimate of 1.1 million unknowingly diabetic Americans on Twitter. Therefore, the successful application of this classfier would identify an estimated 16,000 diabetic and 140,000 prediabetic Americans. Of course, expanding to other English-using Twitter users, and other languages[14] further increases this estimate.

5 Related work

Analysis of social media content for health has been a topic of wide interest (Aramaki et al., 2011; Bian et al., 2012; Prier et al., 2011; Culotta, 2014; Nguyen et al., 2017). Similarly, the literature on detecting user attributes and the effects of those attributes on language use is extensive.

Rao et al. (2010) predict individuals' demographic characteristics of gender, age, and political affiliation based on their tweets. Burger et al. (2011) construct a multilingual dataset of over 100K Twitter accounts, and classify gender better than human annotators, based on account text. Johannsen et al. (2015) study cross-linguistic variation in syntax (part-of-speech and dependency patterns) according to age and gender in online reviews (chosen over tweets for ease of parsing and richer metadata).

Age and gender, while much studied, are not the only available latent characteristics. Mowery et al. (2016) and Vedula and Parthasarathy (2017), for example, predict depressive symptoms in tweet text, a long-term health-variable detection task similar to ours. Similarly, De Choudhury et al. (2013) predict postpartum depression from tweets and Twitter social network structure. Shuai et al. (2016) gather a rich, multi-network feature

---

[14]https://www.npr.org/sections/goatsandsoda/2017/04/05/
522038318/how-diabetes-got-to-be-the-no-1-killer-in-mexico

Correct Label	Predicted Label	Relevance
less-risk	less-risk	chicken waffles tea reading ...
less-risk	at-risk	cake food starving sit sit heart ...
at-risk	less-risk	bacon run cup pack writing rolls parkour ...
at-risk	at-risk	catfish peanut butter pie picnic bland pop ...

Table 4: Examples of relevance displays for words used to assess accounts (most recent tweets first). The red words are relevant to the *less-risk* category, and the blue words to the *at-risk* category, with greater saturation indicating greater relevance.

set to detect social network mental disorders with symptoms such as excessive use of social network sites, measured against gold-data questionnaires. Likewise, Schwartz et al. (2013) predict not only age and gender from the text of Facebook messages, but also the Big Five personality traits (extraversion, emotional stability, agreeableness, conscientiousness, and openness to experience) (Digman, 1990). Moreover, these sometimes-latent user characteristics can inform other classification tasks. For example, Volkova et al. (2013) demonstrate an improvement in the sentiment classification of tweets in a language-independent rule-based model when sentiment vocabulary is adapted for gender-dependent language. Our work continues this direction: here we show that gender information, even when predicted automatically, considerably improves the accuracy of T2DM risk detection.

Much of the previous work on diabetes and weight detection on social media has been at the level of communities. Fried et al. (2014) predict population characteristics such as diabetes and overweight prevalence using location-tagged, food-related tweets. Abbar et al. (2015) analyze correlations between county-level obesity prevalence and food mentions. Again the focus is on predicting dietary choices on a large scale. Relatedly, Eichstaedt et al. (2015) detect heart disease mortality at the county level from tweet text.

There is no known work on detecting individual diabetes risk from social media text. However, Farseev and Chua (2017) capitalize on multiple social media inputs (e.g., a workout tracker) to predict individuals' Body Mass Index category. Wen and Guo (2013) and Kocabey et al. (2017) predict body mass index from images similar to profile pictures, the former from booking photographs and the latter from an internet forum for sharing fitness progress. Of these, only Farseev and Chua (2017) classify solely from text, which is often the only data available from a social media account. Their classification's F_1 is low (17.8) –understandable given the difficulty of this task– which limits its use for realistic T2DM risk prediction. In contrast, our approach obtains a F_1 score that is over 2 times higher, on a task that is arguably more complex.

6 Conclusions

We introduced an approach to the detection of individuals' diabetes risk from their Twitter posts. To this end, we collected a novel dataset linking Twitter activity to a validated, survey-based measure of T2DM risk (Bang et al., 2009). Using this dataset, we proposed the first machine learning approach to predict the T2DM risk of a Twitter account holder using only her tweets. This task is challenging because the data tends to be very sparse, and there are many latent contributing variables (such as genetic predisposition). Our analysis indicates that reducing noise with relevant dictionaries, modeling gender, and modeling posts' temporal recency are valuable in predicting T2DM risk. All in all, our best model achieves an F_1 of 41.9 (vs. the 36.9 "all at risk" baseline and 39.5 of a strong sequence model).

We estimate that if a high-precision variant of this approach were to be deployed at large, e.g., on the public posts of all American Twitter users, it would identify 16,000 diabetic and 140,000 prediabetic Americans that are currently not diagnosed.

Continuing this work, we envision a larger battery of social media-based tools for public-health intervention that focus on the early identification of multiple health risks such as heart disease and various cancers at scale.

7 Release

The system is available as open-source software at

References

Sofiane Abbar, Yelena Mejova, and Ingmar Weber. 2015. You tweet what you eat: Studying food consumption through Twitter. In *Proceedings of the 33rd Annual ACM Conference on Human Factors in Computing Systems*, CHI '15, pages 3197–3206, New York, NY, USA. ACM.

American Diabetes Association. 2013. Economic costs of diabetes in the US in 2012. *Diabetes care*, 36(4):1033–1046.

Eiji Aramaki, Sachiko Maskawa, and Mizuki Morita. 2011. Twitter catches the flu: Detecting influenza epidemics using Twitter. In *Proceedings of the Conference on Empirical Methods in Natural Language Processing*, EMNLP '11, pages 1568–1576, Stroudsburg, PA, USA. Association for Computational Linguistics.

Leila Arras, Franziska Horn, Grégoire Montavon, Klaus-Robert Müller, and Wojciech Samek. 2016. Explaining predictions of non-linear classifiers in NLP. In *Proceedings of the 1st Workshop on Representation Learning for NLP*, pages 1–7.

Leila Arras, Grégoire Montavon, Klaus-Robert Müller, and Wojciech Samek. 2017. Explaining recurrent neural network predictions in sentiment analysis. *EMNLP 2017*, page 159.

Sebastian Bach, Alexander Binder, Grégoire Montavon, Frederick Klauschen, Klaus-Robert Müller, and Wojciech Samek. 2015. On pixel-wise explanations for non-linear classifier decisions by layer-wise relevance propagation. *PloS one*, 10(7):e0130140.

Heejung Bang, Alison M Edwards, Andrew S Bomback, Christie M Ballantyne, David Brillon, Mark A Callahan, Steven M Teutsch, Alvin I Mushlin, and Lisa M Kern. 2009. Development and validation of a patient self-assessment score for diabetes risk. *Annals of internal medicine*, 151(11):775–783.

J. Bian, U. Topaloglu, and F. Yu. 2012. Towards large-scale Twitter mining for drug-related adverse events. In *Proceedings of CIKM Workshop on SHB*, pages 25–32.

Alexander Binder, Sebastian Bach, Gregoire Montavon, Klaus-Robert Müller, and Wojciech Samek. 2016. Layer-wise relevance propagation for deep neural network architectures. In *Information Science and Applications (ICISA) 2016*, pages 913–922. Springer.

Leo Breiman. 1996. Bagging predictors. *Machine learning*, 24(2):123–140.

John D Burger, John Henderson, George Kim, and Guido Zarrella. 2011. Discriminating gender on Twitter. In *Proceedings of the Conference on Empirical Methods in Natural Language Processing*, pages 1301–1309. Association for Computational Linguistics.

Centers for Disease Control and Prevention. 2017. National diabetes statistics report, 2017.

A. Culotta. 2014. Estimating county health statistics with twitter. In *Proceedings of the SIGCHI Conference on Human Factors in Computing Systems*, pages 1335–1344.

Hal Daumé. 2007. Frustratingly easy domain adaptation. In *Proceedings of the 45th Annual Meeting of the Association of Computational Linguistics*, pages 256–263.

Munmun De Choudhury, Scott Counts, and Eric Horvitz. 2013. Predicting postpartum changes in emotion and behavior via social media. In *Proceedings of the SIGCHI Conference on Human Factors in Computing Systems*, pages 3267–3276. ACM.

John M Digman. 1990. Personality structure: Emergence of the five-factor model. *Annual review of psychology*, 41(1):417–440.

Yannis Dimopoulos, Paul Bourret, and Sovan Lek. 1995. Use of some sensitivity criteria for choosing networks with good generalization ability. *Neural Processing Letters*, 2(6):1–4.

Johannes C Eichstaedt, Hansen Andrew Schwartz, Margaret L Kern, Gregory Park, Darwin R Labarthe, Raina M Merchant, Sneha Jha, Megha Agrawal, Lukasz A Dziurzynski, Maarten Sap, et al. 2015. Psychological language on Twitter predicts county-level heart disease mortality. *Psychological science*, 26(2):159–169.

Aleksandr Farseev and Tat-Seng Chua. 2017. TweetFit: Fusing multiple social media and sensor data for wellness profile learning. In *AAAI*, pages 95–101.

D. Fried, M. Surdeanu, S. Kobourov, M. Hingle, and D. Bell. 2014. Analyzing the language of food on social media. In *2014 IEEE International Conference on Big Data (Big Data)*, pages 778–783. IEEE.

Muriel Gevrey, Ioannis Dimopoulos, and Sovan Lek. 2003. Review and comparison of methods to study the contribution of variables in artificial neural network models. *Ecological modelling*, 160(3):249–264.

Mohit Iyyer, Varun Manjunatha, Jordan Boyd-Graber, and Hal Daumé III. 2015. Deep unordered composition rivals syntactic methods for text classification. In *Proceedings of the 53rd Annual Meeting of the Association for Computational Linguistics and the 7th International Joint Conference on Natural Language Processing (Volume 1: Long Papers)*, volume 1, pages 1681–1691.

Anders Johannsen, Dirk Hovy, and Anders Søgaard. 2015. Cross-lingual syntactic variation over age and gender. In *CoNLL*, pages 103–112.

Enes Kocabey, Mustafa Camurcu, Ferda Ofli, Yusuf Aytar, Javier Marin, Antonio Torralba, and Ingmar Weber. 2017. Face-to-BMI: Using computer vision to infer body mass index on social media. In *International AAAI Conference on Web and Social Media*.

Jiwei Li, Xinlei Chen, Eduard Hovy, and Dan Jurafsky. 2015. Visualizing and understanding neural models in NLP. *arXiv preprint arXiv:1506.01066*.

YanFeng Li, Linda S Geiss, Nilka R Burrows, Deborah B Rolka, and Ann Albright. 2013. Awareness of prediabetes — United States, 2005–2010. *Morbidity and Mortality Weekly Report*, 62(11):209–212.

Tomas Mikolov, Kai Chen, Greg Corrado, and Jeffrey Dean. 2013. Efficient estimation of word representations in vector space. In *Proceedings of Workshop at ICLR*.

Danielle Mowery, Albert Park, Mike Conway, and Craig Bryan. 2016. Towards automatically classifying depressive symptoms from Twitter data for population health. In *Proceedings of the Workshop on Computational Modeling of People's Opinions, Personality, and Emotions in Social Media*, pages 182–191.

National Center for Health Statistics. 2017. Health, united states, 2016: with chartbook on long-term trends in health.

Dong Nguyen, Dolf Trieschnigg, A Seza Doğruöz, Rilana Gravel, Mariët Theune, Theo Meder, and Franciska De Jong. 2014. Why gender and age prediction from tweets is hard: Lessons from a crowdsourcing experiment. In *Proceedings of COLING 2014, the 25th International Conference on Computational Linguistics: Technical Papers*, pages 1950–1961.

Thin Nguyen, Mark E Larsen, Bridianne O'Dea, Duc Thanh Nguyen, John Yearwood, Dinh Phung, Svetha Venkatesh, and Helen Christensen. 2017. Kernel-based features for predicting population health indices from geocoded social media data. *Decision Support Systems*.

Olutobi Owoputi, Brendan O'Connor, Chris Dyer, Kevin Gimpel, Nathan Schneider, and Noah A Smith. 2013. Improved part-of-speech tagging for online conversational text with word clusters. In *Proceedings of the Conference of the North American Chapter of the Association for Computational Linguistics: Human Language Technologies*. Association for Computational Linguistics.

Pew Research Center. 2018. Social media fact sheet. http://www.pewinternet.org/fact-sheet/social-media/.

Kyle W. Prier, Matthew S. Smith, Christophe Giraud-Carrier, and Carl L. Hanson. 2011. Identifying health-related topics on Twitter: An exploration of tobacco-related tweets as a test topic. In *Proceedings of the 4th International Conference on Social Computing, Behavioral-cultural Modeling and Prediction*, SBP'11, pages 18–25, Berlin, Heidelberg. Springer-Verlag.

Stephen A. Rains, Melanie D. Hingle, Mihai Surdeanu, Dane Bell, and Stephen Kobourov. 2018. A test of the risk perception attitude framework as a message tailoring strategy to promote diabetes screening. *Health Communication*.

Delip Rao, David Yarowsky, Abhishek Shreevats, and Manaswi Gupta. 2010. Classifying latent user attributes in twitter. In *Proceedings of the 2Nd International Workshop on Search and Mining User-generated Contents*, SMUC '10, pages 37–44, New York, NY, USA. ACM.

William R Rowley, Clement Bezold, Yasemin Arikan, Erin Byrne, and Shannon Krohe. 2017. Diabetes 2030: insights from yesterday, today, and future trends. *Population health management*, 20(1):6–12.

H Andrew Schwartz, Johannes C Eichstaedt, Margaret L Kern, Lukasz Dziurzynski, Stephanie M Ramones, Megha Agrawal, Achal Shah, Michal Kosinski, David Stillwell, Martin EP Seligman, et al. 2013. Personality, gender, and age in the language of social media: The open-vocabulary approach. *PloS one*, 8(9):e73791.

Hong-Han Shuai, Chih-Ya Shen, De-Nian Yang, Yi-Feng Lan, Wang-Chien Lee, Philip S Yu, and Ming-Syan Chen. 2016. Mining online social data for detecting social network mental disorders. In *Proceedings of the 25th International Conference on World Wide Web*, pages 275–285. International World Wide Web Conferences Steering Committee.

Karen Simonyan, Andrea Vedaldi, and Andrew Zisserman. 2013. Deep inside convolutional networks: Visualising image classification models and saliency maps. *arXiv preprint arXiv:1312.6034*.

Nikhita Vedula and Srinivasan Parthasarathy. 2017. Emotional and linguistic cues of depression from social media. In *Proceedings of the 2017 International Conference on Digital Health*, pages 127–136. ACM.

Svitlana Volkova, Theresa Wilson, and David Yarowsky. 2013. Exploring demographic language variations to improve multilingual sentiment analysis in social media. In *EMNLP*, pages 1815–1827.

Sida Wang and Christopher D Manning. 2012. Baselines and bigrams: Simple, good sentiment and topic classification. In *Proceedings of the 50th Annual Meeting of the Association for Computational Linguistics: Short Papers-Volume 2*, pages 90–94. Association for Computational Linguistics.

Lingyun Wen and Guodong Guo. 2013. A computational approach to body mass index prediction from face images. *Image and Vision Computing*, 31(5):392–400.

Treatment Side Effect Prediction from Online User-generated Content

Van Hoang Nguyen   Kazunari Sugiyama   Min-Yen Kan   Kishaloy Halder

School of Computing
National University of Singapore
hoang_van_nguyen@u.nus.edu,   {sugiyama, kanmy, kishaloy}@comp.nus.edu.sg

Abstract

With Health 2.0, patients and caregivers increasingly seek information regarding possible drug side effects during their medical treatments in online health communities. These are helpful platforms for non-professional medical opinions, yet pose risk of being unreliable in quality and insufficient in quantity to cover the wide range of potential drug reactions. Existing approaches which analyze such user-generated content in online forums heavily rely on feature engineering of both documents and users, and often overlook the relationships between posts within a common discussion thread. Inspired by recent advancements, we propose a neural architecture that models the textual content of user-generated documents and user experiences in online communities to predict side effects during treatment. Experimental results show that our proposed architecture outperforms baseline models.

1 Introduction

Seeking medical opinions from online health communities has become commonplace: 71% of age 18–29 (equivalent to 59% of all U.S. adults) reported consulting online health opinion (Fox and Duggan, 2013). These opinions come from an estimated twenty to one hundred thousand health-related websites (Diaz et al., 2002), inclusive of online health communities that network patients with each other to provide information and social support (Johnston et al., 2013). Platforms such as HealthBoards[1] and MedHelp[2] feature users reporting their own health experiences, inclusive of their self-reviewed drugs and medical treatments. Hence, they are valuable sources for researchers (Leyens et al., 2017; Martin-Sanchez and Verspoor, 2014).

Although readers use these platforms to get valuable information about potential drug reactions during treatment, this is not without potentially serious problems. There is lexical variation:

users do refer side effects differently: *"dizziness"* can be expressed as *"giddiness"* or *"my head is spinning"*. More concern is that discussions rarely cover all possible prescribed drugs and their side effects during a treatment, and some topics refer to a condition without mentioning any particular drug. Relying on such information could lead to adverse reactions.

It is important to note that a tool that looks up mentioned drugs' side effects from a static database would not return answers with sufficient coverage. There are also common concerns regarding credibility of user-generated contents – (Impicciatore et al., 1997) have shown that online health information is of variable quality and approached with caution.

Having these caveats in mind though, experienced users can provide valuable expertise. For instance, while reporting expected side effects for a specific treatment, patients with long-term use of certain drugs can be valuable authorities. *E.g.*:

While my experience of 10 years is with Paxil, I expect that Zoloft will be the same. You should definitely feel better within 2 weeks. One way I found to make it easier to sleep was to get lots of exercise. Walk or run or whatever to burn off that anxiety. – User 3690.

This is an answer to a thread asking for expected side effects for depression treatment with Zoloft. User 3690's history of actively discussing about other anti-depressants such as Lexapro and Xanax gives insights in predicting potential drug reactions during the treatment of depression. Table 1 shows that Zoloft (mentioned in the thread) shares many common side effects with the other two anti-depressants: *"changed behavior," "dry mouth,"* and *"sleepiness or unusual drowsiness."*

A method that could differentiate trustworthy user-generated content would be valuable, allowing us to macroscopically harness a large amount of online information that would pave the way to many critical tasks such as digital pharmacovigilance (Salathé, 2016) and disease monitoring (St Louis and Zorlu, 2012). Even on the micro-

---

[1] https://www.healthboards.com/
[2] https://medhelp.ord/

12

Proceedings of the 9th International Workshop on Health Text Mining and Information Analysis (LOUHI 2018), pages 12–21
Brussels, Belgium, October 31, 2018. ©2018 Association for Computational Linguistics

Drugs	Side effects
Lexapro	chills, constipation, cough, decreased appetite, **decreased sexual desire**, **diarrhea**, **dry mouth**, joint pain, muscle ache, tingling feeling, **sleepiness or unusual drowsiness**, unusual dream, **sweating**, ...
Xanax	abdominal or stomach pain, muscle weakness , **changed behavior**, chills, cough, decreased appetite, decreased urine, **diarrhea**, difficult bowel movement, cough, **dry mouth**, tingling feeling, **sleepiness or unusual drowsiness**, slurred speech, sweating, yellow eye...
Zoloft	**changed behavior**, decreased sexual desire, **diarrhea**, **dry mouth**, heartburn, **sleepiness or unusual drowsiness, sweating**...

Table 1: Side effects of anti-depressants.

scopic level of individual posts, such a tool offers users' suggestions for drug reactions and improves the quality of user-generated content.

We address this need in our work. We build a neural architecture that models each post's textual content and its author's experience to predict expected side effects during treatments. Crucially, our supervised neural approach *jointly* learns posts' content and users' experience level within a thread. A key observation we make is that users can be grouped into clusters that share the same expertise or interest in certain drugs, possibly due to their common treatment or medical history. We leverage this expertise by embedding it into a low dimensional vector learned by the model, and subsequently predict side effects that are unmentioned in the discussion. We believe that our model represents trustworthiness more robustly when compared with representations such as a single weights (Li et al., 2016) and traditional drug side effect extraction (Aramaki et al., 2010). Furthermore, inspired by (Halder et al., 2018), we train a cluster-sensitive attention mechanism that allows our model to emphasize varied parts of the post. We also follow general definition of truth discovery and let the model learn a credibility score that is unique to every user and reflective of her trustworthiness. Our experimental results show that integrating the above components outperforms baseline text classification models.

The contributions of our work are summarized as follows:

- We propose a neural network architecture that can capture user expertise, user credibility, individual post's and overall thread's semantic content.

- We formulate the task of side effect prediction during treatment as supervised multi-label classification and apply our proposed method to the task of side effect prediction during treatment.

- We record and analyze the performance of our proposed model through a set of progressively designed experiments. Additionally, we compare the obtained results with traditional text encoding algorithms.

2   Related Work

Our approach learns the representation of posts, threads and users, and then integrates them to apply to the task of drug side effect prediction during treatment. We thus review works on the representation of fundamental objects in online communities, and the discovery of drug side effects.

2.1   Modeling Objects in Online Communities

Post content modeling. In statement credibility prediction, linguistic features of a post are strong indicators for reliability. Stylistic features – *i.e.*, the number of strong/weak modals, conditionals or negations – and affective features – *i.e.*, words that depict an author's attitude and emotion – are adopted to represent a post's content (Mukherjee et al., 2014). Such feature engineering requires a great amount of correlation analysis when applied to a novel problem or dataset. Linguistic features also often fail to fully capture document content, as most do not account for distinctive words in exchange of scalability. Its counter parts, bag of words and per-vocabulary features loosely capture textual content but disregard semantics and suffer scalability with sparsity issues. To address this, state-of-the-art architectures feature complex modeling to model subtle dependencies and rely on word embeddings to address scalability issues, achieving robust results in text classification (Kim, 2014), neural machine translation (Luong et al., 2016), among others.

Inspired by the success of their approaches, we adopt the recurrent neural network architecture (RNN) for post content modeling. Coupled with an attention mechanism, our approach adaptively weights the importance of parts in each post (Luong et al., 2015).

Thread content modeling. Most research working on thread-level modeling usually obtain thread content representation by aggregating each content of its posts (Yang et al., 2014). However, we hypothesize that each post has different contribution to thread content and should be variously weighted to reflect specific factors, such as its author's level of credibility.

13

User ID	Post	Drug mentioned	Aggregated side effects
3690	While my experience of 10 years is with Paxil, I expect that Zoloft will be the same. You should definitely feel better within 2 weeks. One way I found to make it easier to sleep was to get lots of exercize. Walk or run or whatever to burn off that anxiety.	Zoloft	changed behavior, decreased sexual desire, diarrhea, dry mouth, heartburn, sleepiness or unusual drowsiness,...
26521	I've heard of people going "cold turkey" and having withdrawal at 6 months! Please, get in contact with a doctor ASAP! "common symptoms include dizziness, electric shock-like sensations, sweating, nausea, insomnia, tremor, confusion, nightmares and vertigo"		

Table 2: A sample thread, including its list of post–user pairs, mentioned drugs, and side effects.

User modeling. Statement credibility prediction often represents users by a single scalar that indicates their trustworthiness. The intuition is that users who provide trustworthy information frequently will be assigned high reliability scores (Li et al., 2017). Such representation is effective yet insufficient. Recent work have shown that encoding users into high-dimensional embeddings can improve system performance (Yu et al., 2016), which we have adopted in our model.

2.2 Side Effect Discovery

Most drug reaction discovery methods focus on extracting mentioned side effects. A common technique is to apply Named Entity Recognition (NER) and Relation Extraction (RE) systems in a supervised manner. (Sampathkumar et al., 2014) demonstrates its effectiveness in detecting drugs and side effects that appear in a target document (in-context), and predicting if they are related.

However, in our side effect prediction during treatment, our model is required to cover potentially encountered reactions, many of which are not explicitly mentioned in the given post (out-of-context). Hence, we do not identify our task with traditional task of adverse drug side effect extraction (Leaman et al., 2010). Our approach overcomes the limitations of the existing works by modeling user experience, and credibility during post and thread encoding, then subsequently predicting both in- and out-of-context side effects.

3 Preliminaries

Basic Terminologies. To ensure a consistent representation, let us first define some terminology:

- A *drug d* has a set of side effects,
  $S_d = \{s_1, s_2, \ldots, s_k\}$

- A *post p* is the most basic document, containing a sequence of sentences. It is written by a *user u*, and belongs to a *thread t*.

- A *user u* is a member of an online community. She participates in certain threads, *i.e.,* $T_u = \{t_1, t_2, \ldots, t_l\}$ by writing at least one post in each thread. We use the terms *user* and *author*, as well as *user experience* and *user expertise* interchangeably.

- A *thread t* (see Table 2) is an ordered collection of post–user pairs,
  $Q_t = [(p_1, u_1), (p_2, u_2), \ldots, (p_n, u_n)].$
  Every thread discusses the treatment of a particular condition and entails a list of prescribed drugs $D_t = \{d_1, d_2, \ldots, d_m\}$. Hence, every thread has a list of aggregated side effects defined as $S_t = S_{d_1} \cup S_{d_2} \cdots \cup S_{d_m}$, which is also the list of potential side effects experienced during the treatment.

Task Definition. *Drug side effect prediction during treatment* is the task of assigning the most relevant subset of side effects to threads discussing certain treatment, from a large collection of potential side effects. We view the drug side effect prediction problem as a multi-label classification task. In our setting, an instance of item–label is a tuple $(\boldsymbol{x_t}, \boldsymbol{y})$ where $\boldsymbol{x_t}$ is the feature vector of thread t derived from its list of post–user pairs Q_t and $\boldsymbol{y}$ is the side effect label vector *i.e.,* $\boldsymbol{y} \in \{0, 1\}^S$, where S is the number of possible side effect labels. Given training instances, we train our classifier to predict the list of treatment side effects in unseen threads.

Formal Hypothesis. Given a thread t with Q_t, we hypothesize that considering the credibility and experience of user $u \in (p, u) \in Q_t$ improves the quality of feature representation in thread t, resulting in better treatment side effect prediction.

4 Proposed Method

Figure 1 shows the detailed network architecture of our model. It has several components which we shall detail sequentially. Ablation of certain components will serve as baseline systems for comparative evaluation later.

User Expertise Representation (UE): We embed each user $u \in U$ as a vector $\boldsymbol{v}_u$ so that the vector captures user u's experience with certain drugs.

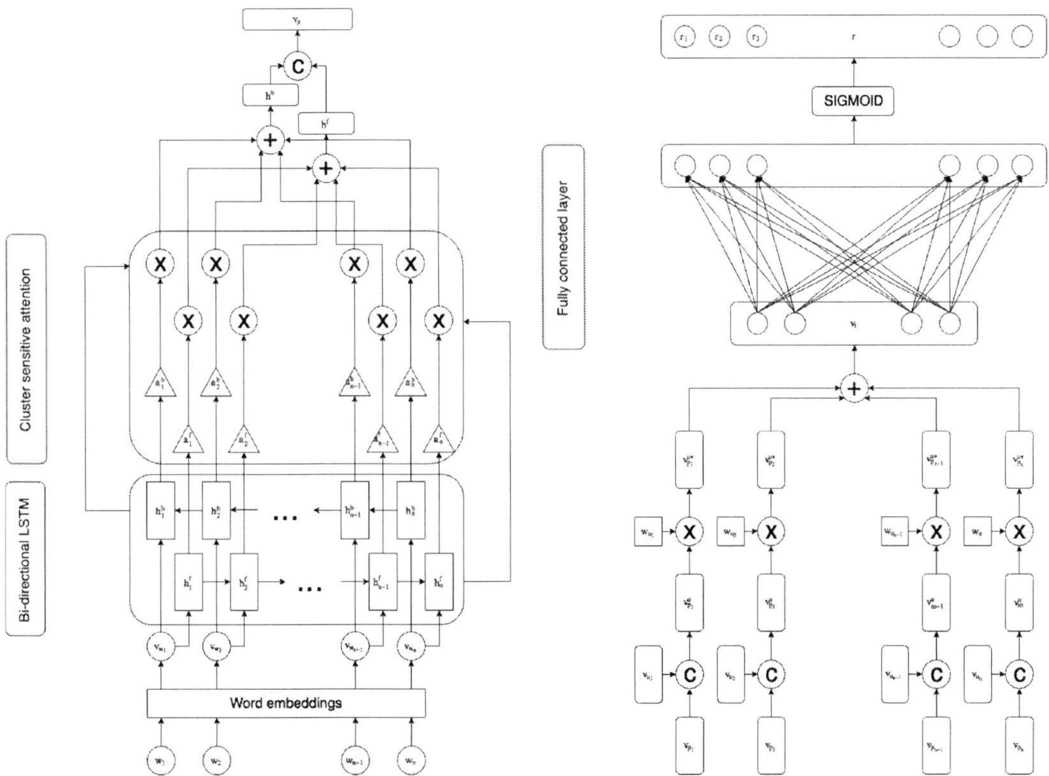

Figure 1: Full model architecture: (l, 1st half) word embeddings and cluster sensitive attention, (r, 2nd half) thread representation and multi label side effect prediction.

As each user u participates in the threads T_u, entailing a list of experienced drugs, we derive user drug experience vector $\boldsymbol{v_u^*} \in \mathbb{R}^{|D|}$ where D is the set of all possible drugs and $v_{u_i}^* = n_{u_i}$ where user u has mentioned i^{th} drug in n_{u_i} threads. We obtain a user drug experience matrix $\boldsymbol{M^*} \in \mathbb{R}^{|U| \times |D|}$ where j^{th} row of $\boldsymbol{M^*}$ denotes user drug experience vector of jth user $u_j \in U$. Since the average number of drugs experienced per user is much fewer than the total number of drugs (see Table 3), M^* suffers from data sparsity and limited scalability. Without dimensionality reduction, the model learns at least $|D|$ parameters for every user, amounting to $|D| \times |U|$ when aggregated for all users. Data sparsity leads to a large number of insufficiently tuned parameters, which significantly increases training time, storage, and reduces the system's robustness.

We apply Principal Component Analysis (PCA) (Jolliffe, 1986) to $\boldsymbol{M^*}$ obtained from training set. Figure 2 shows percentage of variance explained versus number of included principal components (PCs) to determine the number of PCs g. Since our PCA plots do not show added explanation percentage beyond 50 components, we use $g = 50$ com-

ponents, reducing our original $\boldsymbol{M^*} \in \mathbb{R}^{|U| \times |D|}$ to user expertise matrix $\boldsymbol{M} \in \mathbb{R}^{|U| \times g}$.

User Clustering: To model per-user expertise, in a naïve setting, we would train $\approx |U| \times g$ parameters. Given limited data, this is infeasible as it faces sparsity issues. We make a second, key assumption that our set of users U can be grouped into a set of meaning clusters C of size k where $k \ll |U|$. Users within a cluster would have experience with similar drugs, and hence representable using a single vector, reducing the number of learned parameters to $k \times g$.

We apply K-means clustering algorithm (Mac-Queen, 1967) to cluster the users into k groups. To determine the number of clusters k, we analyze the total distance to the nearest centroid versus number of potential clusters in set C – as in Figure 3, where $D(C)$ is defined as follows:

$$D(C) = \frac{\sum_{c \in C} \sum_{u \in c} dist(\boldsymbol{v_c}, \boldsymbol{v_u})}{argmax D(C)}, \quad (1)$$

where $argmax D(C)$ is the maximum total distance obtained when $|C| = 1$.

Since clustering does not gain significant reduction in total distance beyond 100 clusters, we sort

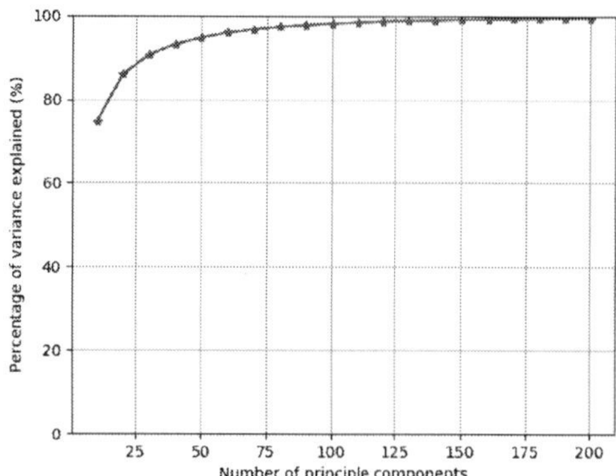

Figure 2: Principal component analysis.

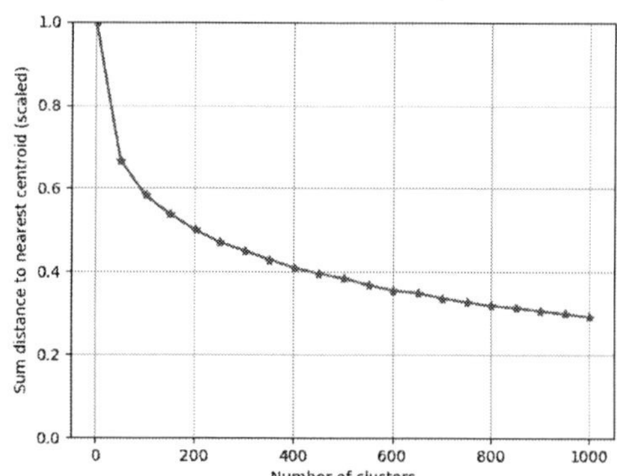

Figure 3: K-means analysis.

each user to a cluster $c \in C$ where $|C| = k = 100$. For each user, we consider the vector of her assigned cluster's centroid to be her expertise vector.

Post Content Encoding: The network takes the content of a thread t as input, which is a list of post–user pairs Q_t. Post p_i of pair $(p_i, u_i) \in Q_t$ consists of a sequence of words $(w_1, \ldots, w_n)$. We seek to represent a post p_i as vector v_p that effectively captures its semantics. We embed each word into a low dimensional vector and transform the post into a sequence of word vectors $\{v_{w_1}, v_{w_2}, \ldots, v_{w_n}\}$. Each word vector is initialized using Google's pre-trained word2vec (Mikolov et al., 2013). Additionally, while each out-of-vocabulary word vector is initialized randomly, we keep it tunable during training to capture domain-specific meanings. Such model adaptation is necessary, as the model needs to learn the embeddings for the drug names, most of which are not included in the pre-trained embeddings but are critical to predict the side effects.

We employ Long-Short Term Memory (LSTM) (Hochreiter and Schmidhuber, 1997) to encode the textual content. A bi-directional LSTM encodes the word vector sequence, outputting two sequences of hidden states: a forward sequence, $H^f = h_1^f, h_2^f, \ldots, h_n^f$ that starts from the beginning of the text; and a backward sequence, $H^b = h_1^b, h_2^b, \ldots, h_n^b$ that starts from the end of the text. For many sequence encoding tasks, knowing both past (left) and future (right) contexts has proven to be beneficial (Dyer et al., 2015). The states h_i^f and h_j^b in the forward and backward sequences are computed as follows:

$$h_i^f = LSTM(h_{i-1}^f, w^i), \quad h_j^b = LSTM(h_{j+1}^b, w^j),$$

where $h_i^f, h_j^b \in \mathbb{R}^e$, and e are the number of encoder units.

# Users	# Threads	Avg. # of words per post	
14,388	99,682	73.65	
Avg. # of posts per thread		Avg. # of threads per user	
8.16		26.21	
# Side effects (SE)		Avg. # of SEs per thread	
1,500		90.47	
# Drugs	Avg. # of drugs per user		
1869	19.72		

Table 3: Dataset statistics.

Cluster-sensitive Attention (CA): Inspired by (Halder et al., 2018), we initialize an attention vector, $v_{a_i} \in \mathbb{R}^e$ for each cluster c_i. Given a forward sequence $H^f = h_1^f, h_2^f, \cdots, h_n^f$ and backward sequence $H^b = h_1^b, h_2^b, \cdots, h_n^b$ of hidden post states p written by user u belonging to cluster c_i, the corresponding w_j weights each hidden state h_j^f and h_j^b of both sequences based on their similarity with the attention vector are:

$$w_{a_j} = \frac{exp(v_{a_i} h_j)}{\sum_{l=1}^{n} exp(v_{a_i} h_j)}. \tag{2}$$

The intuition behind Equation (2), inspired by (Luong et al., 2015), is that hidden states which are similar to the attention vector v_{a_i} should be paid more attention to; hence are weighted higher during document encoding. v_{a_i} is adjusted during training to capture hidden states that are significant in forming the final post representation. w_{a_j} is then used to compute forward and backward weighted feature vectors:

$$h^f = \sum_{j}^{n} w_{a_j} h_j^f, \quad h^b = \sum_{j}^{n} w_{a_j} h_j^b. \tag{3}$$

We concatenate the forward and backward vectors to obtain a single vector, following previous bi-directional RNN practice (Ma and Hovy, 2016).

Thread Content Encoding with Credibility Weights (CW): For every post–user pair (p_i, u_i) of thread t, we first compute post p_i feature vector $\boldsymbol{v}_{p_i}$. It is then concatenated with user u_i's expertise vector $\boldsymbol{v}_{u_i}$ to form post–user complex vector $\boldsymbol{v}_{n_i}^p$. This user-post complex is weighted by a user credibility score w_{u_i}, which is initially randomized and updated while training, to obtain final post–user pair representation $\boldsymbol{v}_{n_i}^{p*}$. This follows the general intuition from the truth discovery literature that users providing high quality answers should assign higher credibility scores, and answers from credible users are more significant. Thus, the thread content representation can be defined as the weighted sum of each post–user complex vector:

$$\boldsymbol{v}_t = \sum_{i=1}^{n} \boldsymbol{v}_{n_i}^{p*} = \sum_{i=1}^{n} w_{u_i} \boldsymbol{v}_{n_i}^p. \tag{4}$$

Multi-label Prediction: We feed the thread content representation $\boldsymbol{v}_t$ through a fully connected layer which outputs can be computed as:

$$\boldsymbol{s}_t = \boldsymbol{W} tanh(\boldsymbol{v}_t) + \boldsymbol{b}, \tag{5}$$

where $\boldsymbol{W}$ and $\boldsymbol{b}$ are weights and biases of the layer. The output vector $\boldsymbol{s}_t \in \mathbb{R}^{|S|}$ is finally passed through a sigmoid activation function, and trained using cross-entropy loss as defined by L:

$$\begin{aligned}
L = \frac{1}{T} \sum_{t=1}^{T} (\mathbf{y_t} \cdot \log(\sigma(\mathbf{s_t})) \\
+ (1 - \mathbf{y_t}) \cdot \log(1 - \sigma(\mathbf{s_t}))) + \lambda \sum_u \mathbf{v_u}^2
\end{aligned} \tag{6}$$

We adopt regularization that penalizes the training loss with the user experience matrix's $L2$ norm by a factor of $\lambda = 0.0065$, obtained via hypertuning. The loss function is differentiable, thus trainable with Adam (Kingma and Ba, 2015). During our gradient-based learning, user credibility score w_{u_i} of user u_i can be updated by calculating $\frac{\partial L}{\partial w_{u_i}}$ by back-propagation:

$$\begin{aligned}
\frac{\partial L}{\partial w_{u_i}} &= \frac{\partial L}{\partial \boldsymbol{s}_t} \frac{\partial \boldsymbol{s}_t}{\partial \boldsymbol{v}_t} \frac{\partial \boldsymbol{v}_t}{\partial w_{u_i}} \\
&= \frac{\partial L}{\partial \boldsymbol{s}_t} \boldsymbol{W} (1 - tanh^2(\boldsymbol{v}_t)) \boldsymbol{v}_{n_i}^p
\end{aligned} \tag{7}$$

5 Experiments

We conduct experiments to validate the effectiveness of our proposed model. In specific, (1) we want to compare our architecture with text encoding baselines, (2) highlight performance improvements incrementally, and (3) evaluate and analyze the obtained results, both at the macroscopic and microscopic levels.

5.1 Baselines

As a competitive baseline from prior work, CNN-KIM (Kim, 2014) constructs a document matrix that incorporates word embeddings, then applies a convolution filter to obtain feature maps. These feature maps are passed through a max-pooling filter to construct a document representation. During prediction, the representation is fed through a fully connected layer. We replace the final softmax layer of the author's model with sigmoid to make it work in a multi-label prediction setting.

The following baselines are used to perform an ablation study of our model.

- **RNN:** We implement a bi-directional LSTM baseline, which is equivalent to our proposed method without CA, UE and CW.

- **Weighted Post Encoder (WPE):** We construct thread representation by summing each of its post–user complex vector weighted by user credibility. This is equivalent to our proposed methodology without CA and UE.

- **Weighted Post Encoder with User Expertise (WPEU):** We concatenate user expertise with post vector to create post–user complex vector. This is equivalent to our proposed method without CA.

5.2 Dataset

We conduct our experiments on the same dataset as (Mukherjee et al., 2014) including 15,000 users and 2.8 million posts extracted from 620,510 HealthBoards[1] threads.

Ground truth possible side effects experienced during treatment are defined as the side effects of drugs mentioned in the discussion. As annotating such amount of posts is expensive, drug side effects are extracted from Mayo Clinic's Drugs and Supplements portal[3] and are used as surrogates for potential reactions of treatments.

5.3 Experimental Settings

We applied a standard natural language preprocessing — Snowball stemming (Porter, 1980) and stop-word elimination — before representation modeling. From the original dataset, we only extract threads that are annotated with drugs and their side effects, along with the lists of contained posts and corresponding users. Table 3 shows the dataset statistics. We divide our data into 10

---

[3] https://www.mayoclinic.org/drugs-supplements

System	Components			Experiment 1			Experiment 2		
	CW	UE	CA	Pre.	Rec.	F_1	Pre.	Rec.	F_1
1. CNN-KIM				0.818	0.677	0.751	0.813	0.503	0.614
2. RNN				0.810	0.657	0.735	0.808	0.484	0.599
3. WPE	✓			**0.873**	0.678	0.773	**0.859**	0.507	0.638
4. WPEU	✓	✓		0.865	0.705	0.781	0.819	0.537	0.643
5. Our model	✓	✓	✓	0.844	**0.730**	**0.793**	0.788	**0.573**	**0.659**

Table 4: Experimental results with both actual (Experiment 1) and Strict (Experiment 2) settings. In the Component columns, "CW", "UE", "CA" denote "Credibility Weights", "User Expertise" and "Cluster Attention module components", respectively.

folds to perform cross-validation (8,1,1 folds for training, validation, and testing respectively). We perform PCA and K-means clustering on training set, using scikitlearn's built-in modules (Pedregosa et al., 2011), with $g = 50$ principal components and $k = 100$ clusters.

For CNN-KIM, we experiment with filters with varying window sizes from 2 to 5, and set the number of feature maps for each filter to 256 and dropout to 0.5. For our proposed model and baseline models using the RNN architecture, when performing post content encoding, we set the number of units in the LSTM cell to 128. Dropout rates of 0.2 and 0.5 are used in our LSTM cells and FC layers, respectively. Cluster attention vectors and user credibility values are initialized with values ranging from -1.0 to 1.0. For each user u, we initialize her expertise vector with the value of v_u obtained in Section 4 and allow training to fine-tune. All models are trained using Tensorflow[4] library.

We conducted two separate experiments:

- **Experiment 1**: We keep the text as-is. Any mentioned drugs are retained inside the thread.

- **Experiment 2**: We remove all mentions of any drug in our drug list. This is a more aggressive experiment which asks the model to predict the treatment's side effects without any mention of the experienced drugs.

6 Results and Evaluation

Table 4 shows the precision, recall, and F_1 obtained by our method and the four baselines.

Macroscopic Analysis: Firstly, all of the three models that apply credibility weighting (CW) – WPE, WPEU and our model – outperform both RNN and CNN baselines in both experiments. Specifically, weighting each post by its author credibility improves the performance of naive post encoder by 6.32%, 2.15% and 3.86% on precision,

recall and F_1 respectively for Experiment 1. Results for Experiment 2 are similar. This demonstrates the effectiveness of accounting for author credibility when encoding thread content, improving side effect prediction.

Improvements by incorporating user experience (UE) are less pronounced. In Experiment 1, adding UE (WPEU vs. WPE) improves recall by 2.65% and 0.8% in F_1. Again, the stricter Experiment 2 shows similar performance trends. On a macro scale, these statistics indicate that our model successfully learns to include more side effects in its prediction, where many are relevant to the ground truth. This is consistent with our hypothesis that considering author experience of each post is effective in predicting out-of-context side effects.

Applying cluster-sensitive attention (CA) in combining RNN's hidden states also improves the performance. In Experiment 1, we observe that adding CA (our model vs. WPEU) also improves recall and F_1, where again, Experiment 2 demonstrates similar but slightly more pronounced performance changes. These indicate that the attention mechanism is more effective when the drugs are present since the drug names in our documents are the phrases that receive greater emphasis.

As settings in Experiment 1 start with more information compared with those in Experiment 2, the task is easier and thus performance is improved (12.7% to 14.15% in F_1). The margin for improvement for Experiment 2 is larger, which explains why absolute score improvements are larger in Experiment 2. When measuring relative improvement, the gains are comparable.

Generally, according to the macroscopic analysis of results in Table 4, we conclude that all of the three components in our proposed architecture, namely, CW, UE, and CA have a positive impact on the overall performance of the model. We observe consistent improvements in F_1 after adding each component is consistent with our stated hypotheses, in both experimental settings.

---

[4] https://www.tensorflow.org/

User ID	Posts	Output side effects					
		CNN-KIM	RNN	WPE	WPEU	Our model	Ground truth
24296 (credibility: 0.11)	[...] **little red rashes all over my body** that resembled vasculitis. [...] I was diagnosed and treated with the "standard treatment" twice, to not much effect), a very stiff neck, really bad brain fog and **confusion**. [...]	diarrhea, skin rash	skin rash	headache, diarrhea, skin rash	headache, diarrhea, unusual tiredness and weakness, dizziness, sleepiness fever, nausea, bad breath	headache, diarrhea, unusual tiredness and weakness, dizziness, sleepiness, fever, nausea, heartburn, belching, indigestion, acid stomac, difficult bowel movement, bad breath, bone joint pain	headache, diarrhea, unusual tiredness and weakness, dizziness, fever, nausea, loss appetite, chills, heartburn, belching, indigestion, acid stomach, confusion, skin rash, weight loss, difficult bowel movement, shakiness
1537 (credibility: 0.32)	[...] now last month my symptoms including joint pains, twitching and tremors and **bug crawling under my skalp sensations reappeared** [...]						
5232 (credibility: 0.36)	[...] I don't know about **cysts in the brain per se** [...]						
16248 (credibility: 0.21)	[...] I've been growing increasingly **sensitive to more foods** over the last year [...] How do you know that you had **damage to your intestines** from Lyme? [...] I'm curious because I am in the process of getting a Lyme work up and my **intestines are messed up**, but all GI tests came back negative.						

Table 5: A sample thread in the test set, mentioning drugs *Flagyl, Tinidazole, Plaquentil,* and *Vitamins.*

User ID	Experienced drugs	Top common experienced side effects
24296	rifampin, **vitamin**, clarithromycin, aciphex, a zithromax, **plaquenil**, **flagyl**, minocycline, levaquin, tetracycline, **tinidazole**, advil	**diarrhea**, bad breath, **headache, heartburn, unusual tiredness and weakness, nausea, fever**
1537	**vitamin**, rocephin, hydroquinone, **plaquenil**, **flagyl**, minocycline, levaquin, **tinidazole**	**diarrhea, skin rash, headache, dizziness, heartburn**, bad breath, **sleepiness**
5232	doxycycline, prozac, **vitamin**, norvasc, tylenol, **flagyl**, questran, biotin, cefuroxime, **plaquenil**	bad breath, **diarrhea, nausea, dizziness, unusual tiredness and weakness**
16248	celexa, prilosec, **vitamin**, rocephin, klonopin, nexium, fumarate, elidel, citrate, prozac	**diarrhea**, sneezing, **nausea**, excessive gas, body pain, loss voice, **heart burn**

Table 6: Experienced drugs and common side effects among users.

Microscopic Analysis: We also analyze our model performance at per-sample level to check whether they are consistent with the macroscopic results. We aim to confirm three hypotheses: (1) Considering author expertise improves prediction on out-of-context side effects. (2) Considering author credibility improves the extraction of both in- and out-of-context side effects from trustworthy users' content. (3) Placing attention on different parts of the document enhances the performance of in-context side effect extraction. Tables 5 and 6 show a sample testing thread, its users' commonly experienced drugs, and its side effects.

We observe that CNN-KIM and the simple, RNN-based post encoding can capture side effects that are mentioned both directly (*e.g.*, "*skin rash*") as well as indirectly (*e.g.*, "*diarrhea*"), but fail to capture the remaining symptoms, many of which are out-of-context.

Considering User 1537's credibility shows performance improvements. In her posts, User 1537 indirectly refers to "*headache*" by mentioning "*bug crawling under my skalp sensations*". The calculated higher credibility score weights User 1537 experiences with "*sleepiness*" higher in the WPEU (CW + UE) baseline prediction, which is correct. These observations are consistent with our hypothesis about user credibility.

User experience is effective in predicting out-of-context symptoms. In the illustrated sample training set, all of the four users have experience with similar drugs with common side effects such as "*unusual tiredness and weakness*", "*nausea*", and "*fever*". As "*bad breath*" is also a shared side effect, it is comprehensible that the model outputs "*bad breath*". Nonetheless, it is intuitive for the model to pick up such commonness among users and compute relevant results. These observations are consistent with our hypothesis on user experience.

Finally, the model with CA can learn different parts of the documents. Especially for User 16248's posts that mentioned digestive problems, hidden states encode phrases such as "*increasingly sensitive to more foods*", and "*damage to your intestines*" receive higher attention, resulting in the prediction of "*heartburn*", "*belching*", "*indigestion*", "*acid stomach*", and "*difficult bowel movement*". This functionality is consistent with our original purpose and expectation for adding attention to the post encoder architecture.

7 Conclusion

We have addressed the importance of user experience and credibility in modeling thread contents of online communities, specifically through the task of drug side effect prediction during treatment. We suggest a subset of side effects relevant to the mentioned treatment in the given discussion, taking into account the each post content and its author expertise in certain treatments. Mainstream models for online communities fail to fully capture post content semantically and user experience with previous drugs.

We model users' expertise by examining their experience with different drugs, then group users with similar experience into clusters that share a common experience vector representation. Experimental results show that our proposed thread content encoder outperforms state-of-the-art document encoders, and that our neural components play a significant role in improving task performance.

We believe that our model is adaptable to other domains. We aim to use it for downstream application in online health community such as credibility analysis and thread recommendation in the future.

References

Eiji Aramaki, Yasuhide Miura, Masatsugu Tonoike, Tomoko Ohkuma, Hiroshi Masuichi, Kayo Waki, and Kazuhiko Ohe. 2010. Extraction of Adverse Drug Fffects from Clinical Records. *MedInfo 2010*, 160:739–743.

Joseph A. Diaz, Rebecca A. Griffith, James J. Ng, Steven E. Reinert, Peter D. Friedmann, and Anne W. Moulton. 2002. Patients' Use of the Internet for Medical Information. *Journal of General Internal Medicine*, 17(3):180–185.

Chris Dyer, Miguel Ballesteros, Wang Ling, Austin Matthews, and Noah A Smith. 2015. Transition-Based Dependency Parsing with Stack Long Short-Term Memory. In *Proc. of the 53rd Annual Meeting of the Association for Computational Linguistics (ACL 2015)*, pages 334–343.

Susannah Fox and Maeve Duggan. 2013. Health Online 2013. *Health*, 2013:1–55.

Kishaloy Halder, Lahari Poddar, and Min-Yen Kan. 2018. Cold Start Thread Recommendation as Extreme Multi-label Classification. In *Proc. of the Workshop on Extreme Multilabel Classification for Social Media co-located with the Web Conference (WWW'18 Companion)*, pages 1911–1918.

Sepp Hochreiter and Jürgen Schmidhuber. 1997. Long Short-Term Memory. *Neural Computation*, 9(8):1735–1780.

Piero Impicciatore, Chiara Pandolfini, Nicola Casella, and Maurizio Bonati. 1997. Reliability of Health Information for the Public on the World Wide Web: Systematic Survey of Advice on Managing Fever in Children at Home. *BMJ*, 314(7098):1875.

Allen C. Johnston, James L. Worrell, Paul M. Di Gangi, and Molly Wasko. 2013. Online Health Communities: an Assessment of the Influence of Participation on Patient Empowerment Outcomes. *Information Technology & People*, 26(2):213–235.

Ian T Jolliffe. 1986. Principal Component Analysis and Factor Analysis. *Statistical Methods in Medical Research*, 1(1):115–128.

Yoon Kim. 2014. Convolutional Neural Networks for Sentence Classification. In *Proc. of the 2014 Conference on Empirical Methods on Natural Language Processing (EMNLP 2014)*, pages 1746–1751.

Diederik P Kingma and Jimmy Ba. 2015. Adam: A Method for Stochastic Optimization. In *Proc. of the 3rd International Conference for Learning Representations (ICLR2015)*.

Robert Leaman, Laura Wojtulewicz, Ryan Sullivan, Annie Skariah, Jian Yang, and Graciela Gonzalez. 2010. Towards internet-age pharmacovigilance: extracting adverse drug reactions from user posts to health-related social networks. In *Proc. of the 2010 Workshop on Biomedical Natural Language Processing (BioNLP 2010)*, pages 117–125.

Lada Leyens, Matthias Reumann, Nuria Malats, and Angela Brand. 2017. Use of Big Data for Drug Development and for Public and Personal Health and Care. *Genetic Epidemiology*, 41(1):51–60.

Yaliang Li, Nan Du, Chaochun Liu, Yusheng Xie, Wei Fan, Qi Li, Jing Gao, and Huan Sun. 2017. Reliable Medical Diagnosis from Crowdsourcing: Discover Trustworthy Answers from Non-Experts. In *Proc. of the 10th ACM International Conference on Web Search and Data Mining (WSDM 2017)*, pages 253–261.

Yaliang Li, Jing Gao, Chuishi Meng, Qi Li, Lu Su, Bo Zhao, Wei Fan, and Jiawei Han. 2016. A survey on truth discovery. *ACM SIGKDD Explorations Newsletter*, 17(2):1–16.

Minh-Thang Luong, Quoc V Le, Ilya Sutskever, Oriol Vinyals, and Lukasz Kaiser. 2016. Multi-task Sequence to Sequence Learning. In *Proc. of the 4th International Conference for Learning Representations (ICLR2016)*.

Minh-Thang Luong, Hieu Pham, and Christopher D Manning. 2015. Effective Approaches to Attention-based Neural Machine Translation. In *Proc. of the 2015 Conference on Empirical Methods in Natural Language Processing (EMNLP 2015)*, pages 1412–1421.

Xuezhe Ma and Eduard Hovy. 2016. End-to-end Sequence Labeling via Bi-directional LSTM-CNNs-CRF. In *Proc. of the 54th Annual Meeting of the Association for Computational Linguistics (ACL 2016)*, pages 1064–1074.

J. MacQueen. 1967. Some Methods for Classification and Analysis of Multivariate Observations. In *Proc. of the 5th Berkeley Symposium on Mathmatical Statistics and Probability*, pages 281–297.

F. Martin-Sanchez and K. Verspoor. 2014. Big data in medicine is driving big changes. *Yearbook of Medical Informatics*, 9(1):14–20.

Tomas Mikolov, Ilya Sutskever, Kai Chen, Greg S Corrado, and Jeff Dean. 2013. Distributed representations of words and phrases and their compositionality. In *In Proc. of the Advances in Neural Information Processing Systems (NIPS 2013)*, pages 3111–3119.

Subhabrata Mukherjee, Gerhard Weikum, and Cristian Danescu-Niculescu-Mizil. 2014. People on Drugs: Credibility of User Statements in Health Communities. In *Proc. of the 20th ACM SIGKDD International Conference on Knowledge Discovery and Data Mining (KDD'14)*, pages 65–74.

F. Pedregosa, G. Varoquaux, A. Gramfort, V. Michel, B. Thirion, O. Grisel, M. Blondel, P. Prettenhofer, R. Weiss, V. Dubourg, J. Vanderplas, A. Passos, D. Cournapeau, M. Brucher, M. Perrot, and E. Duchesnay. 2011. Scikit-learn: Machine Learning in Python. *Journal of Machine Learning Research (JMLR)*, 12(2011):2825–2830.

M. F. Porter. 1980. An Algorithm for Suffix Stripping. *Program*, 14(3):130–137.

Marcel Salathé. 2016. Digital Pharmacovigilance and Disease Surveillance: Combining Traditional and Big-Data Systems for Better Public Health. *The Journal of Infectious Diseases*, 214(suppl_4):S399–S403.

Hariprasad Sampathkumar, Xue-Wen Chen, and Bo Luo. 2014. Mining Adverse Drug Reactions from Online Healthcare Forums using Hidden Markov Model. *BMC Medical Informatics and Decision Making*, 14(1):91–108.

Connie St Louis and Gozde Zorlu. 2012. Can Twitter Predict Disease Outbreaks? *BMJ: British Medical Journal (Online)*, 344(e2353).

Diyi Yang, Mario Piergallini, Iris Howley, and Carolyn Rose. 2014. Forum Thread Recommendation for Massive Open Online Courses. In *Proc. of the 7th International Conference on Educational Data Mining (EDM 2014)*, pages 257–260.

Yang Yu, Xiaojun Wan, and Xinjie Zhou. 2016. User Embedding for Scholarly Microblog Recommendation. In *Proc. of the 54th Annual Meeting of the Association for Computational Linguistics (ACL 2016)*, pages 449–453.

Revisiting neural relation classification in clinical notes with external information

Simon Šuster, Madhumita Sushil and **Walter Daelemans**
Computational Linguistics & Psycholinguistics Research Center,
University of Antwerp, Belgium
`firstname.lastname@uantwerpen.be`

Abstract

Recently, segment convolutional neural networks have been proposed for end-to-end relation extraction in the clinical domain, achieving results comparable to or outperforming the approaches with heavy manual feature engineering. In this paper, we analyze the errors made by the neural classifier based on confusion matrices, and then investigate three simple extensions to overcome its limitations. We find that including ontological association between drugs and problems, and data-induced association between medical concepts does not reliably improve the performance, but that large gains are obtained by the incorporation of semantic classes to capture relation triggers.

1 Introduction

The extraction of relations from clinical notes is a fundamental clinical NLP task, crucial to support automated health care systems and to enable secondary use of clinical notes for research (Wang et al., 2017). In clinical relation extraction, the 2010 i2b2/VA challenge dataset has been by far the most widely used. Three categories of relations are annotated in discharge summaries: those between medical treatments and problems (**TrP**)[1], between tests and problems (**TeP**)[2] and between pairs of problems (**PP**)[3] (Uzuner et al., 2011). Many systems participating in the shared task used carefully crafted syntactic and semantic features, sometimes in combination with rules (Grouin et al., 2010; Rink et al., 2011). Recently, neural network approaches have been applied to this task, where they serve as feature extractors, with a softmax layer for classification. In this case, human-engineered or external features are usually not included. Two examples

on which we base our work are Sahu et al. (2016) and Luo et al. (2017), who achieve results similar to or better than the best-scoring approaches participating in the i2b2 challenge. They use convolutional neural networks, in which a convolutional unit processes a piece of text segment (SegCNN) in a sliding window manner, and then applies a max-pooling operation to provide the hidden features. In Sahu et al. (2016), the unit of text is simply a sentence, and the CNN constructs a global representation. On the other hand, Luo et al. (2017) argue that since multiple relations can occur in a single sentence, one representation is not sufficient. Therefore, they break the sentence into segments, so the encoding and the pooling operations apply to one segment at a time. Each sentence consists of five segments: tokens preceding the first concept c_1; c_1 itself; tokens between c_1 and c_2; concept c_2; and the tokens following it. This idea is related to *dynamic pooling*, known from previous event extraction work on the ACE 2005 dataset (Chen et al., 2015). More generally, the extension of neural networks with background information have been studied, inter alia, for text categorization, natural language inference, and entity and event extraction (K. M. et al., 2018; Yang and Mitchell, 2017).

In our work, we aim to boost the performance of a SegCNN classifier by first identifying its weakest points in a confusion matrix analysis, and then addressing these with external linguistic and domain features. We observe as much as a 6 point improvement in % F1 by a simple addition of semantic classes; a modest improvement with PMI features for PP relations; and no effect when adding association information between drugs and problems. We make the code, which is a modification of Luo et al. (2017)'s implementation of segment convolutional neural networks, available at `https://github.com/SimonSuster/seg_cnn`.

---

[1] Tr[A|C|I|NA|W]P: treatment {<u>a</u>dministered for, <u>c</u>auses, <u>i</u>mproves, <u>n</u>ot <u>a</u>dministered because of, <u>w</u>orsens} a problem.

[2] Te[C|R]P: test {<u>c</u>onducted for, <u>r</u>evealed} a problem.

[3] PIP: problem <u>i</u>ndicates a medical problem.

Proceedings of the 9th International Workshop on Health Text Mining and Information Analysis (LOUHI 2018), pages 22–28
Brussels, Belgium, October 31, 2018. ©2018 Association for Computational Linguistics

$g\backslash s$	None	TrAP	TrCP	TrIP	TrNAP	TrWP
None	**980**	86	15	3	7	0
TrAP	139	**423**	5	3	3	0
TrCP	48	27	**69**	0	0	0
TrIP	11	12	1	**16**	0	0
TrNAP	11	24	3	0	**7**	0
TrWP	11	16	5	4	1	**4**

(a) TrP relations.

$g\backslash s$	None	TeCP	TeRP
None	**575**	17	294
TeCP	41	**52**	36
TeRP	89	9	**612**

(b) TeP relations.

$g\backslash s$	None	PIP
None	**2544**	135
PIP	122	**343**

(c) PP relations.

Table 1: Confusion matrices for different relation categories of the base SegCNN. The first diagonal represents the number of correctly classified relations, and is shown in bold. The colored cells highlight low sensitivity (blue), hallucinating relations (green) and confusable relations (orange).

2 Analysis of limitations

To better understand the limitations of a SegCNN extractor, we analyze its results with confusion matrices. In Table 1, we use color coding to point to three types of challenges: a) **poor sensitivity** (blue cells), which are errors due to the classifier's conservativeness in proclaiming a relation; b) **"hallucinating" relations** (green), which are precision errors where relations should not be identified; and c) **confusable relations** (orange), where we see that the TrCP relation is often classified as TrAP (27/69 times), and similarly for the other treatment-problem relations. This is especially true for the less frequent relations TrNAP and TrWP, where the correct predictions are outnumbered by the cases wrongly predicted as TrAP. The TrAP predictions by the system account for the most mistakes. We can see from the number of a) and b) errors on the TrP relations—76% of all mistakes made by the model—that identifying the presence of a relation is more challenging than type classification of relations, cf. Rink et al. (2011). Similar observations can be made about the test-problem relations. For example, TeCP is frequently confused with TeRP (36), and the TeRP type is often hallucinated (294). Overall, determining the presence of a relation is more difficult than discriminating between TeCP and TeRP as 91% of mistakes are only due to detection. This number is higher here than for TrP relations since we are dealing with a smaller number of relation types, which causes less confusion in class assignment. For problem-problem relations, the matrix shows the model is somewhat more likely to predict the relation spuriously than to miss the relation.

In a qualitative analysis, we find that relations are often unrecognized in sentences with several (coordinated) concepts:

(1)   *she also had climbing bilirubin [. . .] and was started on zosyn$_{tr}$ for suspected biliary obstruction and ascending cholangitis$_{pr}$ coverage .* (gold: TrAP)

Relations can be hallucinated especially when two concepts may seem to be associated, but the knowledge of syntax or the domain tells us they are not:

(2)   *the patient was treated with tylenol orally$_{tr}$ as well as ativan for anxiety$_{pr}$ that she had about going home* (gold: none)

Here, medical knowledge of compatibility between drugs and problems could help, e.g. that tylenol is not indicated for anxiety, but ativan is. In the following example, the classifier wrongly predicts TeCP, although there is a clear cue for the correct relation TeRP in the predicate ("found"):

(3)   *during initial evaluation$_{te}$ for a coronary artery bypass graft , 80% to 90% of the right coronary artery stenosis$_{pr}$ was found*

3 Addressing the limitations

To deal with poor sensitivity and hallucinated relations mentioned above, we introduce simple domain knowledge in the form of association between a pair of concepts. We collect the association information either from an ontology (§ 3.1) or induce it from the data (§ 3.2). To increase the discriminatory power of the extractor to differentiate between the relations, we incorporate a semantic class feature which could give the classifier an explicit cue about the presence of a relation (§ 3.3).

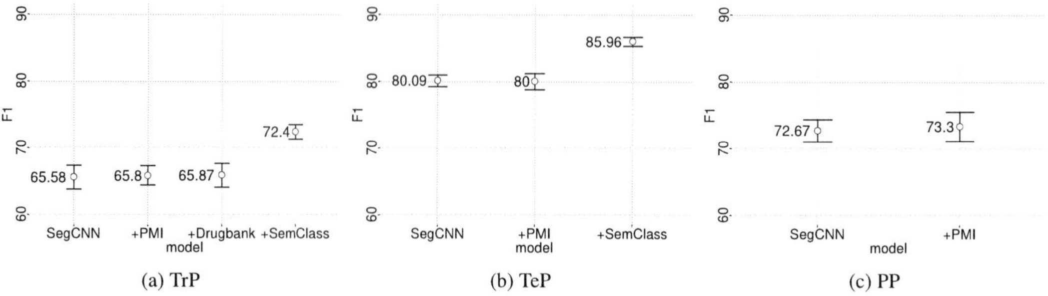

Figure 1: Results per relation category in percentage F1. The reported scores are averaged over 20 runs, and the 95% confidence intervals are shown.

3.1 Drug-problem association (Drugbank)

We use Drugbank (Wishart et al., 2017) to obtain a compatibility score between a drug treatment and a problem. We create a mapping from all drug names, synonyms and product names, to their indications. We also extract a mapping between drugs and their adverse reactions. In this way, we obtain 71,683 drug names, 3108 indications and 1163 adverse reactions. If there is a match for an observed treatment-problem pair in the drug-indication mapping, we simply assign a value of 1 (and scale it, as explained in Appendix) and -1 otherwise. Consider the example where we consider creating a relation between $neurontin_{tr}$ and $seizure\ history_{pr}$. In the indication for neurontin from Drugbank, seizures are mentioned as a possible medical problem, so this type of information could serve as background evidence for the classifier. The adverse drug effects represent a separate feature and are included in the same way. Due to low coverage of the drug-problem features for the treatment-problem concept pairs in the data (416 pairs are found, out of 7699), we also investigate a more general, data-induced approach, described next.

3.2 Concept-concept association (PMI)

We obtain association scores for concept pairs in all relation types by estimating a pointwise mutual information (PMI) model on a large corpus. We use the MIMIC-III corpus (Johnson et al., 2016) to compute the PMI for the co-occurring concepts. We first recognize clinical concepts in MIMIC-III using CLAMP (Soysal et al., 2017), and use Ucto (Van Gompel et al., 2012) for preprocessing. We then collect the counts, where two concepts are taken as co-occurring if they are mentioned in the same sentence, irrespective of the ordering. If found, we remove any determiners and pronouns. The concept type identified by CLAMP is appended to its mention. For a concept pair in our data, we perform a type-sensitive and order-insensitive lookup. In case of no match, we back-off by gradually removing up to two leftmost tokens. We find that the coverage lies between 68–82% depending on the relation category and the dataset split, and that the highest coverage applies for PP relations. The concept-concept association for relation extraction has been studied previously by Demner-Fushman et al. (2010) and de Bruijn et al. (2011), who used Medline® as the resource, whereas we achieved better results and coverage on the development set with MIMIC-III than Medline®.

3.3 Semantic classes

The semantic classes can provide cues about the relation types present in the sentence and facilitate distinguishing between different TrP and TeP relations[4]. We obtain the classes with WordNet (Miller, 1995) and an online thesaurus[5]. This was a manual process, in which we looked up the synonyms for all relation type names. For the seven TrP and TeP relation types, a hundred lexical triggers were obtained in total. For example, {*show, reveal, display...*} belong to the "revealing" class indicative of the TeRP relation. Lexical triggers are matched to their semantic classes if they occur in the non-concept sentence segments. We find that for TrP relations, matching only with the middle segment works best, but for TeP, the preceding, middle and succeeding segments work best.

---

[4]We do not use semantic classes for PP since there is only one relation type, PIP.

[5]en.oxforddictionaries.com

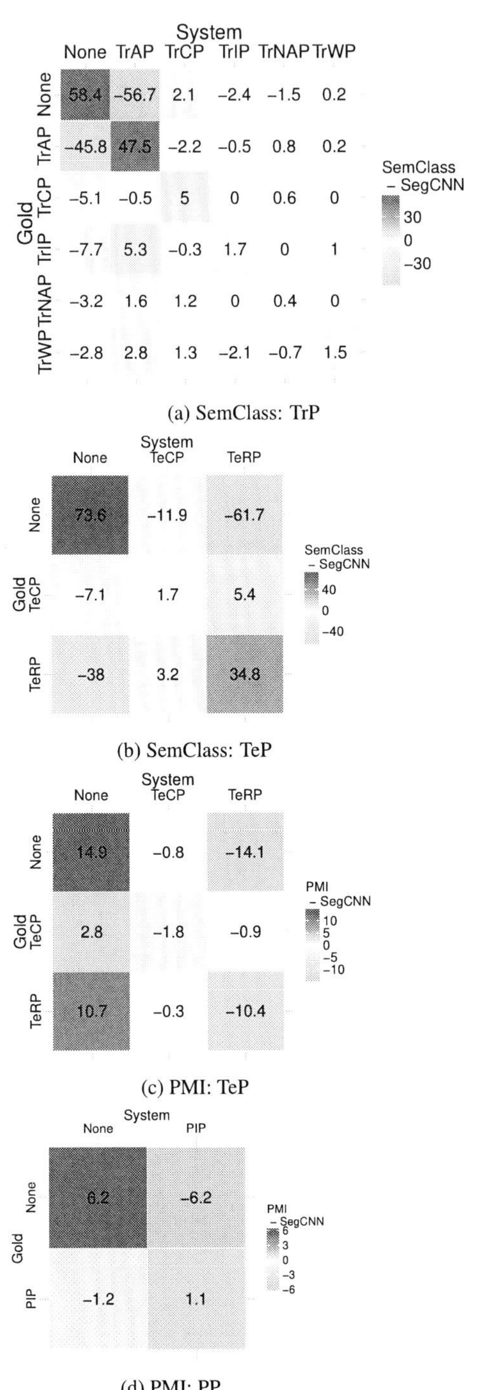

(a) SemClass: TrP

(b) SemClass: TeP

(c) PMI: TeP

(d) PMI: PP

Figure 2: A comparison of counts between a SegCNN and a model using either semantic classes or PMI features, for different relation categories.

4 Results

In our experiments, we use different data splits from those used in Luo et al. (2017) to increase the size of the training part and to also create a development set. The details, including the experimental setting, can be found in the Appendix. For the results using the vanilla SegCNN, we retrain the original models by Luo et al. (2017) and report their performance on our data splits. This gives us the results which are a few points lower on TrP and TeP relations, but also few points higher on PP relations, than the results reported in their paper.

We show the results in Figure 1, where % F1 is reported for different relation categories. Overall, the highest scores are achieved on TeP relations. The addition of semantic classes helps the most, with an improvement of almost 7 points over SegCNN for TrP, and 6 points for TeP relations. We think the advantage comes from the fact that the relation triggers are represented explicitly as the input to the classifier, whereas in the case of the base SegCNN, the classifier can only rely on a dense vectorial representation, which captures the trigger words more fuzzily. The contribution of the association features is less pronounced. The drug-problem (SemClass) and concept-concept (PMI) features have a small positive effect for TrP relations, with PMI working best (+0.5) for PP relations, where the coverage is the highest.

We now have a detailed look at the effect of the individual features. For this, we contrast the confusion matrix obtained from the base SegCNN with the confusion matrix of an extended model, where these matrices represent counts averaged over 20 runs. We obtain a new, contrasted matrix by subtracting the SegCNN matrix from that of the extended model, and display it as a heat map. An extension works well when the counts in the first diagonal are positive, and all the remaining counts are negative. In Figures 2a and 2b, we see an increase in correct classifications for semantic class features across all relation types, which speaks about the generality of this feature. The sensitivity for all relations has also increased (first column) as there are fewer true relations that remained unidentified. However, the counts of the less frequent relations (TrIP, TrNAP and TrWP) have shifted to incorrect relations (note the pale-red cells in the lower left corner of 2a). The improvements are the most obvious for the most frequent relations (TrAP and TeRP), with a clear increase in sensitivity, and a reduction in the number of unrelated (None) concepts classified as either TrAP or TeRP. The confusion matrix comparison for the problem-problem association (PMI) feature is shown in Figures 2c and 2d.[6] For TeP relations, we see that the addition of this feature type helps in reducing the number of hallucinated relations (first row), but at the expense of sensitivity—note that several relations are left unidentified (the counts in the TeCP and TeRP in the first column increased). A slight positive effect of PMI features can be seen for the PP relation, where the model becomes less prone to proclaim unrelated concepts as related (first row). Based on these figures, we can conclude that the PMI feature helps in deciding whether a pair of concepts should be linked with a relation or not, but does not have sufficient power to distinguish between different relations.

In conclusion, results show that the SegCNN model often misses, hallucinates or confuses relations, and that including semantic classes for relation triggers helps for different relation types.

Acknowledgments

We would like to thank the anonymous reviewers for their useful comments. This research was carried out within the Accumulate strategic basic research project, funded by the government agency Flanders Innovation & Entrepreneurship (VLAIO) [grant number 150056].

References

Berry de Bruijn, Colin Cherry, Svetlana Kiritchenko, Joel Martin, and Xiaodan Zhu. 2011. Machine-learned solutions for three stages of clinical information extraction: the state of the art at i2b2 2010. *Journal of the American Medical Informatics Association*, 18(5):557–562.

Yubo Chen, Liheng Xu, Kang Liu, Daojian Zeng, and Jun Zhao. 2015. Event extraction via dynamic multi-pooling convolutional neural networks. In *ACL-IJCNLP*.

Dina Demner-Fushman, Emilia Apostolova, R Islamaj Dogan, et al. 2010. NLM's system description for the fourth i2b2/VA challenge. In *Proceedings of the 2010 i2b2/VA Workshop on Challenges in Natural Language Processing for Clinical Data. Boston, MA, USA: i2b2*.

Cyril Grouin, Asma Ben Abacha, Delphine Bernhard, Bruno Cartoni, Louise Deleger, Brigitte Grau, Anne-Laure Ligozat, Anne-Lyse Minard, Sophie Rosset, and Pierre Zweigenbaum. 2010. CARAMBA: concept, assertion, and relation annotation using machine-learning based approaches. In *i2b2 Medication Extraction Challenge Workshop*.

Kai Hakala, Suwisa Kaewphan, Tapio Salakoski, and Filip Ginter. 2016. Syntactic analyses and named entity recognition for pubmed and pubmed central — up-to-the-minute. In *Proceedings of the 15th Workshop on Biomedical Natural Language Processing*, pages 102–107. Association for Computational Linguistics.

Alistair EW Johnson, Tom J Pollard, Lu Shen, Liwei H Lehman, Mengling Feng, Mohammad Ghassemi, Benjamin Moody, Peter Szolovits, Leo Anthony Celi, and Roger G Mark. 2016. MIMIC-III, a freely accessible critical care database. *Scientific data*, 3:.

Annervaz K. M., Somnath Basu Roy Chowdhury, and Ambedkar Dukkipati. 2018. Learning beyond datasets: Knowledge graph augmented neural networks for natural language processing. In *Proceedings of the 2018 Conference of the North American Chapter of the Association for Computational Linguistics: Human Language Technologies, Volume 1 (Long Papers)*, pages 313–322. Association for Computational Linguistics.

Yuan Luo, Yu Cheng, Özlem Uzuner, Peter Szolovits, and Justin Starren. 2017. Segment convolutional neural networks (Seg-CNNs) for classifying relations in clinical notes. *Journal of the American Medical Informatics Association*, 25(1):93–98.

Tomas Mikolov, Kai Chen, Greg Corrado, and Jeffrey Dean. 2013. Efficient estimation of word representations in vector space. In *ICLR Workshop Papers*.

George A. Miller. 1995. WordNet: A Lexical Database for English. *Communications of the ACM*, 38(11).

Bryan Rink, Sanda Harabagiu, and Kirk Roberts. 2011. Automatic extraction of relations between medical concepts in clinical texts. *Journal of the American Medical Informatics Association*, 18(5):594–600.

Sunil Sahu, Ashish Anand, Krishnadev Oruganty, and Mahanandeeshwar Gattu. 2016. Relation extraction from clinical texts using domain invariant convolutional neural network. In *Proceedings of the 15th Workshop on Biomedical Natural Language Processing*, pages 206–215. Association for Computational Linguistics.

Ergin Soysal, Jingqi Wang, Min Jiang, Yonghui Wu, Serguei Pakhomov, Hongfang Liu, and Hua Xu. 2017. CLAMP — a toolkit for efficiently building customized clinical natural language processing pipelines. *Journal of the American Medical Informatics Association*.

Özlem Uzuner, Brett R South, Shuying Shen, and Scott L DuVall. 2011. 2010 i2b2/va challenge on concepts, assertions, and relations in clinical text. *Journal of the American Medical Informatics Association*, 18(5):552–556.

---

[6]We include the remaining TrP matrix and the matrices for the Drugbank model in Appendix.

	# documents	# TrP	# TeP	# PP
train	272 (64%)	2220	2233	1413
dev.	68 (16%)	587	485	325
test	86 (20%)	846	839	465

Table 2: Data statistics.

Maarten Van Gompel, Ko van der Sloot, and Antal van den Bosch. 2012. Ucto: Unicode Tokeniser. Technical report, Tilburg Centre for Cognition and Communication, Tilburg University and Radboud Centre for Language Studies, Radboud University Nijmegen.

Yanshan Wang, Liwei Wang, Majid Rastegar-Mojarad, Sungrim Moon, Feichen Shen, Naveed Afzal, Sijia Liu, Yuqun Zeng, Saeed Mehrabi, Sunghwan Sohn, et al. 2017. Clinical information extraction applications: A literature review. *Journal of biomedical informatics*.

David S Wishart, Yannick D Feunang, An C Guo, Elvis J Lo, Ana Marcu, Jason R Grant, Tanvir Sajed, Daniel Johnson, Carin Li, Zinat Sayeeda, et al. 2017. Drugbank 5.0: a major update to the drugbank database for 2018. *Nucleic acids research*, 46(D1):D1074–D1082.

Bishan Yang and Tom Mitchell. 2017. Leveraging knowledge bases in LSTMs for improving machine reading. In *Proceedings of the 55th Annual Meeting of the Association for Computational Linguistics (Volume 1: Long Papers)*, volume 1, pages 1436–1446.

A   Supplemental Material

A.1   Experimental setup

Luo et al. (2017) used a part of the i2b2/VA dataset that is no longer available to those requesting the dataset. We therefore only have 170 documents for training and 256 documents for testing. Since our goal is to build an accurate relation extractor, we re-balance the dataset by increasing the size of the training corpus, reducing the size of the test set and creating a small development set. The sizes of the final splits are shown in Table 2. In all our experiments, we use the gold-standard concept annotations, and train one classifier per relation category.

Hyper-parameters   We use the same set of hyper-parameters as Luo et al. (2017), except that we turn off the drop out on the final layer of the classifier network, which harmed the performance in our experiments on the development set. We also noticed that scaling of the added features positively affected the results, so we tuned the scaling factor as well.

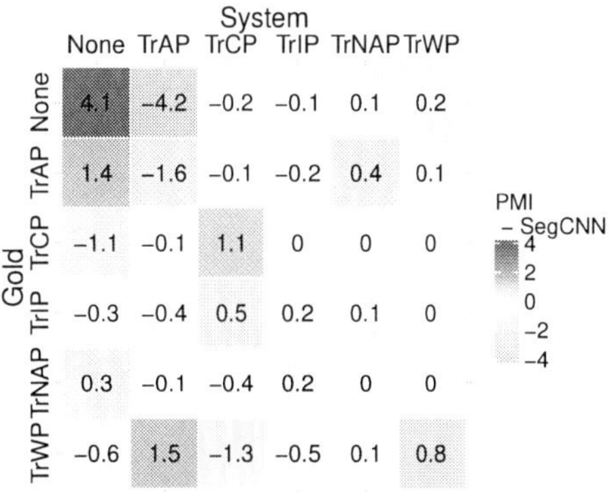

Figure 3: A comparison of counts between a base SegCNN and a model extended with PMI features, for different relation categories.

Embeddings   We trained the word embeddings on a combination of PubMed abstracts, open-access PMC articles (Hakala et al., 2016) and MIMIC-III intensive care notes (Johnson et al., 2016), all segmented and tokenized, totaling around 9 billion tokens. We induce the embeddings using word2vec's CBOW model (Mikolov et al., 2013) and the default parameters, except for dimensionality, which we set to 200 for TrP relations, 500 for TeP and 400 for PP relations, as in Luo et al. (2017).

A.2   Supplementary results

The additional results from a contrastive confusion matrix analysis are shown in Figure 3 for the PMI extension, and in Figure 4 for the model with the added drug-treatment association feature.

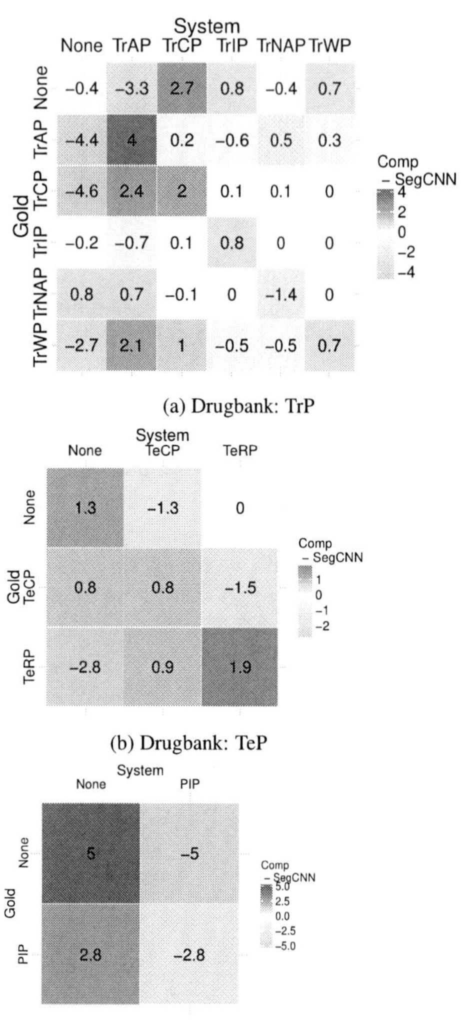

(a) Drugbank: TrP

(b) Drugbank: TeP

(c) Drugbank: PP

Figure 4: A comparison of counts between a base SegCNN and a model extended with Drugbank features, for different relation categories.

Supervised Machine Learning for Extractive Query Based Summarisation of Biomedical Data

Mandeep Kaur
Macquarie University
Sydney, Australia
mandeep-kaur.mandeep-kaur@
students.mq.edu.au

Diego Mollá
Macquarie University
Sydney, Australia
diego.molla-aliod@mq.edu.au

Abstract

The automation of text summarisation of biomedical publications is a pressing need due to the plethora of information available online. This paper explores the impact of several supervised machine learning approaches for extracting multi-document summaries for given queries. In particular, we compare classification and regression approaches for query-based extractive summarisation using data provided by the BioASQ Challenge. We tackled the problem of annotating sentences for training classification systems and show that a simple annotation approach outperforms regression-based summarisation.

1 Introduction

Text summarisation is a task of abridgement full text into a compact version while preserving the crucial information of the original text that is relevant to a user. The continuous increase of volume of digital text over the internet has reached such tremendous magnitude that a plethora of on-line text is available in regard to a topic. Consequently, manual skimming of text faces paramount obstacles like information overload (Das and Martins, 2007). This problem is particularly important for medical practitioners who need to analyse all the relevant information to diagnose and determine the best course of action for a particular patient. For example, there are cases in which medical practitioners fail to pursue answers to their queries (Ely et al., 2005). Moreover, manually searching the information is an extremely time-consuming and expensive task. Therefore, there is a strong motivation for building text processing systems that can automate some of the processes involved in this practice.

Our focus is to perform query-focused summarisation, also known as user-focused summarisation, of biomedical publications, by extracting and summarising the content relevant to the query given by the practitioner. The extraction system used in our experiments takes into account a specific query written as a question in plain English and tries to identify the information within a set of retrieved documents that is relevant to the query. Motivated by the success of machine learning in automatic text summarisation, we address the task of automatic query-based summarisation of biomedical text by using supervised machine learning techniques. We generate summaries by identifying the most significant content from the input text within the context of a query and generating a final summary by utilising that content.

In addition, this research also deals with a burning issue of availability of annotated corpora for supervised learning. In computational linguistics, labelled corpora are used to train machine learning algorithms and assess the performance of automatic summarisation methods. The employment of annotated corpora to the field of summarisation dates back to the late 1960s. These annotations typically consist of human-produced summaries, and it is not trivial to determine how to convert this information into the specific annotations required for supervised machine learning approaches to summarisation. Getting this data manually labelled is quite expensive and time-consuming; automatic annotation of data is still an active research question.

The contributions of this paper include:

1. A comparison of supervised approaches to query-focused extractive text summarisation of biomedical data.

2. A comparison of annotation approaches for classification-based approaches to query-focused extractive summarisation of biomedical data.

29

Proceedings of the 9th International Workshop on Health Text Mining and Information Analysis (LOUHI 2018), pages 29–37
Brussels, Belgium, October 31, 2018. ©2018 Association for Computational Linguistics

The rest of the paper is organised as follows. Section 2 provides a brief review of related work on the topic of extractive summarisation, with references to systems using biomedical text. Section 3 discusses the BioASQ Challenge, whose data are used in our experiments, and how it relates to query-focused summarisation. Section 4 presents the details of our summarisation framework. Section 5 discusses various annotation approaches used to train classifiers for supervised machine learning. Section 6 illustrates the results of our experiments for regression and classification approaches, along with an analysis of the output from our classification models using different annotation approaches. Finally, Section 7 concludes the paper with remarks on our future direction.

2 Related Work

Text summarisation has a rich background of research algorithms starting form late 1950's. The earliest works on text summarisation used sentence extraction as a primary component of a text summarisation system and the classic extractive approaches applied to extract summaries used statistical features for selecting significant content from the source text. The text features utilised by these approaches were based on bag-of-words (BOW) approaches. BOW models including word frequency and tf-idf are the most frequently used methods to discover the important content (Wu et al., 2008). More recently, word embeddings generated by deep learning approaches have also been shown to be useful for text summarisation (Malakasiotis et al., 2015; Mollá, 2017).

In recent years, the main focus of research in the summarisation field has been directed towards the application of machine learning to generate better summaries. Popular features such as multiple words, noun phrases, main verbs, named entities and word embeddings (Barzilay and Elhadad, 1997; Filatova and Hatzivassiloglou, 2004; Harabagiu and Lacatusu, 2002; Malakasiotis et al., 2015; Mollá, 2017) have been heavily exploited for summarisation.

In contrast to other domains, research on automatic text processing in the medical domain is still very much in its infancy. In the recent past, there has been steady ongoing research in biomedical text processing (Zweigenbaum et al., 2007). Factors such as the requirement of large volume of data, highly complex domain-specific terminologies and domain-specific format, and typology of questions (Athenikos and Han, 2010) makes it complex to process biomedical text. Most of the researchers working on summarisation for the medical domain apply the same kinds of techniques developed in other domains.

Three main supervised machine learning approaches have been used for text summarisation: classification, regression, and learning to rank.

Classification: The concept of summarising text by using supervised classification approaches was pioneered by Kupiec et al. (1995). They categorised each sentence as worthy of extraction or not by a classification function, using a Naïve Bayes classifier. In this classification approach the sentences are treated individually. At first, most machine learning systems assumed feature independence and relied on Naïve Bayes methods (Das and Martins, 2007). However, later models shifted the focus towards breaking the assumption that features are independent of each other (Lin and Chin-Yew, 1999).

Classification approaches have also been applied for summarisation of biomedical text. A work proposed by Chuang and Yang (2000) used decision trees and Naïve Bayes classifiers to train the summariser to extract important sentence segments based on feature vectors in order to generate a final summary. Other work by Sarkar (2009) and Sarkar et al. (2011) applied classification techniques to extractive summarisation by classifying individual sentences. The features used were term frequency, sentence similarity to document title, position of sentence, presence of domain specific cue phrases, presence of novel terms, and sentence length.

Regression: Regression approaches for summarisation try to fit the predicted score of a sentence as close as possible to the target score instead of labelling the sentences. An early work using regression for summarisation is by Ouyang et al. (2011) using support vector regression (SVR). Support vector regression (SVR) has also been used in conjunction with other techniques like integer linear programming (ILP) for generating summaries (Galanis et al., 2012) and has achieved state-of-the-art results in comparison to other competitive extractive summarisers.

A system named FastSum (Schilder and Kondadadi, 2008) used regression SVM for training their data set by using the least computationally expensive NLP techniques to generate the summary. The system used a set of clusters as input data and simple pre-processing was performed on the sentences. A comparison of this system with MEAD (Radev et al., 2000) showed that it is more than 4 times faster than MEAD.

Some of the recent work on biomedical data (Malakasiotis et al., 2015) used BioASQ data which is the data used in this paper. As in this paper, their work addressed the task of multi-document query focused summarisation. They used SVR to assign relevance scores to the sentences of the given relevant abstracts, and an alternative greedy strategy to select the most relevant sentences avoiding redundant ones.

A system by Mollá (2017) also experimented using BioASQ data in conjunction with SVR. The feature set used was based on Malakasiotis et al. (2015). In addition to SVR, Mollá (2017) used other regression approaches with deep learning architectures including convolutional neural networks (CNNs) and long-short term memory networks (LSTMs).

Learning to rank: Learning to rank transforms the task into a simple problem of ranking extracts from an original text. Given sentences with labelled importance scores, it is possible to get learning to rank models to train a model capable of assigning high rank to the most important sentences.

Ranking SVMs are the most commonly used approaches for learning to rank. When comparing SVMs and ranking SVMs to model the relevance of sentences to queries, Wang et al. (2007) show that ranking SVMs outperform standard SVMs on a small test collection. Learning to rank has also been applied to the summarisation of XML documents with a goal of learning how to best combine the sentence features such that within each document, summary sentences get higher scores than non-summary ones (Amini et al., 2007).

Another significant work done in this category uses ranking SVM to combine features for extractive query focused multi-document summarisation (Shen and Li, 2011). In order to do that, a graph-based method was proposed for training data generation by utilizing the sentence relationships and a cost sensitive loss was introduced to improve the robustness of learning. The method outperformed

Query: Name synonym of Acrokeratosis paraneoplastica.

Exact answer: Bazex syndrome.

Ideal answer: Acrokeratosis paraneoplastic (Bazex syndrome) is a rare, but distinctive paraneoplastic dermatosis characterized by erythematosquamous lesions located at the acral sites and is most commonly associated with carcinomas of the upper aerodigestive tract.

Figure 1: Example of query, exact answer, and ideal answer from the BioASQ 5b Phase B shared task.

the baseline strategies.

We are not aware of any work on biomedical summarisation using learning to rank techniques.

3 The BioASQ Challenge

We utilised a biomedical corpus provided by the BioASQ Challenge[1]. The BioASQ Challenge organises shared tasks on aspects related to biomedical semantic indexing and question answering (Tsatsaronis et al., 2015). One of the tasks, Task B, focuses on question answering, and Phase B of Task B asks participants to respond to a query by providing "exact answers" and "ideal answers". Whereas the exact answers are the usual output of a factoid question answering system, the ideal answers contain additional text such as explanations and justifications, and can be viewed as examples of query-focused summarisation. Figure 1 shows an example of a question, its exact answer, and its ideal answer, as provided in the training set of BioASQ 5b.

In the BioASQ data set each question contains, among other information, the text of the question, the question type, and a list of source documents. The list of documents has been extracted manually by annotators and are relevant to the query. They can be viewed as the ideal output of a text retrieval system and are used as the input data of our experiments. The training data set contains a total of 1306 questions.

4 Summarisation Model

Our system performs query-focused extractive summarisation of biomedical data, and our model

---

[1] http://bioasq.org/

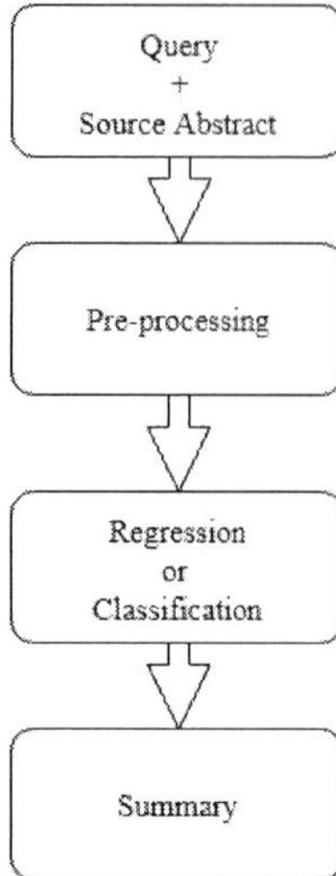

Figure 2: The overall summarisation model.

is trained with data from the BioASQ 5b Challenge. We follow a three-stage summarisation model for the generation of the summaries. In the first stage, the question and input text are pre-processed and transformed to an intermediate representation. In the second stage, each sentence in the input is assigned an importance score or label depending on the approach applied. Finally, in the final stage, the n most highly ranked sentences are selected to generate a summary. Figure 2 outlines the summarisation model.

4.1 Pre-processing

Pre-processing refers to the first stage of the model. First, the data are partitioned into training and testing using 10-fold cross validation. After partitioning the data, the sentences and questions are vectorised by computing the tf-idf of their words.

We also incorporate a technique that compares sentences with the associated queries. In particular, we compute the cosine similarity of each candidate (S_i) sentence with the associated query (Q_i), using the tf-idf vector representations for each:

$$Sim(S_i, Q_i) = \frac{S_i \cdot Q_i}{\|S_i\| \|Q_i\|}$$

4.2 Approaches for Extracting Summaries

Regression and classification-based techniques are used for generating a summary for a given query. To enable the comparison of all techniques, we have used a common feature set. In our case the feature set used is:

1. tf-idf vector of the candidate sentence.

2. Cosine similarity between the tf-idf vector of the question and the tf-idf vector of the candidate sentence.

Since the intent of this work is to compare the performance of regression and classification approaches, and not to obtain the best possible results, the feature set used is fairly simple and is commonly used on the most popular supervised approaches for query-based extractive summarisation.

For the regression approaches, each sentence of the training data is annotated with the F1 ROUGE-SU4 score of the sentence compared to the target summary. ROUGE-SU4 considers skip bigrams with a maximum distance of 4 words between the words of each skip bigram (Lin, 2004). This measure has also been found to correlate well with human judgements in extractive summarisation. Other systems have used ROUGE for annotating data and its application has been proved useful, e.g the system by Galanis et al. (2012); Peyrard and Eckle-Kohler (2016). We use Support Vector Regression (SVR), which has performed well in past regression approaches to summarisation.

For the classification approaches, we use the standard two-class labelling approach where class 1 indicates sentences that are selected for the final summary, and class 0 indicates sentences that are not selected. We use Support Vector Machine (SVM), which has performed well in many other classification problems.

5 Data Annotation for Classification

Supervised machine learning requires annotated training data to generate summaries. Often the summary annotations consist of sample reference summaries, but it is not straightforward to translate this information into the target labels 1 and 0 for classification. Although many researchers attempted to tackle this issue by manually selecting the summary-worthy sentences for their experiments (Ulrich et al., 2008), manual annotation consumes a considerable amount of time.

We have experimented with several approaches to determine when to assign a label 1 or 0 to an input sentence for the training procedure. As mentioned above, the inherent annotation of the sentences for the regression approach is based on their ROUGE score. Whereas it is straightforward to use ROUGE for the regression approach, we need to convert the ROUGE score into a binary value for classification. We experimented with various thresholds, and compared with a more complex approach based on Marcu (1999)'s work.

5.1 ROUGE Annotation with Thresholds

We tried two thresholds to define the labels for both the summary and the non-summary classes so that, if the ROUGE-SU4 score of the sentence is above the threshold, the sentence is labelled 1. Otherwise the sentence is labelled 0. This is done for every sentence associated with a query.

Firstly, we experimented by labelling the three highest SU4 scoring sentences as summary (i.e. label 1) for each query in the data. Secondly, we tried a threshold of 0.1. We labelled the sentence as 1 if its SU4 score is higher than 0.1 and labelled the rest as 0.

5.2 Marcu Annotation

In addition to the above-mentioned ROUGE annotation approaches, we also experimented with a greedy approach proposed by Marcu (1999) that we call the Marcu annotation. The motivation behind using this approach for our experiments is that it takes into account the similarity between the taget abstract and the entire set of sentences selected for the summary.

This method, instead of selecting sentences which are identical to those in the abstract, eliminates sentences which do not appear to be similar to ones in the abstract. The rationale of the methodology is that, if the similarity between the document and its target abstract does not decrease when a sentence is removed from the document, then we can say that the sentence is not relevant to the target abstract (Marcu, 1999). This elimination process continues while the similarity does not decrease as we remove sentences.

The original algorithm by Marcu (1999) is divided into two parts: generating the core extract and cleaning-up the core extract. The first part of the algorithm results in an extract through which important sentences in the text can be identified and annotated. In the second part, some cosmetic procedures are performed to the generated extract. In this second clean-up step Marcu employed some heuristics to further reduce the set of sentences.

We only implemented the first part of the algorithm. There are two reasons for not implementing the second part of the algorithm. Firstly, some of the heuristics require knowledge of the rhetorical structure of the source to be able to apply them. This information was not available, and could not be easily obtained. In addition, for some of the heuristics, the details were insufficient to know exactly how to implement them.

Algorithm 1 shows the algorithm for generating the extract. The input to the algorithm is a reference abstract and input text to summarise. In step 1, the input text is broken into sentences. Step 2 then pre-processes the abstract and text. Pre-processing involves tokenising all the information into words and then performing stemming and removing stop words. We use NLTK for steps 1 and 2 in contrast to Marcu (1999)'s approach, who used a shallow clause boundary and discourse marker identification (CB-DM-I) algorithm for this task. This algorithm is more complex and considers the information related to various textual units to perform pre-processing.

Initially, we assume the extract to be the whole text (step 3 in Algorithm 1).

Steps 4 and 5 can be explained as follows: If we delete from E a sentence S that is totally distinct from the abstract A, we obtain a new extract $E \backslash S$ whose similarity with A is higher than that of E. We therefore apply a greedy approach and repeatedly delete sentences from E so that at each step the resulting extract has maximum similarity with the abstract. We eventually reach a state where we can no longer delete sentences without decreasing the similarity of E with the abstract. The resulting E at this stage is considered the extract that we are looking for.

The similarity operator $Sim(X, Y)$ is the cosine similarity between the tf-idf of X and Y.

6 Evaluation and Results

We evaluated all our approaches automatically using the ROUGE evaluation tool (Lin, 2004). Our system-generated summaries are all evaluated by comparing them with the associated gold standard summaries which are the BioASQ ideal answers in our case.

Figure 3 shows the results of the regression and the following classification approaches described in Section 5:

Data:
Abstract (A): The reference summary.
Text (T): Input text to summarise.
Result:
Extract (E): A set of sentences from text
which has maximum similarity to abstract

1   $T_1, \cdots T_n$ = sentences from T
2   Stem and delete stop words from
     $A, T_1, \cdots T_n$
3   $E = T$
4   $S = \mathrm{argmax}_{S' \in E}\, Sim(E \backslash S', A)$
5   **while** $Sim(E, A) < Sim(E \backslash S, A)$ **do**
     $E = E \backslash S$
     $S = \mathrm{argmax}_{S' \in E}\, Sim(E \backslash S', A)$
end

Algorithm 1: Marcu's greedy approach for the generation of a core extract.

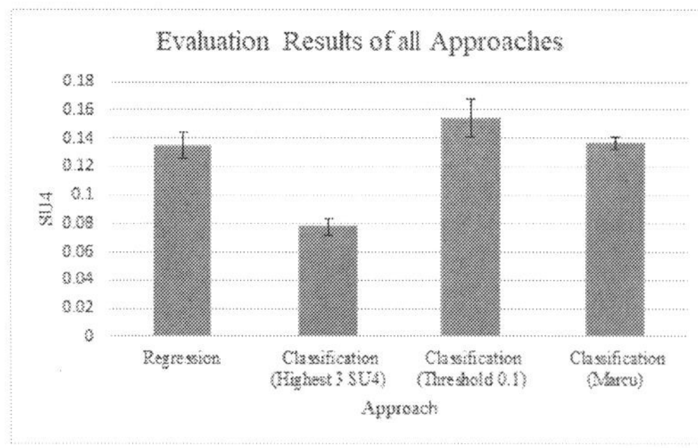

Figure 3: Comparison of the results of the regression and three classification approaches. The results show the mean of 10-fold cross-validation, and the error bars show the standard deviation.

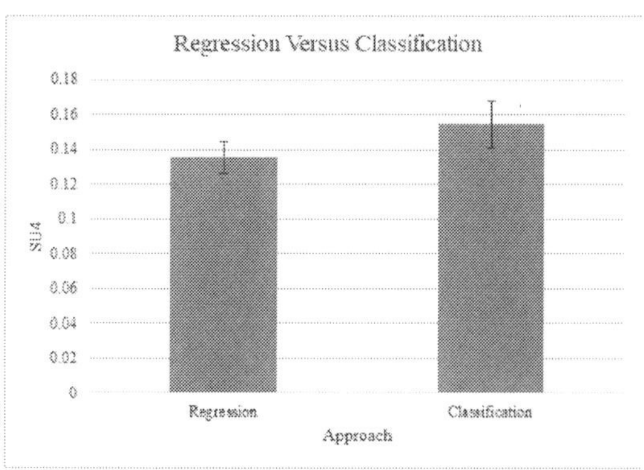

Figure 4: Comparison of classification (with 0.1 threshold) and regression according to their ROUGE-SU4 (error bars refer to standard deviation) in 10-fold cross-validation.

1. Label with 1 the three sentences with highest ROUGE score per question.

2. Label with 1 all sentences with ROUGE score higher than 0.1.

3. Label with 1 the sentences annotated according to Algorithm 1.

6.1 Regression Versus Classification

To produce comparable results, we kept preprocessing, feature extraction and number of sentences (3 sentences) in the final summary constant. The same data partition into training and testing was used in all cases.

Figure 4 compares the F1 ROUGE-SU4 scores of regression and the best classification approach. We can observe that the average SU4 score of the classification approach is higher than the score of the regression approach. The classification approach mentioned in Figure 4 is the one with threshold 0.1. The standard deviation for both approaches is indicated by the error bars.

To have a more precise evaluation, we analyse the variation of SU4 at each cross-validation fold for each approach to see whether classification is performing better than regression at every fold of cross-validation. In Figure fig:10folds, the variation of the SU4 score over each of the 10 folds for both techniques is shown and classification SU4 can be seen on the higher side for all the folds except for the last one.

6.2 Comparing Annotation Approaches

Figure 6 shows F1 ROUGE-SU4 scores of all of the classification approaches: (i) using three sen-

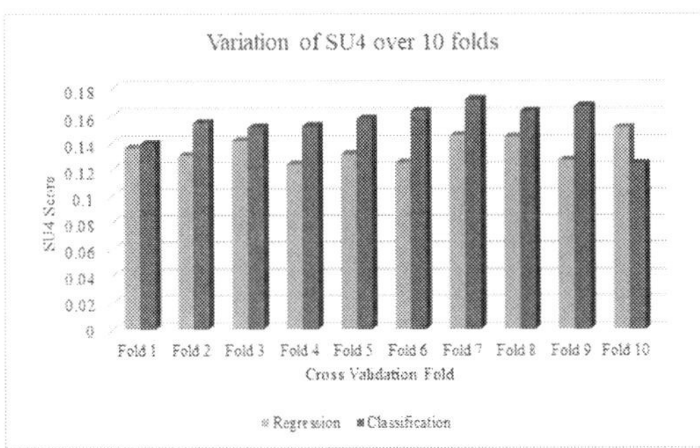

Figure 5: ROUGE SU4 variation over 10-Fold cross-validation for classification and regression.

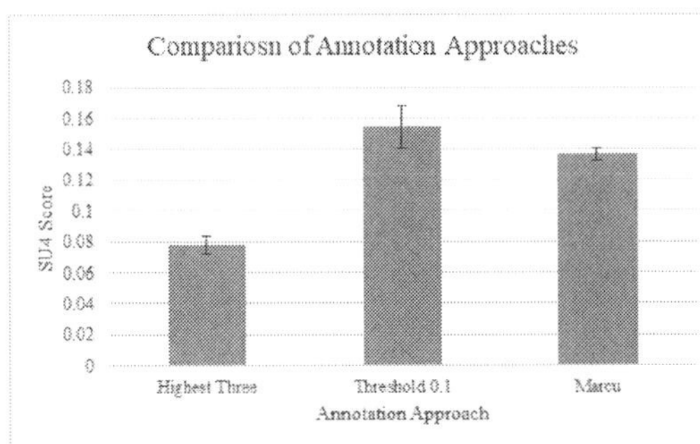

Figure 6: Comparison of various annotation approaches (error bars refers to standard deviation)

tences with highest SU4 as summary class, (ii) use threshold 0.1, and (iii) use the approach based on Marcu (1999)'s work.

The second approach (i.e. with threshold 0.1) can be seen as outperforming all the other approaches. In contrast, the first approach produces the lowest SU4 score among all the three. Whereas Marcu's approach is better than the approach with the highest three, it is outperformed by the approach with threshold of 0.1. The standard deviations for all of the approaches through 10-fold cross-validation are also presented as error bars in Figure 6.

6.3 Comparison with Ouyang et al.

A similar work performed by Ouyang et al. (2011) reported better results for regression than for classification in their experiments. They used different evaluation data, different features, and different approaches. In particular, they used data provided by the Document Understanding Conferences (DUC), and their annotation approach used two thresholds. They positively annotated the sentences with ROUGE score higher than 0.7 and

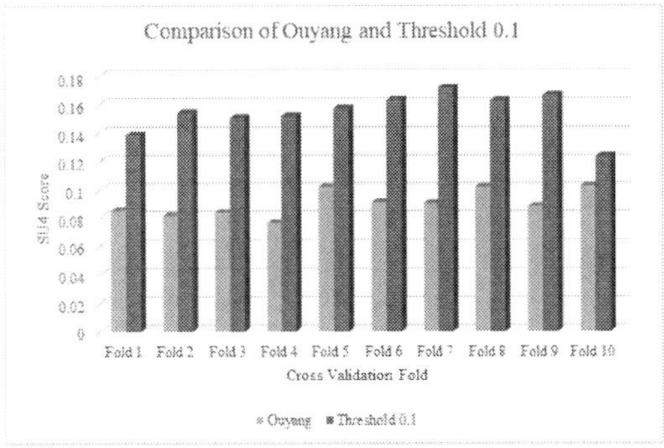

Figure 7: Classification with Ouyang et al. and our annotation approach (0.1 as threshold).

negatively annotated those with score lesser than 0.3. Apparently, sentences with score between 0.3 and 0.7 were not used in their experiments.

We therefore replicated their annotation approach using the BioASQ data set and our features so that we could compare with our other experiments and obtained an average ROUGE-S4 of 0.09. This is lower than the results of our regression approach.

Our results are therefore compatible with the results provided by Ouyang et al. (2011) when we use their annotation approach for classification. We can consequently conclude that classification can deliver better results than regression, but we need to be careful with the approach used to annotate the training sentences.

Figure 7 provides a comparison of our best performing annotation with Ouyang et al. (2011)'s approach by showing the variation of SU4 over all cross-validation folds.

The results reported in this paper are not directly comparable with the official results of the BioASQ runs for two reasons. First of all, the system implemented in this paper uses the entire source summaries as input. In contrast, systems participating in BioASQ can use additional information about what snippets from the source summaries are most relevant. Second, Mollá (2017) observed that the results of cross-validation with the training data gave much poorer results than the results evaluated using the BioASQ test set and the BioASQ evaluation scripts. Of the runs submitted by Mollá (2017), only the one labelled RNN used as input the full summaries without information about relevant snippets. The average of ROUGE-SU4 across all batches was 0.435. However, our (unpublished) experiments revealed that cross-validation of the same system achieved a ROUGE-SU4 of 0.144. This is lower than our best results using classification reported in this paper.

7 Conclusions

We have presented a comparison of two supervised machine learning techniques for extractive query focused summarisation. In addition, we have also explored the difficult phase of annotating data for classification approaches for summarisation, drawing a comparison among several annotation techniques.

To evaluate the model for both approaches, we have conducted an automatic evaluation and compared the performance of our system against human generated systems by using ROUGE. A series of experiments have been conducted by labelling data by different mechanisms for classification-based approaches.

Our experiments revealed that classification performs better than regression when a threshold of 0.1 SU4 is applied for annotating data.

When comparing the different annotation techniques for the classification approach, we observed a considerable difference between the results when using threshold 0.1, using the highest three SU4 scoring sentences, or using other annotation techniques such as the ones by Marcu (1999) and Ouyang et al. (2011).

As part of future work, we plan to conduct further experiments to determine the best annotation techniques for classification-based approaches. In particular, we plan to explore the impact of the second part of Marcu's greedy approach to see any improvement in results, along with utilising ROUGE as a similarity measure instead of cosine similarity to generate the extract. In addition, we will explore automatic approaches to determine the best thresholds. We empirically tried several thresholds and observed that 0.1 improved results but ideally this part would be done automatically.

We also plan to conduct an analysis of experiments by using learning to rank approaches. This type of learning algorithms may help improve performance.

References

Massih R. Amini, Anastasios Tombros, Nicolas Usunier, and Mounia Lalmas. 2007. Learning-based summarisation of XML documents. *Information Retrieval*, 10(3):233–255.

Sofia J Athenikos and Hyoil Han. 2010. Biomedical question answering: A survey. *Computer Methods and Programs in Biomedicine*, 99(1):1–24.

Regina Barzilay and Michael Elhadad. 1997. Using lexical chains for text summarization. In *Proceedings of the ACL/EACL 1997 Workshop on Intelligent Scalable Text Summarization*, pages 10–17.

Wesley T. Chuang and Jihoon Yang. 2000. Extracting sentence segments for text summarization: A machine learning approach. In *Proceedings of the 23rd annual international ACM SIGIR conference on Research and development in information retrieval - SIGIR '00*, pages 152–159, New York, New York, USA. ACM Press.

Dipanjan Das and André F.T. Martins. 2007. A survey on automatic text summarization. Technical report, CMU.

John W Ely, Jerome A Osheroff, M Lee Chambliss, Mark H Ebell, and Marcy E Rosenbaum. 2005. Answering physicians' clinical questions: Obstacles and potential solutions. *J Am Med Inform Assoc.*, 12(2):217–224.

Elena Filatova and Vasileios Hatzivassiloglou. 2004. Event-based extractive summarization. *Proceedings of ACL Workshop on Summarization*, pages 104–111.

Dimitrios Galanis, Gerasimos Lampouras, and Ion Androutsopoulos. 2012. Extractive multi-document summarization with integer linear programming and support vector regression. In *Proceedings of COLING 2012*.

Sanda M. Harabagiu and Finley Lacatusu. 2002. Generating single and multi-document summaries with GISTEXTER. *Proceedings of the Workshop on Automatic Summarization*, 11:30—-38.

Julian M. Kupiec, Jan Pedersen, and Francine Chen. 1995. A trainable document summarizer. In *Proceedings of the 18th annual international ACM SIGIR conference on Research and development in information retrieval - SIGIR '95*, pages 68–73, New York, New York, USA. ACM Press.

Chin-Yew Lin. 2004. ROUGE: A package for automatic evaluation of summaries. In *ACL Workshop on Tech Summarisation Branches Out*.

Chin-Yew Lin and Chin-Yew. 1999. Training a selection function for extraction. In *Proceedings of the eighth international conference on Information and knowledge management - CIKM '99*, pages 55–62, New York, New York, USA. ACM Press.

Prodromos Malakasiotis, Emmanouil Archontakis, and Ion Androutsopoulos. 2015. Biomedical question-focused multi-document summarization: ILSP and AUEB at BioASQ3. In *CLEF 2015 Working Notes*.

Daniel Marcu. 1999. The automatic construction of large-scale corpora for summarization research. In *Proceedings of the 22nd annual international ACM SIGIR conference on Research and development in information retrieval*, pages 137–144. ACM.

Diego Mollá. 2017. Macquarie University at BioASQ 5b — query-based summarisation techniques for selecting the ideal answers. In *Proc. BioNLP2017*.

You Ouyang, Wenjie Li, Sujian Li, and Qin Lu. 2011. Applying regression models to query-focused multi-document summarization. *Information Processing and Management*, 47(2):227–237.

Maxime Peyrard and Judith Eckle-Kohler. 2016. Optimizing an approximation of ROUGE — a problem-reduction approach to extractive multi-document summarization. In *Proceedings of the 54th Annual Meeting of the Association for Computational Linguistics (Volume 1: Long Papers)*, volume 1, pages 1825–1836, Stroudsburg, PA, USA. Association for Computational Linguistics.

Dragomir R. Radev, Hongyan Jing, and Malgorzata Budzikowska. 2000. Centroid-based summarization of multiple documents. In *NAACL-ANLP 2000 Workshop on Automatic summarization -*, volume 4, pages 21–30, Morristown, NJ, USA. Association for Computational Linguistics.

Kamal Sarkar. 2009. Using domain knowledge for text summarization in medical domain. *International Journal of Recent Trends in Engineering (ACEEE)*, 1(1):6.

Kamal Sarkar, Mita Nasipuri, and Suranjan Ghose. 2011. Using machine learning for medical document summarization. *International Journal of Database Theory and Application International Journal of Database Theory and Application*, 4(1):31–48.

Frank Schilder and Ravikumar Kondadadi. 2008. FastSum: fast and accurate query-based multi-document summarization. In *Proceedings of the 46th Annual Meeting of the Association for Computational Linguistics on Human Language Technologies: Short Papers*, pages 205–208. Association for Computational Linguistics.

Chao Shen and Tao Li. 2011. Learning to rank for query-focused multi-document summarization. In *2011 IEEE 11th International Conference on Data Mining*, pages 626–634. IEEE.

George Tsatsaronis, Georgios Balikas, Prodromos Malakasiotis, Ioannis Partalas, Matthias Zschunke, Michael R Alvers, Dirk Weissenborn, Anastasia Krithara, Sergios Petridis, Dimitris Polychronopoulos, Yannis Almirantis, John Pavlopoulos, Nicolas Baskiotis, Patrick Gallinari, Thierry Artiéres, Axel-Cyrille Ngonga Ngomo, Norman Heino, Eric Gaussier, Liliana Barrio-Alvers, Michael Schroeder, Ion Androutsopoulos, and Georgios Paliouras. 2015. An overview of the BIOASQ large-scale biomedical semantic indexing and question answering competition. *BMC Bioinformatics*, 16(1):138.

Jan Ulrich, Gabriel Murray, and Giuseppe Carenini. 2008. A publicly available annotated corpus for supervised email summarization. In *Proc. AAAI Email-2008 Workshop*, pages 77–82.

Changhu Wang, Feng Jing, Lei Zhang, and Hong-Jiang Zhang. 2007. Learning query-biased web page summarization. In *Proceedings of the sixteenth ACM conference on Conference on information and knowledge management - CIKM '07*, page 555, New York, New York, USA. ACM Press.

Ho Chung Wu, Robert Wing Pong Luk, Kam Fai Wong, and Kui Lam Kwok. 2008. Interpreting TF-IDF term weights as making relevance decisions. *ACM Transactions on Information Systems*, 26(3):1–37.

Pierre Zweigenbaum, Dina Demner-Fushman, Hong Yu, and Kevin B Cohen. 2007. Frontiers of biomedical text mining: current progress. *Briefings in Bioinformatics*, 8(5):358–375.

Comparing CNN and LSTM character-level embeddings in BiLSTM-CRF models for chemical and disease named entity recognition

Zenan Zhai, Dat Quoc Nguyen and **Karin Verspoor**
School of Computing and Information Systems
The University of Melbourne, Australia
`zenanz@student.unimelb.edu.au`, {`dqnguyen`, `karin.verspoor`}`@unimelb.edu.au`

Abstract

We compare the use of LSTM-based and CNN-based character-level word embeddings in BiLSTM-CRF models to approach chemical and disease named entity recognition (NER) tasks. Empirical results over the BioCreative V CDR corpus show that the use of either type of character-level word embeddings in conjunction with the BiLSTM-CRF models leads to comparable state-of-the-art performance. However, the models using CNN-based character-level word embeddings have a computational performance advantage, increasing training time over word-based models by 25% while the LSTM-based character-level word embeddings more than double the required training time.

1 Introduction

Bi-directional Long-Short Term Memory Conditional Random Field models (BiLSTM-CRF), in which a BiLSTM is coupled with a CRF layer to connect output tags, have been shown to achieve state-of-art performance in sequence tagging tasks including part of speech (POS) tagging, chunking, and NER (Huang et al., 2015). The combination of word embeddings and character-level word embeddings has been explored in this context, with Ma and Hovy (2016) using Convolutional Neural Networks (CNNs) to construct character-level word embeddings and Lample et al. (2016) applying LSTM networks. This work showed that the use of character-level word embeddings improves the performance of the models, by contributing the ability to recognize unseen words.

Biomedical Named Entity Recognition (BNER) is a vital initial step for information extraction tasks in the biomedical domain, including the Chemical-Disease Relationship (CDR) extraction task where both chemical and disease entities must be identified (Li et al., 2016). Character-level word embeddings could be particularly significant in this context, given that new entity names are frequently created, and may follow consistent patterns including productive morphology such as common prefixes (e.g., *di-*) or suffixes (e.g., *-ase*). Features that capture word-internal characteristics have been shown to be effective for BNER tasks in CRF models (Klinger et al., 2008).

Lyu et al. (2017) applied a BiLSTM-CRF model with LSTM-based character-level word embeddings to a gene and protein NER task, demonstrating state-of-art performance that outperformed traditional feature-based models. Luo et al. (2018) further improved on this result on a chemical NER task by adding an attention layer between the BiLSTM and CRF layers (Att-BiLSTM-CRF).

In an experiment by Reimers and Gurevych (2017b), optimal hyper-parameters for LSTM networks in sequence tagging tasks were explored, with the finding that incorporation of character-level word embeddings significantly improved performance on NER tasks on general datasets including CoNLL 2003 (Tjong Kim Sang and De Meulder, 2003). However, the choice of CNN-based (Ma and Hovy, 2016) or LSTM-based character-level word embeddings (Lample et al., 2016) did not affect the performance significantly. Since the CNN has fewer parameters to train than BiLSTM network, it is better in terms of training efficiency, and was recommended as the preferred approach.

In this paper, we implement and compare models with each type of word embedding to generate empirical results for the tasks of chemical and disease NER, using the BioCreative V CDR corpus (Li et al., 2016). These BNER categories are the most searched entities in the biomedical literature (Islamaj Dogan et al., 2009), and hence particularly important to study.

The results show that models with CNN-based character-level word embeddings achieve state-of-the-art results comparable to LSTM-based character-level word embeddings, while having

38

Proceedings of the 9th International Workshop on Health Text Mining and Information Analysis (LOUHI 2018), pages 38–43
Brussels, Belgium, October 31, 2018. ©2018 Association for Computational Linguistics

the advantage of reduced training complexity, demonstrating that the prior results also hold for the BNER task.

2 Experimental methodology

This section presents our empirical approach to comparing state-of-the-art neural network models for chemical and disease NER.

2.1 Dataset

In our experiments, we use the BioCreative V CDR corpus (Li et al., 2016). This corpus provides a set of 1000 manually-annotated abstracts (9193 sentences) for training and development, and another set of 500 manually-annotated abstracts (4840 sentences) for test. In particular, we used a pre-processed version of the CDR corpus from Luo et al. (2018),[1] which provides predicted POS-, chunking- and gazetteer-based tags:

- POS and chunking tags are predicted by the GENIA tagger (Tsuruoka et al., 2005).[2]

- Gazetteer tags are encoded in BIO tagging scheme based on matching to the external Jochem chemical dictionary (Hettne et al., 2009).

Following Luo et al. (2018), we randomly sample 10% from the set of 1000 abstracts for development, and use the remaining for training.

2.2 Models

We use the following BiLSTM-CRF-based sequence labeling models:

- Baseline BiLSTM model (Schuster and Paliwal, 1997; Hochreiter and Schmidhuber, 1997) which uses a softmax layer to predict NER labels of input words.

- BiLSTM-CRF (Huang et al., 2015) extends the BiLSTM model with a CRF layer which allows the model to use sentence-level tag information for sequence prediction.

- BiLSTM-CRF + CNN-char (Ma and Hovy, 2016) extends the BiLSTM-CRF model with character-level word embeddings. For each word, its character-level word embedding is derived by applying a CNN to the character sequence in the word.

Hyper-para.	Value
Optimizer	Nadam
Mini-batch size	32
Clipping	$\tau = 1$
Dropout	[0.25, 0.25]

Table 1: Fixed hyper-parameter configurations.

CNN-based		LSTM-based	
Hyper-para.	**Value**	**Hyper-para.**	**Value**
charEmbedSize	30	charEmbedSize	30
Window size	3	BiLSTM layer	1
# of filters	30	LSTM size	25
# of Params.	2,730	# of Params.	11,200

Table 2: Hyper-parameters for learning character-level word embedding. "charEmbedSize" and "# of Params." denote the vector size of character embeddings and the total number of parameters, respectively.

- BiLSTM-CRF + LSTM-char also extends the BiLSTM-CRF model with character-level word embeddings which are derived by applying a BiLSTM to the character sequence in each word (Lample et al., 2016).

Following Luo et al. (2018), we also consider the impact of extra features including syntactic features such as POS and chunking tags, and a chemical term feature based on matching to an external gazetteer. Figure 1 illustrates the general BiLSTM-CRF model architecture with character-level word embeddings and additional features, while Figure 2 illustrates CNN-based and LSTM-based architectures for learning the character-level word embeddings.

2.3 Implementation details

We used a well-known implementation of BiLSTM-CRF-based models from Reimers and Gurevych (2017b).[3] We used the training set to learn model parameters, the development set to select optimal hyper-parameters, and the test set to report final results. Here, we tune the model hyper-parameters using the performance across both NER categories ("Overall") on the development set.

We employed pre-trained 50-dimensional word vectors from Luo et al. (2018). These pre-trained vectors were derived by training the Word2Vec skip-gram model (Mikolov et al., 2013) on a large text collection of 2 million MEDLINE abstracts.

---

[1] https://github.com/lingluodlut/Att-ChemdNER

[2] http://www.nactem.ac.uk/GENIA/tagger

[3] https://github.com/UKPLab/emnlp2017-bilstm-cnn-crf

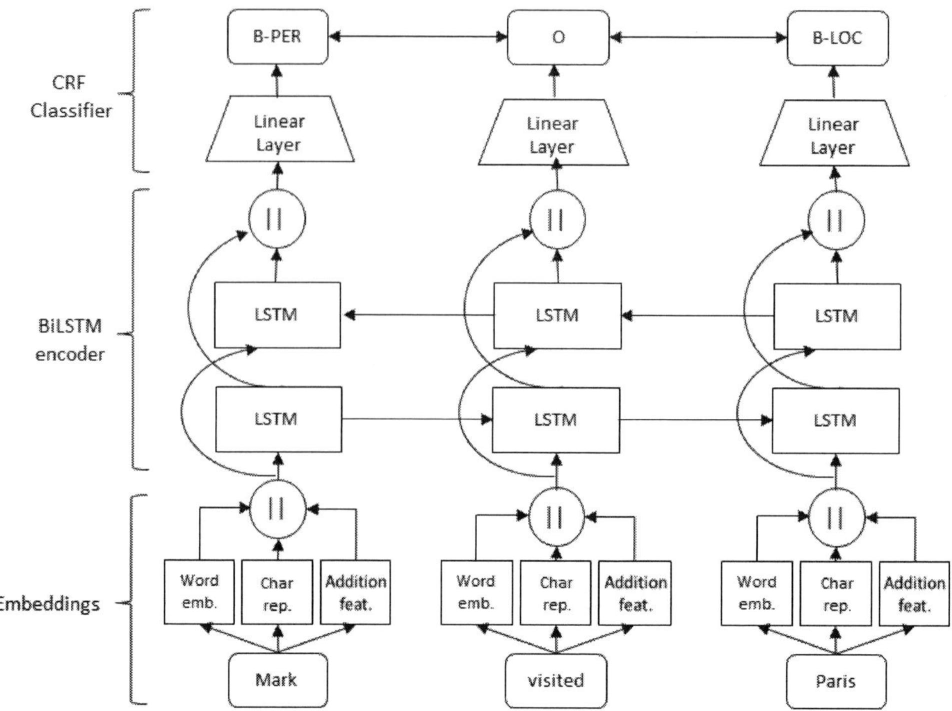

Figure 1: Architecture of BiLSTM-CRF models with character-level word representations and additional features. This figure is adapted from Reimers and Gurevych (2017a).

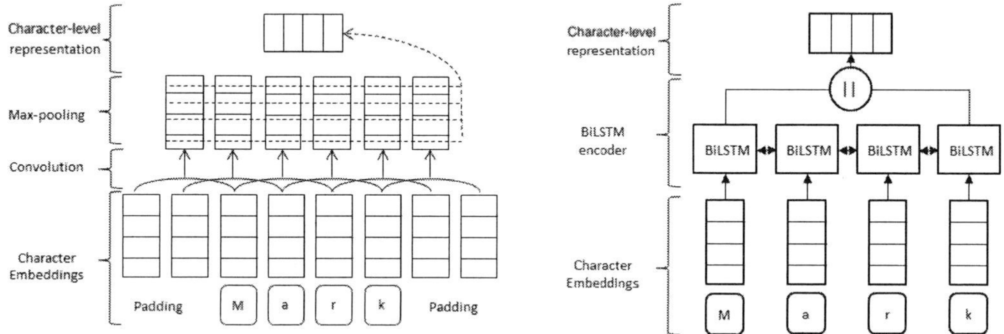

(CNN-based character-level word representation)   (LSTM-based character-level word representation)

Figure 2: Character-level word representations. This figure is also adapted from Reimers and Gurevych (2017a).

Reimers and Gurevych (2017b) showed that the BiLSTM-CRF model achieved best performance with 2 BiLSTM layers. Therefore, in our experiment, we only evaluated models up to 2 stacked BiLSTM layers. The size of LSTM hidden states in each layer was selected from [100, 150, 200, 250]. We achieved the highest F1 score on the development set when using 250-dimensional LSTM hidden states for all models.

By default, each of the additional features (POS, chunking tags, gazetteer match tag) was incorpo-rated into the model via a 10-dimensional embedding. Other hyper parameters were also fixed as in Reimers and Gurevych (2017b) during initialization. See tables 1 and 2 for more details.

In the training process, we used the score on development set to assess model improvement. Early stopping was applied if there was no improvement after 10 epochs. The threshold for a word that was not in the word embedding vocabulary to be added into the embedding was set to 5. The average training time for each epoch was also recorded.

Model	Chemical			Disease			Overall		
	P	R	F_1	P	R	F_1	P	R	F_1
BiLSTM	87.48	91.61	89.50	78.22	83.54	80.80	83.26	87.97	85.55
BiLSTM + CNN-char	90.65	90.70	90.67	79.34	82.66	80.97	85.44	87.07	86.25
BiLSTM + LSTM-char	90.47	91.64	**91.05**	79.43	83.97	**81.64**	85.37	88.18	**86.76**
BiLSTM-CRF	90.75	90.96	90.86	80.74	83.75	82.21	86.15	87.71	86.92
BiLSTM-CRF + CNN-char	91.64	92.24	**91.94**	81.42	84.67	**83.01**	86.95	88.83	**87.88**
BiLSTM-CRF + LSTM-char	92.08	91.79	**91.94**	81.48	84.22	82.83	87.20	88.38	87.79
BiLSTM-CRF$_{+Gazetteer}$	92.26	91.01	91.63	81.87	82.19	82.03	87.53	87.03	87.28
BiLSTM-CRF$_{+Gazetteer}$+ CNN-char	92.62	92.03	**92.32**	80.72	85.28	**82.94**	87.07	88.99	**88.02**
BiLSTM-CRF$_{+Gazetteer}$ + LSTM-char	92.11	92.33	92.22	82.13	83.66	82.89	87.57	88.42	87.99
Att-BiLSTM-CRF (LSTM-char) (Luo et al., 2018)	92.88	91.07	91.96	-	-	-	-	-	-
Att-BiLSTM-CRF$_{POS+Chunking+Gazetteer}$ (LSTM-char)	93.49	91.68	**92.57**	-	-	-	-	-	-
TaggerOne (Leaman and Lu, 2016) [♠]	94.2	88.8	**91.4**	85.2	80.2	**82.6**	-	-	-
tmChem (Leaman et al., 2015) [♠]	93.2	84.0	88.4	-	-	-	-	-	-
Dnorm (Leaman et al., 2013) [♠]	-	-	-	82.0	79.5	80.7	-	-	-

Table 3: Results (in %) on the test set. [♠] denotes results reported on a 950/50 training/development split rather than our 900/100 split. As indicated, Att-BiLSTM-CRF used LSTM-char word embeddings.

3 Main results

3.1 Baseline results

Table 3 presents our empirical results. The first three rows show the performance of baseline models without the CRF layer, the next three rows show the performance of BiLSTM-CRF models without additional features, and then the next three rows show the results for BiLSTM-CRF models with additional gazetteer features.

As the empirical results in Table 3 show, the model with CNN character-level embeddings (CNN-char) and the model with LSTM character-level embeddings (LSTM-char) achieved similar overall F1 scores (87.88% and 87.79%, respectively), outperforming BiLSTM-CRF by approximately 1% in absolute terms. In particular, on chemical NER, both BiLSTM-CRF-based models with character-level word embeddings obtained the same F1 score (91.94%), while on disease NER the model with CNN-char obtained slightly higher performance (83.01%) than the model with LSTM-char (82.83%). All models with the CRF layer outperformed their respective baseline BiLSTM models in F1 scores for all entity categories.

3.2 Effect of additional features

When incorporating additional POS and chunking features into three baseline BiLSTM-CRF-based models, we found that no performance improvement based on the baseline models was observed.

On chemical NER, the additional gazetteer feature improved the baseline BiLSTM-CRF by about 0.8% while it only improved the baselines BiLSTM-CRF + CNN-char and BiLSTM-CRF + LSTM-char by about 0.3%, thus clearly indicating that character-level word embeddings can capture unseen word information. Considering both NER categories together ("Overall"), the best performance was also obtained when the gazetteer feature was added, reaching overall F1 scores of 88.02% and 87.99%, respectively, for the two CNN-based and LSTM-based character-level embedding models.

3.3 Comparison with prior work

The performance comparison between our BiLSTM-CRF-based models and other machine learning approaches to the two studied NER tasks is also shown in Table 3. The pattern of chemical NER outperforming disease NER is consistent across all tools.

The Att-BiLSTM-CRF model (Luo et al., 2018) used a BiLSTM-CRF model with LSTM character-level word embedding and an additional attention layer. It achieved an F1 score of 91.96% on chemical NER without additional features. The positive effect of a gazetteer feature was also observed in their results; the model with syntactic and gazetteer features reached an F1 score of 92.57%. Note that the datasets used in this paper might not be exactly the same as ours due to random sampling.

The last three rows of Table 3 show the results presented in Leaman and Lu (2016), where 950 of the abstracts were used for training and 50 for development (cf. our 900/100 split). Dnorm (Leaman et al., 2013) is a model based on pairwise learning to rank on disease name normalization, which achieved F1 score of 80.7% on disease

NER. The tmChem (Leaman et al., 2015) is based on CRF; using numerous hand-crafted features it reached an F1 score of 88.4% on chemical entities. As a semi-Markov model with a richer set of features for NER tasks, TaggerOne (Leaman and Lu, 2016) achieved F1 score of 91.4% and 82.6% on chemical and disease entities, respectively.

Compared to previous non-deep-learning methods using CRFs, the BiLSTM-CRF models have significant advantage on F1 score of both chemical and disease entities, primarily due to improvement on recall.

3.4 Discussion

In our experiment on the effect of additional features, we found that syntactic features such as POS and chunking information did not have clear positive effect on the performance. In contrast, the match/partial match between words and entries in the chemical gazetteer is a good indicator for the presence of chemical entities. Since the Jochem dictionary contains only chemical entities, it is not surprising that the performance on diseases was not substantially impacted by adding the gazetteer feature, although some small variations in performance can be observed, likely due to changed influences from neighboring terms.

The empirical results shown that models using either CNN-char or LSTM-char achieve a similar overall F1 score on chemical and disease NER. The results are further comparable with other state-of-the-art models. This indicates that these character-level models have sufficient complexity to learn the generalizable morphological and lexical patterns in biomedical named entity terms.

On the other hand, as shown by the substantial differences in the number of parameters in Table 2, CNN (LeCun et al., 1989) has the advantage of reduced training complexity as compared to the LSTM models (Hochreiter and Schmidhuber, 1997) under similar experimental settings. In our experimental environment, the execution time of the model with LSTM-char increased 115% relative to the baseline BiLSTM-CRF model, while it only increased by 25% for with CNN-char, as detailed in Table 4. Therefore, consistent with prior results on general NER, we conclude that CNN-based embeddings are preferable to LSTM-based embeddings for BNER.

We analyzed the error cases of the CNN-char and LSTM-char models without additional features: 3326 and 3271 words were incorrectly predicted using CNN-char and LSTM-char, respectively, with 2138 mistakes in common. In errors which only was made by one of the two models, we found that CNN-char made more false positive predictions and fewer false negative predictions, while LSTM-char made approximately an even number of the two kinds of false predictions.

The relationship between the length of words and these errors was also explored. For words less than 20 characters in length, the distribution of errors is almost identical for the two models. However, for longer words, the model with LSTM-char tends to make more mistakes. This supports prior observations that LSTM can be difficult to apply to long sequences of input (Bradbury et al., 2017). In approximately 50% of error cases, the word length is short, less than 5 characters. Short biomedical named entities are usually abbreviations and tend to be out-of-vocabulary terms, and are therefore particularly difficult for the character-level word embedding models to capture (Habibi et al., 2017).

4 Conclusion

We compared the performance of BiLSTM-CRF models with CNN-based and LSTM-based character-level word embeddings for biomedical named entity recognition. We confirmed previously published results on chemical and disease NER that demonstrate that character-level embeddings are helpful. We further show empirically, generalizing prior results for general NER to the biomedical context, that there is little difference between the two approaches: both types of character-level word embeddings achieved identical F1 score on the chemical NER task, and similar performance on disease NER (with CNN-char showing a slight performance advantage). However, the CNN embeddings show a substantial advantage in reduced training complexity.

Model	Avg. Runtime per Epoch (seconds)	Δ
BiLSTM-CRF	106	0
+ CNN-char	**134**	**+26%**
+ LSTM-char	229	+115%

Table 4: Training time of best performing models (2 BiLSTM layers and 250 LSTM units), computed on a Intel Core i5 2.9 GHz PC.

Acknowledgments

This work was supported by the ARC Discovery Project DP150101550 and ARC Linkage Project LP160101469.

References

James Bradbury, Stephen Merity, Caiming Xiong, and Richard Socher. 2017. Quasi-Recurrent Neural Networks. In *Proceedings of the 2017 International Conference on Learning Representations*.

Maryam Habibi, Leon Weber, Mariana Neves, David Luis Wiegandt, and Ulf Leser. 2017. Deep learning with word embeddings improves biomedical named entity recognition. *Bioinformatics*, 33(14):i37–i48.

Kristina M. Hettne, Rob H. Stierum, Martijn J. Schuemie, Peter J. M. Hendriksen, Bob J. A. Schijvenaars, Erik M. van Mulligen, Jos Kleinjans, and Jan A. Kors. 2009. A dictionary to identify small molecules and drugs in free text. *Bioinformatics*, 25(22):2983–2991.

Sepp Hochreiter and Jürgen Schmidhuber. 1997. Long Short-Term Memory. *Neural Computation*, 9(8):1735–1780.

Zhiheng Huang, Wei Xu, and Kai Yu. 2015. Bidirectional LSTM-CRF models for sequence tagging. *arXiv preprint*, arXiv:1508.01991.

Rezarta Islamaj Dogan, G. Craig Murray, Aurélie Névéol, and Zhiyong Lu. 2009. Understanding PubMed user search behavior through log analysis. *Database*, 2009:bap018.

Roman Klinger, Corinna Kolářik, Juliane Fluck, Martin Hofmann-Apitius, and Christoph M. Friedrich. 2008. Detection of IUPAC and IUPAC-like chemical names. *Bioinformatics*, 24(13):i268–i276.

Guillaume Lample, Miguel Ballesteros, Sandeep Subramanian, Kazuya Kawakami, and Chris Dyer. 2016. Neural Architectures for Named Entity Recognition. In *Proceedings of the 2016 Conference of the North American Chapter of the Association for Computational Linguistics: Human Language Technologies*, pages 260–270.

Robert Leaman, Rezarta Islamaj Doan, and Zhiyong Lu. 2013. DNorm: disease name normalization with pairwise learning to rank. *Bioinformatics*, 29(22):2909–2917.

Robert Leaman and Zhiyong Lu. 2016. TaggerOne: joint named entity recognition and normalization with semi-Markov Models. *Bioinformatics*, 32(18):2839–2846.

Robert Leaman, Chih-Hsuan Wei, and Zhiyong Lu. 2015. tmChem: a high performance approach for chemical named entity recognition and normalization. *Journal of Cheminformatics*, 7(1):S3.

Y. LeCun, B. Boser, J. S. Denker, D. Henderson, R. E. Howard, W. Hubbard, and L. D. Jackel. 1989. Backpropagation Applied to Handwritten Zip Code Recognition. *Neural Comput.*, 1(4):541–551.

Jiao Li, Yueping Sun, Robin J. Johnson, Daniela Sciaky, Chih-Hsuan Wei, Robert Leaman, Allan Peter Davis, Carolyn J. Mattingly, Thomas C. Wiegers, and Zhiyong Lu. 2016. BioCreative V CDR task corpus: a resource for chemical disease relation extraction. *Database*, 2016:baw068.

Ling Luo, Zhihao Yang, Pei Yang, Yin Zhang, Lei Wang, Hongfei Lin, and Jian Wang. 2018. An attention-based BiLSTM-CRF approach to document-level chemical named entity recognition. *Bioinformatics*, 34(8):1381–1388.

Chen Lyu, Bo Chen, Yafeng Ren, and Donghong Ji. 2017. Long short-term memory RNN for biomedical named entity recognition. *BMC Bioinformatics*, 18(1):462.

Xuezhe Ma and Eduard Hovy. 2016. End-to-end Sequence Labeling via Bi-directional LSTM-CNNs-CRF. In *Proceedings of the 54th Annual Meeting of the Association for Computational Linguistics (Volume 1: Long Papers)*, pages 1064–1074.

Tomas Mikolov, Ilya Sutskever, Kai Chen, Greg S Corrado, and Jeff Dean. 2013. Distributed Representations of Words and Phrases and their Compositionality. In *Advances in Neural Information Processing Systems 26*, pages 3111–3119.

Nils Reimers and Iryna Gurevych. 2017a. Optimal Hyperparameters for Deep LSTM-Networks for Sequence Labeling Tasks. *arXiv preprint*, arXiv:1707.06799.

Nils Reimers and Iryna Gurevych. 2017b. Reporting Score Distributions Makes a Difference: Performance Study of LSTM-networks for Sequence Tagging. In *Proceedings of the 2017 Conference on Empirical Methods in Natural Language Processing*, pages 338–348.

M. Schuster and K.K. Paliwal. 1997. Bidirectional Recurrent Neural Networks. *IEEE Transactions on Signal Processing*, 45(11):2673–2681.

Erik F. Tjong Kim Sang and Fien De Meulder. 2003. Introduction to the CoNLL-2003 Shared Task: Language-Independent Named Entity Recognition. In *Proceedings of the Seventh Conference on Natural Language Learning at HLT-NAACL 2003*, pages 142–147.

Yoshimasa Tsuruoka, Yuka Tateishi, Jin-Dong Kim, Tomoko Ohta, John McNaught, Sophia Ananiadou, and Jun'ichi Tsujii. 2005. Developing a Robust Part-of-Speech Tagger for Biomedical Text. In *Advances in Informatics*, pages 382–392.

Deep learning for language understanding of mental health concepts derived from Cognitive Behavioural Therapy

Lina Rojas-Barahona[1], Bo-Hsiang Tseng[1], Yinpei Dai[1], Clare Mansfield[2]
Osman Ramadan[1], Stefan Ultes[1], Michael Crawford [3] and
Milica Gašić[1]
[1] University of Cambridge, [2] CM Insight, [3] Imperial College London
`mg436@cam.ac.uk`

Abstract

In recent years, we have seen deep learning and distributed representations of words and sentences make impact on a number of natural language processing tasks, such as similarity, entailment and sentiment analysis. Here we introduce a new task: understanding of mental health concepts derived from Cognitive Behavioural Therapy (CBT). We define a mental health ontology based on the CBT principles, annotate a large corpus where this phenomena is exhibited and perform understanding using deep learning and distributed representations. Our results show that the performance of deep learning models combined with word embeddings or sentence embeddings significantly outperform non-deep-learning models in this difficult task. This understanding module will be an essential component of a statistical dialogue system delivering therapy.

1 Introduction

Promotion of mental well-being is at the core of the action plan on mental health 2013–2020 of the World Health Organisation (WHO) (World Health Organization, 2013) and of the European Pact on Mental Health and Well-being of the European Union (EU high-level conference: Together for Mental Health and Well-being, 2008). The biggest potential breakthrough in fighting mental illness would lie in finding tools for early detection and preventive intervention (Insel and Scholnick, 2006). The WHO action plan stresses the importance of health policies and programmes that not only meet the need of people affected by mental disorders but also protect mental well-being. The emphasis is on early evidence-based non-pharmacological intervention, avoiding institutionalisation and medicalisation. What is particularly important for successful intervention is the frequency with which the therapy can be accessed (Hansen et al., 2002). This

gives automated systems a huge advantage over conventional therapies, as they can be used continuously with marginal extra cost. Health assistants that can deliver therapy, have gained great interest in recent years (Bickmore et al., 2005; Fitzpatrick et al., 2017). These systems however are largely based on hand-crafted rules. On the other hand, the main research effort in statistical approaches to conversational systems has focused on limited-domain information seeking dialogues (Schatzmann et al., 2006; Geist and Pietquin, 2011; Gasic and Young, 2014; Fatemi et al., 2016; Li et al., 2016; Williams et al., 2017).

In this paper we introduce a new task: understanding of mental health concepts derived from Cognitive Behavioural Therapy (CBT). We present an ontology that is formulated according to Cognitive Behavioural Therapy principles. We label a high quality mental health corpus, which exhibits targeted psychological phenomena. We use the whole unlabelled dataset to train distributed representations of words and sentences. We then investigate two approaches for classifying the user input according to the defined ontology. The first model involves a convolutional neural network (CNN) operating over distributed words representations. The second involves a gated recurrent network (GRU) operating over distributed representation of sentences. Our models perform significantly better than chance and for instances with a large number of data they reach the inter-annotator agreement. This understanding module will be an essential component of a statistical dialogue system delivering therapy.

The paper is organised as follows. In Section 2 we give a brief background of the statistical approach to dialogue modelling, focusing on dialogue ontology and natural language understanding. In Section 3 we review related work in the area of automated mental-health assistants. The sections that

44

Proceedings of the 9th International Workshop on Health Text Mining and Information Analysis (LOUHI 2018), pages 44–54
Brussels, Belgium, October 31, 2018. ©2018 Association for Computational Linguistics

follow represent the main contribution of this work: a CBT ontology in Section 4, a labelled dataset in Section 5, and models for language understanding in Section 6. We present the results in Section 7 and our conclusion in Section 8.

2 Background

A dialogue system can be treated as a trainable statistical model suitable for goal-oriented information seeking dialogues (Young, 2002). In these dialogues, the user has a clear goal that he or she is trying to achieve and this involves extracting particular information from a back-end database. A structured representation of the database, the *ontology* is a central element of a dialogue system. It defines the concepts that the dialogue system can understand and talk about. Another critical component is the natural language understanding unit, which takes textual user input and detects presence of the ontology concepts in the text.

2.1 Dialogue ontology

Statistical approaches to dialogue modelling have been applied to relatively simple domains. These systems interface databases of up to 1000 entities where each entity has up to 20 properties, i.e. *slots* (Cuayáhuitl, 2009). There has been a significant amount of work in spoken language understanding focused on exploiting large knowledge graphs in order to improve coverage (Tür et al., 2012; Heck et al., 2013). Despite these efforts, little work has been done on mental health ontologies for supporting cognitive behavioural therapy on dialogue systems. Available medical ontologies follow a symptom-treatment categorisation and are not suitable for dialogue or natural language understanding (Bluhm, 2017; Hofmann, 2014; Wang et al., 2018).

2.2 Natural language understanding

Within a dialogue system, a natural language understanding unit extracts meaning from user sentences. Both classification (Mairesse et al., 2009) and sequence-to-sequence (Yao et al., 2014; Mesnil et al., 2015) models have been applied to address this task.

Deep learning architectures that exploit distributed word-vector representations have been successfully applied to different tasks in natural language understanding, such as semantic role labelling, semantic parsing, spoken language understanding, sentiment analysis or dialogue belief tracking (Collobert et al., 2011; Kim, 2014; Kalchbrenner et al., 2014; Le and Mikolov, 2014a; Rojas Barahona et al., 2016; Mrkšić et al., 2017).

In this work we consider understanding of mental health concepts of as a classification task. To facilitate this process, we use distributed representations.

3 Related work

The aim of building an automated therapist has been around since the first time researchers attempted to build a dialogue system (Weizenbaum, 1966). Automated health advice systems built to date typically rely on expert coded rules and have limited conversational capabilities (Rojas-Barahona and Giorgino, 2009; Vardoulakis et al., 2012; Ring et al., 2013; Riccardi, 2014; DeVault et al., 2014; Ring et al., 2016). One particular system that we would like to highlight is an affectively aware virtual therapist (Ring et al., 2016). This system is based on Cognitive Behavioural Therapy and the system behaviour is scripted using VoiceXML. There is no language understanding: the agent simply asks questions and the user selects answers from a given list. The agent is however able to interpret hand gestures, posture shifts, and facial expressions. Another notable system (DeVault et al., 2014) has a multi-modal perception unit which captures and analyses user behaviour for both behavioural understanding and interaction. The measurements contribute to the indicator analysis of affect, gesture, emotion and engagement. Again, no statistical language understanding takes place and the behaviour of the system is scripted. The system does not provide therapy to the user but is rather a tool that can support healthcare decisions (by human healthcare professionals).

The Stanford Woebot chat-bot proposed by (Fitzpatrick et al., 2017) is designed for delivering CBT to young adults with depression and anxiety. It has been shown that the interaction with this chat-bot can significantly reduce the symptoms of depression when compared to a group of people directed to a read a CBT manual. The conversational agent appears to be effective in engaging the users. However, the understanding component of Woebot has not been fully described. The dialogue decisions are based on decision trees. At each node, the user is expected to choose one of several predefined responses. Limited language understanding was introduced at specific points in the tree to determine routing to subsequent conversational nodes. Still, one of the main deficiencies reported by the trial participants in (Fitzpatrick et al., 2017) was the inability to converse naturally. Here we address this problem by performing statistical natural language understanding.

4 CBT ontology

To define the ontology we draw from principles of Cognitive Behavioural Therapy (CBT). This is one of the best studied psychotherapeutic interventions, and the most widely used psychological treatment for mental disorders in Britain (Bhasi et al., 2013). There is evidence that CBT is more effective than other forms of psychotherapy (Tolin, 2010). Unlike other, longer-term, forms of therapy such as psychoanalysis, CBT can have a positive effect on the client within a few sessions. Also, due to it being highly structured, it is more easily amenable by computer interpretation. This is why we adopted CBT as the basis of our work.

Cognitive Behavioural Therapy is derived from Cognitive Therapy model theory (Beck, 1976; Beck et al., 1979), which postulates that our emotions and behaviour are influenced by the way we think and by how we make sense of the world. The idea is that, if the client changes the way he or she thinks about their problem, this will in turn change the way he or she feels, and behaves.

A major underlying principle of CBT is the idea of cognitive distortions, and the value in challenging them. In CBT, clients are helped to test their assumptions and views of the world in order to check if they fit with reality. When clients learn that their perceptions and interpretations are distorted or unhelpful they then work on correcting them. Within the realm of cognitive distortion, CBT identifies a number of specific self-defeating thought processes, or thinking errors. There is a core of around 10 to 15 thinking errors, with their exact titles having some fluidity. A strong component of CBT is teaching clients to be able to recognize and identify the thinking errors themselves, and ultimately discard the negative thought processes and 're-think' their problems.

We consider the main analytical step in this therapy: an adequate decoding of these 'thinking error' concepts, and the identification of the key emotion(s) and the situational context of a particular problem. Therefore, our ontology consists of *think-ing errors*, *emotions*, and *situations*.

4.1 Thinking errors

Notwithstanding slight variations in number and terminology, the list of *thinking errors* is fairly well standardised in the CBT literature. We present one such list in Table 1. However, it is important to note that there is a fair degree of overlap between different *thinking errors*, for example, between *Jumping to Negative Conclusions* and *Fortune Telling*, or between *Disqualifying the Positives* and *Mental Filtering*. In addition, within the data used – and as is likely to be the case in any data of spontaneous expressions of psychological upset – a single problem can exhibit several *thinking errors* simultaneously. Thus, the situation is much more challenging than in simple information-seeking dialogues, where ontologies are typically clearly defined and there is no or very little overlap between concepts.

4.2 Emotions

In addition to *thinking errors*, we define a set of *emotions*. We mainly focus on negative emotions, relevant to people in psychological distress. In CBT, emotions tend to be divided into positive and negative, or helpful/healthy and unhelpful/ unhealthy emotions (Branch and Willson, 2010). The set of emotions for this work evolved over time in the early days of annotation. Although we initally agreed to focus on 'unhealthy' emotions, as defined by CBT, there seemed also to be a place for the 'healthy' emotion *Grief/sadness*. Overall, the list of emotions used was drawn from a number of sources, including CBT literature, the annotators' own knowledge of what they work with in psychological therapy, and the common emotions that were seen emerging from the data early on in the process. Note that more than one emotion might be expressed within an individual problem – for example *Depression* and *Loneliness*. The list of *emotions* is given in Table 2.

4.3 Situations

While our main emphasis was on *thinking errors* and *emotions*, we also defined a small set of *situations*. The list of *situations* again evolved during the early days of annotation, with a longer original list being reduced down, for simplicity. Again, it is possible for more than one situation (for example *Work* and *Relationships*) to apply to a single problem. The considered *situations* are given in Table 3.

Thinking Error	Frequency	Exhibited by...
Black and white (or all or nothing) thinking	20.82%	Only seeing things in absolutes No shades of grey
Blaming	8.05%	Holding others responsible for your pain Not seeking to understand your own responsibility in situation
Catastrophising	11.87%	Magnifying a (sometimes minor) negative event into potential disaster
Comparing	3.27%	Making dissatisfied comparison of self versus others
Disqualifying the positive	6.15%	Dismissing/discounting positive aspects of a situation or experience
Emotional reasoning	13.31%	Assuming feelings represent fact.
Fortune telling	25.70%	Predicting how things will be, unduly negatively
Jumping to negative conclusions	44.16%	Anticipating something will turn out badly, with little evidence to support it
Labelling	10.51%	Using negative, sometimes highly coloured, language to describe self or other Ignoring complexity of people
Low frustration tolerance "I can't bear it"	16.03%	Assuming something is intolerable, rather than difficult to tolerate or a temporary discomfort
Inflexibility "should/need/ought"	8.08%	Having rigid beliefs about how things or people must or ought to be
Mental filtering	5.50%	Focusing on the negative Filtering out all positive aspects of a situation
Mind-reading	14.60%	Assuming others think negative things or have negative motives and intentions
Over-generalising	12.69%	Generalising negatively, using words like always, nobody, never, etc
Personalising	5.85%	Interpreting events as being related to you personally and overlooking other factors

Table 1: Taxonomy for *thinking errors* and how they are exhibited.

Emotion	Frequency	Exhibited by ...
Anger (/frustration)	14.76%	Feelings of frustration, annoyance, irritation, resentment, fury, outrage
Anxiety	63.12%	Any expression of fear, worry or anxiety
Depression	20.72%	Feeling down, hopeless, joyless, negative about self and/or life in general
Grief/sadness	5.70%	Feeling sad, upset, bereft in relation to a major loss
Guilt	3.37%	Feeling blameworthy for a wrongdoing or something not done
Hurt	19.88%	Feeling wounded and/or badly treated
Jealousy	3.12%	Antagonistic feeling towards another either wish to be like or to have what they have
Loneliness	7.41%	Feeling of alone-ness, isolation, friendlessness, not understood by anyone
Shame	5.68%	Feeling distress, humiliation, disgrace in relation to own behaviour or feelings

Table 2: Taxonomy for *emotions* and how they are exhibited.

Situation	Frequency
Bereavement	2.65%
Existential	21.93%
Health	10.61%
Relationships	67.58%
School/College	8.28%
Work	6.10%
Other	5.53%

Table 3: Taxonomy for *situations*.

5   The corpus

The corpus consists of 500K written posts that users anonymously posted on the Koko platform[1]. This platform is based on the peer-to-peer therapy proposed by (Morris et al., 2015). In this set-up, a user anonymously posts their problem (referred to as the *problem*) and is prompted to consider their most negative take on the problem (referred to as the *negative take*). Subsequently, peers post responses that attempt to offer a re-think and give a more positive angle on the problem. When first developed, this peer-to-peer framework was shown to be more efficacious than expressive writing, an intervention that is known to improve physical and

[1] https://itskoko.com/

Figure 1: An example of an annotated Koko post.

emotional well-being (Morris et al., 2015). Since then, the app developed by Koko has collected a very large number of posts and associated responses. Initially, any first-time Koko user would be given a short introductory tutorial in the art of 're-thinking'/'re-framing' problems (based on CBT principles), before being able to use the platform. This however changed over time, as the age of the users decreased, and a different tutorial, emphasizing empathy and optimism, was used (less CBT-based than the 're-thinking'). Most of the data annotated in this study was drawn from the earlier phase. Figure 1 gives an annotated post example.

5.1 Annotation

A subset of posts was annotated by two psychological therapists using a web annotation tool that we developed. The annotation tool allowed annotators to have a quick view of the posts, showing up to 50 posts per page, to navigate through posts, to check pending posts and to annotate them by adding or removing *thinking errors*, *emotions* and *situations*. All annotations were stored in a MySQL database.

Initially 1000 posts were analysed. These were used to define the ontology. Then 4035 posts were labelled with *thinking errors*, *emotions* and *situations*. It takes an experienced psychological therapist about one minute to annotate one post. Note that the same post can exhibit multiple *thinking errors*, *emotions* and *situations*, which makes the whole process more complex. We randomly selected 50 posts and calculated the inter-annotator agreement. The inter-annotator agreement was calculated using a contingency table for thinking error, emotion and situation, showing agreement and disagreement between the two annotators. Then, Cohen's kappa was calculated discounting the possibility that the agreement may happen by chance. The result is shown in Table 4. The main reason for the low agreement in *thinking errors* (61%) is

Concept	Thinking error	Situation	Emotion
Kappa	0.61 ± 0.09	0.92 ± 0.08	0.90 ± 0.07

Table 4: Cohen's kappa with a 95% confidence interval

due to the unbounded number of *thinking errors* per post. In other words, the annotators typically have three or four *thinking errors* in common but one of them might have detected one or two more. Still, the agreement is much higher than chance, so we think that while challenging, it is possible to build a classifier for this task. The distributions of labelled posts with multiple sub-categories for three super-categories are shown in Figure 2

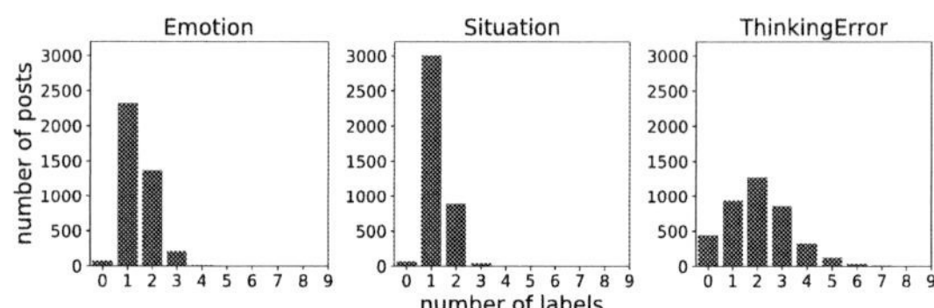

Figure 2: Distribution of posts for each category.

6 Deep learning model

6.1 Distributed representations

The task of decoding *thinking errors* and *emotions* is closely related to the task of sentiment analysis. In sentiment analysis we are concerned with positive or negative sentiment expressed in a sentence. Detecting thinking errors or emotions could be perceived as detecting different kinds of negative sentiment. Distributed representations of words, sentences and documents have gained success in sentiment detection and similarity tasks (Le and Mikolov, 2014a; Maas et al., 2011; Kiros et al., 2015). A key advantage of these representations is that they can be obtained in an unsupervised manner, thus allowing exploitation of large amounts of unlabelled data. This is precisely what we have in our set-up, where only a small portion of our posts is labelled.

We utilise GloVe (Pennington et al., 2014) word vectors, which have previously achieved competitive results in a similarity task. We train the word vectors on the whole dataset and then use a convolutional neural network (CNN) to extract features from posts where words are represented as vectors.

We also consider distributed representation of sentences. A particularly competitive model is the skip-thought model, which is obtained from an encoder-decoder model that tries to reconstruct the surrounding sentences of an encoded passage

(Kiros et al., 2015). On similarity tasks it outper-foms the simpler doc2vec model (Le and Mikolov, 2014a). An approach that represents vectors by weighted averages of word vectors and then mod-ifies them using PCA and SVD outperforms skip-thought vectors (Arora et al., 2017). This method however does not do well on a sentiment analysis task due to down-weighting of words like "not". As these often appear in our corpus, we chose skip-thought vectors for investigation here.

The skip-thought model allows a dense repre-sentation of the utterance. We train skip-thought vectors using the method described in (Kiros et al., 2015). The automatically generated post shown in Fig 3 demonstrates that skip-thought vectors can convey the sentiment well in accordance to context. We then train a gated recurrent unit (GRU) network using the skip-thoughts as input.

> i 'm so depressed . i 'm worthless . No one likes me
> i 'm try being nice but . No light at every point i 'm
> unpopular and i 'm a <NUM> year old potato . my
> most negative take is that i 'll never know how to be
> as socially as a quiet girl. i will stop talking to how
> fragile is and be any ways of normal people .

Figure 3: An example of a generated post using skip-thought vectors initialised with "I'm so depressed".

6.2 Convolutional neural network model

The convolutional neural network (CNN) model is preferred over a recurrent neural network (RNN) model, because the posts are generally too long for an RNN to maintain memory over words. The convolutional neural network (CNN) used in this work is inspired by (Kim, 2014) and operates over pre-trained GloVe embeddings of dimensionality d. As shown in Fig 4, the network has two inputs, one for the *problem* and the other for the *negative take*. These are represented as two tensors. A convolu-tional operation involves a filter $\mathbf{w} \in \mathcal{R}^{ld}$ which is applied to l words to produce the feature map. Then, a max-pooling operation is applied to pro-duce two vectors: $\mathbf{p}$ for *problem* and $\mathbf{n}$ for *negative take*. The reason for this is that the *negative take* is usually a summary of the post, carrying stronger sentiment (see Figure 1). We use a gating mecha-nism to combine $\mathbf{p}$ and $\mathbf{n}$ as follows:

$$\mathbf{g} = \sigma(\mathbf{W}_p\mathbf{p} + \mathbf{W}_n\mathbf{n} + \mathbf{b}) \tag{1}$$

$$\mathbf{h} = \mathbf{g} \odot \mathbf{p} + (1 - \mathbf{g}) \odot \mathbf{n} \tag{2}$$

Here, σ is the sigmoid function, $\mathbf{W}_p$, $\mathbf{W}_n$ and $\mathbf{W}$ are weight matrices, $\mathbf{b}$ is a bias term, $\mathbf{1}$ is a vector of ones, $\odot$ is the element-wise product, and $\mathbf{g}$ is

the output of the gating mechanism. The extracted feature $\mathbf{h}$ is then processed with a one-layer fully-connected neural network (FNN) to perform binary classification. The model is illustrated in Fig 4.

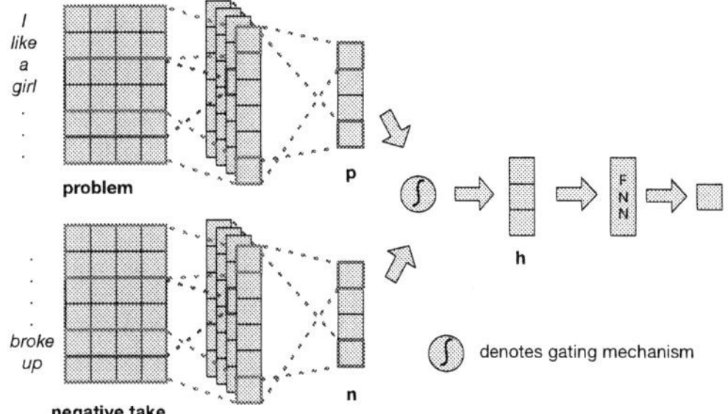

Figure 4: CNN with gating mechanism.

6.3 Gated recurrent unit model

We use the gated recurrent unit (GRU) model to process skip-thought sentence vectors, for two rea-sons. First, most posts contain less than 5 sentences, so a recurrent neural network is more suitable than a convolutional neural network. Second, since our corpus only comprises very limited labelled data, a GRU should perform better than a long short-term memory (LSTM) network as it has less parameters.

Denote each post as $P = \{s_1, s_2, ..., s_t, ...\}$, where s_t is the tth sentence in post P. First, we use an already trained GRU to extract skip-thought em-beddings $\mathbf{e_t}$ from the sentences s_t. Then, taking the sequence of sentence vectors $\{\mathbf{e_1}, \mathbf{e_2}, ..., \mathbf{e_t}, ...\}$ as input, another GRU is used as follows:

$$\mathbf{z_t} = \sigma(\mathbf{W}_z\mathbf{h}_{t-1} + \mathbf{U}_z\mathbf{e_t} + \mathbf{b_z}) \tag{3}$$

$$\mathbf{r_t} = \sigma(\mathbf{W}_r\mathbf{h}_{t-1} + \mathbf{U}_r\mathbf{e_t} + \mathbf{b_r}) \tag{4}$$

$$\tilde{\mathbf{h}_t} = \tanh(\mathbf{W}(\mathbf{r_t} \odot \mathbf{h}_{t-1}) + \mathbf{U}\mathbf{e_t} + \mathbf{b_h}) \tag{5}$$

$$\mathbf{h_t} = \mathbf{z_t} \odot \mathbf{h}_{t-1} + (1 - \mathbf{z_t}) \odot \tilde{\mathbf{h}_t} \tag{6}$$

$\mathbf{W}_z, \mathbf{U}_z, \mathbf{W}_r, \mathbf{U}_r, \mathbf{W}, \mathbf{U}$ are recurrent weight ma-trices, $\mathbf{b_z}, \mathbf{b_r}, \mathbf{b_h}$ are bias terms, $\odot$ is the element-wise dot product, and σ is the sigmoid function.

Finally, the last hidden state $\mathbf{h}_T$ is fed into a FNN with one hidden layer of the same size as input. The model is illustrated in Fig 5.

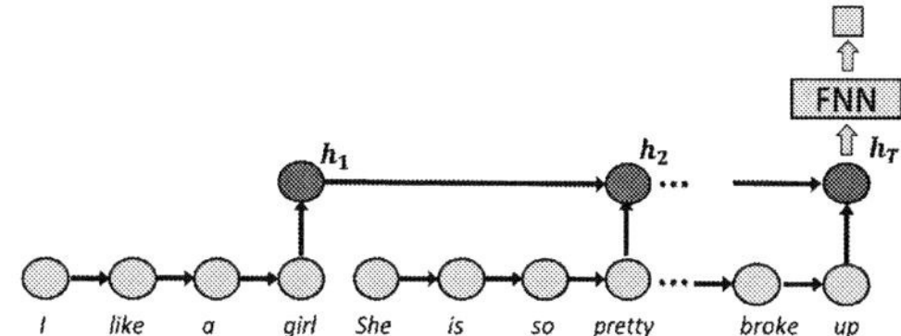

Figure 5: GRU with skip-thought vectors.

6.4 Training set-up

We first train 100 and 300 dimensions for both GloVe embeddings and skip-thought embeddings using the same mechanism as in (Pennington et al., 2014; Kiros et al., 2015). In some posts the length of sentences is very large, so we bound the length at 50 words. We do not treat the *problem* separately from the *negative take* as the GRU will anyway put more importance on the information that comes last. We split the labelled data in a 8 : 1 : 1 ratio for training, validation and testing in a 10-fold cross validation for both GRU and CNN training. A distinct network is trained for each concept, i. e. one for *thinking errors*, one for *emotions* and one for *situations*. The hidden size of the FNN is 150.

To tackle the data bias problem, we utilise oversampling. Different ratios (1:1, 1:3, 1:5, 1:7) of positive and negative samples are explored.

We used filter windows of 2, 3, and 4 with 50 feature maps for the CNN model. For the GRU model, the hidden size is set at 150, so that both models have comparable number of parameters. Mini-batches of size 24 are used and gradients are clipped with maximum norm 5. We initialise the learning rate as 0.001 with a decay rate of 0.986 every 10 steps. The non-recurrent weights with a truncated normal distribution $(0, 0.01)$, and the recurrent weights with orthogonal initialisation (Saxe et al., 2013). To overcome over-fitting, we employ dropout with rate 0.8 and $l2$-normalisation. Both models were trained with Adam algorithm and implemented in Tensorflow (Girija, 2016).

7 Results

7.1 Baselines

For rule-based models, we chose a chance classifier and a majority classifier, where all the posts are treated as positive examples for each class. In addition, we trained two non-deep-learning models, the logistic regression (LR) model and the Support Vector Machine (SVM). Both of them take the bag-of-words feature as input and implemented in sklearn (Pedregosa et al., 2011). For completeness, we also trained 100 and 300 dimensions PV-DM document embeddings (Le and Mikolov, 2014b) as the distributed representations of the posts using the *gensim* toolkit (Řehůřek and Sojka, 2010), and employ FNNs to do the classification, the hidden size is set as 800 to ensure parameters of all deep learning models comparable. All the baseline models are trained with the same set-up as described in section 6.4.

7.2 Analysis

Table 5 gives the average F1 scores and the average F1 scores weighted with the frequency of CBT labels for all models under the oversampling ratio 1:1. It shows that GloVe word vectors with CNN achieves the best performance both in 100 and 300 dimensions.

Model	AVG. F1	Weighted AVG F1
Chance	0.203±0.008	0.337±0.008
Majority	0.24±0.000	0.432±0.000
LR-BOW	0.330±0.011	0.479±0.008
SVM-BOW	0.403±0.000	0.536±0.000
FNN-DocVec-100d	0.339±0.006	0.502±0.005
FNN-DocVec-300d	0.349±0.007	0.508±0.005
GRU-SkipThought-100d	0.401±0.005	0.558±0.004
GRU-SkipThought-300d	0.423±0.005	0.570±0.004
CNN-GloVe-100d	**0.443 ± 0.007**	0.576±0.005
CNN-GloVe-300d	0.442 ± 0.007	**0.578 ± 0.006**

Table 5: F1 scores for all models with 1:1 oversampling

Table 6 shows the F1-measure of the compared models that detect *thinking errors*, *emotions* and *situations* under the 1 : 1 oversampling ratio. We only include the results of the best performing models, SVMs, CNNs and GRUs, due to limited space. The results show that both models outperform SVM-BOW in larger embedding dimensions. Although SVM-BOW is comparable to 100 dimensional GRU-Skip-thought in terms on average F1, in all other cases CNN-GloVe and GRU-Skip-thought overshadow SVM-BOW. We also find that CNN-GloVe on average works better than GRU-Skip-thought, which is expected as the space of words is smaller in comparison to the space of sentences so the word vectors can be more accurately trained. While the CNN operating on 100 dimensional word vectors is comparable to the CNN operating on 300 dimensional word vectors, the GRU-Skip-thought tends to be worse on 100 dimensional skip-thoughts, suggesting that sentence vectors generally need to be of a higher dimension to represent the meaning more accurately than word vectors.

Table 7 shows a more detailed analysis of the 300 dimensional CNN-GloVe performance, where both precision and recall are presented, indicating that oversampling mechanism can help overcome the data bias problem. To illustrate the capabilities of this model, we give samples of two posts and their predicted and true labels in Figure 6, which shows that our model discerns the classes reasonably well even in some difficult cases.

	Freq. Num.	SVM-BOW	100d		300d	
			CNN-Glove	GRU-Skip-thought	CNN-Glove	GRU-Skip-thought
Emotion						
Anxiety	2547	0.798±0.000	0.805±0.003	0.805±0.002	0.805±0.006	**0.816 ± 0.002**
Depression	836	0.564±0.000	0.605±0.003	0.568±0.001	**0.611 ± 0.008**	0.578±0.005
Hurt	802	0.448±0.000	0.505±0.007	0.483±0.003	**0.506 ± 0.005**	0.496±0.006
Anger	595	0.375±0.001	0.389±0.009	0.384±0.007	0.383±0.004	**0.425 ± 0.007**
Loneliness	299	0.558±0.000	0.495±0.008	0.445±0.007	**0.549 ± 0.009**	0.457±0.005
Grief	230	0.433±0.005	0.462±0.010	0.373±0.008	**0.462 ± 0.008**	0.382±0.005
Shame	229	0.220±0.000	**0.304 ± 0.011**	0.243±0.004	0.277±0.007	0.254±0.004
Jealousy	126	0.217±0.000	**0.228 ± 0.012**	0.159±0.004	0.216±0.005	0.216±0.009
Guilt	136	0.252±0.000	**0.295 ± 0.012**	0.186±0.007	0.279±0.014	0.225±0.008
AVG. F1 score for **Emotion**		0.429±0.001	**0.454 ± 0.008**	0.405±0.005	**0.454 ± 0.007**	0.428±0.006
Situation						
Relationships	2727	0.861±0.000	0.871±0.003	0.886±0.001	0.878±0.006	**0.889 ± 0.003**
Existential	885	0.556±0.000	0.591±0.002	**0.600 ± 0.005**	0.594±0.007	0.599±0.006
Health	428	0.476±0.000	**0.589 ± 0.003**	0.555±0.005	0.585±0.008	0.587±0.006
School_College	334	0.633±0.000	0.670±0.004	0.641±0.003	0.673±0.009	**0.680 ± 0.002**
Other	223	0.196±0.001	0.255±0.011	0.241±0.008	0.256±0.005	**0.281 ± 0.006**
Work	246	0.651±0.000	**0.663 ± 0.004**	0.572±0.006	0.661±0.011	0.639±0.006
Bereavement	107	0.602±0.000	0.637±0.021	0.402±0.024	**0.639 ± 0.021**	0.493±0.011
AVG. F1 score for **Situation**		0.568±0.000	0.611±0.007	0.557±0.007	**0.612 ± 0.010**	0.595±0.006
Thinking Error						
Jumping_to_negative_conclusions	1782	0.590±0.000	0.696±0.004	0.685±0.004	**0.703 ± 0.005**	0.687±0.002
Fortune_telling	1037	0.458±0.000	**0.595 ± 0.002**	0.558±0.004	0.585±0.006	0.564±0.005
Black_and_white	840	0.395±0.000	0.431±0.002	0.437±0.004	0.432±0.003	**0.441 ± 0.003**
Low_frustration_tolerance	647	0.318±0.000	0.322±0.007	0.330±0.003	0.313±0.005	**0.336 ± 0.001**
Catastrophising	479	0.352±0.000	**0.375 ± 0.002**	0.358±0.005	0.371±0.004	0.364±0.003
Mind-reading	589	0.360±0.000	0.404±0.005	0.353±0.011	**0.419 ± 0.006**	0.356±0.007
Labelling	424	0.399±0.001	0.453±0.007	0.335±0.004	**0.462 ± 0.004**	0.373±0.002
Emotional_reasoning	537	0.290±0.000	**0.319 ± 0.007**	0.285±0.005	0.306±0.006	0.293±0.008
Over-generalising	512	0.405±0.001	0.405±0.006	0.375±0.004	**0.418 ± 0.008**	0.389±0.004
Inflexibility	326	0.202±0.001	0.203±0.014	0.188±0.007	**0.218 ± 0.003**	0.208±0.005
Blaming	325	0.209±0.001	**0.304 ± 0.007**	0.264±0.002	0.277±0.003	0.274±0.004
Disqualifying_the_positive	248	0.146±0.000	0.194±0.007	0.176±0.005	0.187±0.003	**0.195 ± 0.005**
Mental_filtering	222	0.088±0.000	0.142±0.007	0.150±0.001	0.141±0.002	**0.155 ± 0.003**
Personalising	236	0.212±0.000	0.230±0.012	0.220±0.005	**0.236 ± 0.004**	0.221±0.005
Comparing	132	0.242±0.000	**0.289 ± 0.014**	0.177±0.008	0.255±0.009	0.227±0.007
AVG. F1 score for **Thinking Error**		0.311±0.000	**0.358 ± 0.007**	0.326±0.005	0.355±0.0050	0.339±0.004
AVG. F1 score		0.403±0.000	0.443±0.007	0.401±0.005	0.442±0.007	0.423±0.005
AVG. F1 score weighted with Freq.		0.536±0.000	0.576±0.005	0.558±0.004	**0.578±0.006**	0.570±0.004

Table 6: F1 score of the models trained with embeddings with dimensionality of 300 and 100 respectively.

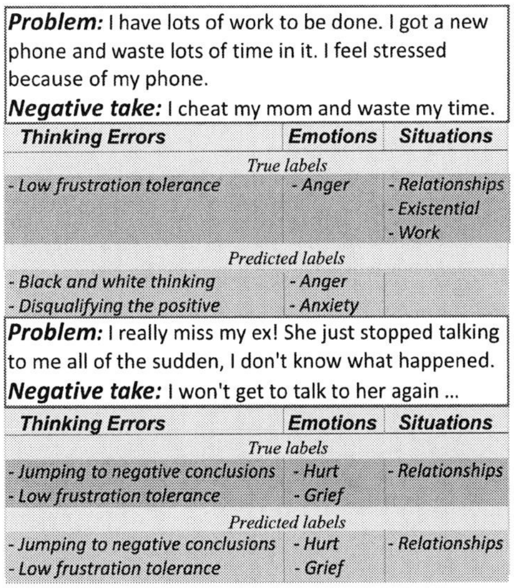

While oversampling is essential for both models, GRU-Skip-thought is less sensitive to lower oversampling ratios, suggesting that skip-thoughts can already capture sentiment on the sentence level. Therefore, including only a limited ratio of positive samples is sufficient to train the classifier. Instead, models using word vectors need more positive data to learn sentence sentiment features.

8 Conclusion

We presented an ontology based on the principles of Cognitive Behavioural Therapy. We then annotated data that exhibits psychological problems and computed the inter-annotator agreement.

We found that classifying *thinking errors* is a difficult task as suggested by the low inter-annotator agreement. We trained GloVe word embeddings and skip-thought embeddings on 500K posts in an unsupervised fashion and generated distributed representations both of words and of sentences. We

Problem: I have lots of work to be done. I got a new phone and waste lots of time in it. I feel stressed because of my phone.		
Negative take: I cheat my mom and waste my time.		
Thinking Errors	**Emotions**	**Situations**
True labels		
- Low frustration tolerance	- Anger	- Relationships - Existential - Work
Predicted labels		
- Black and white thinking	- Anger	
- Disqualifying the positive	- Anxiety	
Problem: I really miss my ex! She just stopped talking to me all of the sudden, I don't know what happened.		
Negative take: I won't get to talk to her again ...		
Thinking Errors	**Emotions**	**Situations**
True labels		
- Jumping to negative conclusions	- Hurt	- Relationships
- Low frustration tolerance	- Grief	
Predicted labels		
- Jumping to negative conclusions	- Hurt	- Relationships
- Low frustration tolerance	- Grief	

Figure 6: predictions of posts by 300 dim CNN-GloVe

Figure 7 gives the comparative performance of two models under different oversampling ratios.

label	precision	recall	F1 score	accuracy
Anxiety	0.739±0.007	0.884±0.005	0.805±0.006	0.729±0.012
Depression	0.538±0.010	0.708±0.005	0.611±0.008	0.813±0.010
Hurt	0.428±0.005	0.620±0.004	0.506±0.005	0.763±0.011
Anger	0.313±0.005	0.491±0.000	0.383±0.004	0.769±0.012
Loneliness	0.479±0.010	0.643±0.008	0.549±0.009	0.923±0.006
Grief	0.437±0.013	0.490±0.000	0.462±0.008	0.937±0.005
Shame	0.219±0.008	0.378±0.004	0.277±0.007	0.891±0.007
Jealousy	0.170±0.002	0.296±0.012	0.216±0.005	0.935±0.006
Guilt	0.221±0.014	0.378±0.008	0.279±0.014	0.936±0.008
Relationships	0.847±0.005	0.912±0.007	0.878±0.006	0.829±0.011
Existential	0.516±0.008	0.700±0.004	0.594±0.007	0.789±0.009
Health	0.520±0.010	0.668±0.005	0.585±0.008	0.900±0.006
School_College	0.570±0.009	0.821±0.008	0.673±0.009	0.934±0.004
Other	0.209±0.004	0.331±0.007	0.256±0.005	0.894±0.007
Work	0.601±0.015	0.733±0.006	0.661±0.011	0.955±0.003
Bereavement	0.567±0.029	0.733±0.008	0.639±0.021	0.979±0.002
Jumping_to_negative_conclusions	0.643±0.005	0.775±0.004	0.703±0.005	0.711±0.009
Fortune_telling	0.486±0.006	0.737±0.004	0.585±0.006	0.733±0.010
Black_and_white	0.330±0.003	0.625±0.003	0.432±0.003	0.657±0.011
Low_frustration_tolerance	0.222±0.005	0.531±0.002	0.313±0.005	0.631±0.028
Catastrophising	0.291±0.005	0.509±0.000	0.371±0.004	0.796±0.012
Mind-reading	0.343±0.008	0.540±0.002	0.419±0.006	0.783±0.014
Labelling	0.376±0.004	0.597±0.003	0.462±0.004	0.853±0.007
Emotional_reasoning	0.241±0.006	0.417±0.004	0.306±0.006	0.748±0.017
Over-generalising	0.337±0.009	0.548±0.002	0.418±0.008	0.808±0.014
Inflexibility	0.162±0.002	0.336±0.006	0.218±0.003	0.807±0.012
Blaming	0.218±0.002	0.381±0.005	0.277±0.003	0.841±0.009
Disqualifying_the_positive	0.125±0.002	0.365±0.008	0.187±0.003	0.808±0.016
Mental_filtering	0.087±0.001	0.386±0.009	0.141±0.002	0.741±0.026
Personalising	0.179±0.003	0.345±0.007	0.236±0.004	0.871±0.009
Comparing	0.257±0.009	0.253±0.009	0.255±0.009	0.952±0.003

Table 7: Precision, recall, F1 score and accuracy for 300 dim CNN-GloVe with oversampling ratio 1:1

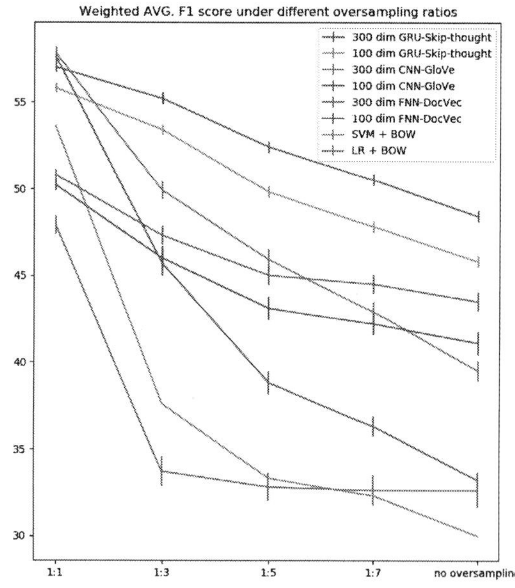

Figure 7: Weighted AVG. F1 for different models

then used the GloVe word vectors as input to a CNN and the skip-thought sentence vectors as input to a GRU. The results suggest that both models significantly outperform a chance classifier for all *thinking errors*, *emotions* and *situations* with CNN-GloVe on average achieving better results.

Areas of future investigation include richer dis-tributed representations, or a fusion of distributed representations from word-level, sentence-level and document-level, to acquire more powerful semantic features. We also plan to extend the current ontology with its focus on *thinking errors*, *emotions* and *situations* to include a much lager number of concepts. The development of a statistical system delivering therapy will moreover require further research on other modules of a dialogue system.

Acknowledgements

This work was funded by EPSRC project Natural speech Automated Utility for Mental health (NAUM), award reference EP/P017746/1. The authors would also like to thank anonymous reviewers for their valuable comments. The code is available at `https://github.com/YinpeiDai/NAUM`

References

S. Arora, Y. Liang, and T. Ma. 2017. A simple but tough-to-beat baseline for sentence embedding. In *ICLR*.

A.T. Beck. 1976. *Cognitive Therapy and the Emotional Disorders*. New York, International Universities Press.

A.T. Beck, J. Rush, B. Shaw, and G Emery. 1979. *Cognitive Therapy of Depression*. New York, Guildford Press.

Charissa Bhasi, Rohanna Cawdron, Melissa Clapp, Jeremy Clarke, Mike Crawford, Lorna Farquharson, Elizabeth Hancock, Miranda Heneghan, Rachel Marsh, and Lucy Palmer. 2013. Second Round of the National Audit of Psychological Therapies for Anxiety and Depression (NAPT).

Timothy Bickmore, Amanda Gruber, and Rosalind Picard. 2005. Establishing the computer–patient working alliance in automated health behavior change interventions. *Patient education and counseling*, 59(1):21–30.

Robyn Bluhm. 2017. The need for new ontologies in psychiatry. *Philosophical Explorations*, 20(2):146–159.

R. Branch and R. Willson. 2010. *Cognitive Behavioural Therapy for Dummies*. Wiley.

Ronan Collobert, Jason Weston, Léon Bottou, Michael Karlen, Koray Kavukcuoglu, and Pavel Kuksa. 2011. Natural language processing (almost) from scratch. *Journal of Machine Learning Research*, 12(Aug):2493–2537.

Heriberto Cuayáhuitl. 2009. *Hierarchical reinforcement learning for spoken dialogue systems*. Ph.D. thesis, University of Edinburgh, Edinburgh.

D DeVault, R Artstein, G Ben, T Dey, E Fast, A Gainer, K Georgila, J Gratch, A Hartholt, M Lhommet, G Lucas, S Marsella, F Morbini, A Nazarian, S Scherer, G Stratou, A Suri, D Traum, R Wood, Y Xu, A Rizzo, and L-P Morency. 2014. Simsensei kiosk: A virtual human interviewer for healthcare decision support. In *International Conference on Autonomous Agents and Multiagent Systems*.

EU high-level conference: Together for Mental Health and Well-being. 2008. European Pact on Mental Health and Well-being.

Mehdi Fatemi, Layla El Asri, Hannes Schulz, Jing He, and Kaheer Suleman. 2016. Policy networks with two-stage training for dialogue systems. In *Proceedings of SIGDIAL*.

Kathleen Kara Fitzpatrick, Alison Darcy, and Molly Vierhile. 2017. Delivering cognitive behavior therapy to young adults with symptoms of depression and anxiety using a fully automated conversational agent (woebot): a randomized controlled trial. *JMIR mental health*, 4(2).

M. Gasic and S. Young. 2014. Gaussian processes for pomdp-based dialogue manager optimization. *Audio, Speech, and Language Processing, IEEE/ACM Transactions on*, 22(1):28–40.

M Geist and O Pietquin. 2011. Managing Uncertainty within the KTD Framework. In *Proceedings of the Workshop on Active Learning and Experimental Design*, Sardinia (Italy).

Sanjay Surendranath Girija. 2016. Tensorflow: Large-scale machine learning on heterogeneous distributed systems.

Nathan B. Hansen, Michael J. Lambert, and Evan M. Forman. 2002. The psychotherapy dose-response effect and its implications for treatment delivery services. *Clinical Psychology: Science and Practice*, 9(3):329–343.

Larry P Heck, Dilek Hakkani-Tür, and Gökhan Tür. 2013. Leveraging knowledge graphs for web-scale unsupervised semantic parsing. In *Proceedings of Interspeech*, pages 1594–1598.

Stefan Hofmann. 2014. Toward a cognitive-behavioral classification system for mental disorders. *Behavior Therapy*, 45(4):576 – 587.

TR Insel and EM Scholnick. 2006. Cure therapeutics and strategic prevention: raising the bar for mental health research. *Molecular Psychiatry*, 11(1):11-17.

Nal Kalchbrenner, Edward Grefenstette, and Phil Blunsom. 2014. A convolutional neural network for modelling sentences. *arXiv preprint arXiv:1404.2188*.

Yoon Kim. 2014. Convolutional neural networks for sentence classification. *arXiv preprint arXiv:1408.5882*.

R. Kiros, Y. Zhu, R. Salakhutdinov, R.S. Zemel, A. Torralba, R. Urtasun, and S. Fidler. 2015. Skip-thought vectors. *NIPS*.

Quoc Le and Tomas Mikolov. 2014a. Distributed representations of sentences and documents. In *Proceedings of the 31st International Conference on International Conference on Machine Learning - Volume 32*, ICML'14, pages II–1188–II–1196. JMLR.org.

Quoc Le and Tomas Mikolov. 2014b. Distributed representations of sentences and documents. In *International Conference on Machine Learning*, pages 1188–1196.

Jiwei Li, Will Monroe, Alan Ritter, and Dan Jurafsky. 2016. Deep reinforcement learning for dialogue generation. In *Proceedings of EMNLP*.

Andrew L Maas, Raymond E Daly, Peter T Pham, Dan Huang, Andrew Y Ng, and Christopher Potts. 2011. Learning word vectors for sentiment analysis. In *Proceedings of the 49th annual meeting of the association for computational linguistics: Human language technologies-volume 1*, pages 142–150. Association for Computational Linguistics.

F. Mairesse, M. Gašić, F. Jurčíček, S. Keizer, B. Thomson, K. Yu, and S. Young. 2009. Spoken language understanding from unaligned data using discriminative classification models. In *Proceedings of ICASSP*.

Grégoire Mesnil, Yann Dauphin, Kaisheng Yao, Yoshua Bengio, Li Deng, Dilek Hakkani-Tur, Xiaodong He, Larry Heck, Gokhan Tur, Dong Yu, and Geoffrey Zweig. 2015. Using recurrent neural networks for slot filling in spoken language understanding. *IEEE Transactions on Audio, Speech, and Language Processing*, 23(3):530–539.

RR Morris, Schueller SM, and Picard RW. 2015. Efficacy of a Web-Based, Crowdsourced Peer-To-Peer Cognitive Reappraisal Platform for Depression: Randomized Controlled Trial. *J Med Internet Res*, 17(3).

Nikola Mrkšić, Diarmuid Ó Séaghdha, Tsung-Hsien Wen, Blaise Thomson, and Steve Young. 2017. Neural belief tracker: Data-driven dialogue state tracking. In *Proceedings of the 55th Annual Meeting of the Association for Computational Linguistics (Volume 1: Long Papers)*, pages 1777–1788. Association for Computational Linguistics.

F. Pedregosa, G. Varoquaux, A. Gramfort, V. Michel, B. Thirion, O. Grisel, M. Blondel, P. Prettenhofer, R. Weiss, V. Dubourg, J. Vanderplas, A. Passos, D. Cournapeau, M. Brucher, M. Perrot, and E. Duchesnay. 2011. Scikit-learn: Machine learning in Python. *Journal of Machine Learning Research*, 12:2825–2830.

Jeffrey Pennington, Richard Socher, and Christopher D. Manning. 2014. Glove: Global vectors for word representation. In *Empirical Methods in Natural Language Processing (EMNLP)*, pages 1532–1543.

Radim Řehůřek and Petr Sojka. 2010. Software Framework for Topic Modelling with Large Corpora. In *Proceedings of the LREC 2010 Workshop on New Challenges for NLP Frameworks*, pages 45–50, Valletta, Malta. ELRA. `http://is.muni.cz/publication/884893/en`.

Giuseppe Riccardi. 2014. Towards healthcare personal agents. In *Proceedings of the 2014 Workshop on Roadmapping the Future of Multimodal Interaction Research Including Business Opportunities and Challenges*, RFMIR '14, pages 53–56, New York, NY, USA. ACM.

Lazlo Ring, Barbara Barry, Kathleen Totzke, and Timothy Bickmore. 2013. Addressing loneliness and isolation in older adults: Proactive affective agents provide better support. In *Proceedings of the 2013 Humaine Association Conference on Affective Computing and Intelligent Interaction*, ACII '13, pages 61–66, Washington, DC, USA. IEEE Computer Society.

Lazlo Ring, Timothy Bickmore, and Paola Pedrelli. 2016. An affectively aware virtual therapist for depression counseling. In *CHI 2016 Computing and Mental Health Workshop*.

Lina M. Rojas Barahona, M. Gasic, N. Mrkšić, P-H Su, S. Ultes, T-H Wen, and S. Young. 2016. Exploiting

sentence and context representations in deep neural models for spoken language understanding. In *Proceedings of COLING 2016, the 26th International Conference on Computational Linguistics: Technical Papers*, pages 258–267, Osaka, Japan. The COLING 2016 Organizing Committee.

Lina Maria Rojas-Barahona and Toni Giorgino. 2009. Adaptable dialog architecture and runtime engine (adarte): A framework for rapid prototyping of health dialog systems. *I. J. Medical Informatics*, 78(Supplement-1):S56–S68.

Andrew M Saxe, James L McClelland, and Surya Ganguli. 2013. Exact solutions to the nonlinear dynamics of learning in deep linear neural networks. *arXiv preprint arXiv:1312.6120*.

J Schatzmann, K Weilhammer, MN Stuttle, and S Young. 2006. A Survey of Statistical User Simulation Techniques for Reinforcement-Learning of Dialogue Management Strategies. *KER*, 21(2):97–126.

D.F. Tolin. 2010. Is cognitivebehavioral therapy more effective than other therapies? A meta-analytic review. *Clinical Psychology Review*, 30:710–720.

Gökhan Tür, Minwoo Jeong, Ye-Yi Wang, Dilek Hakkani-Tür, and Larry P Heck. 2012. Exploiting the semantic web for unsupervised natural language semantic parsing. In *Proceedings of Interspeech*.

L.P. Vardoulakis, L. Ring, B. Barry, C. Sidner, and T. Bickmore. 2012. Designing relational agents as long term social companions for older adults. In Yukiko Nakano, Michael Neff, Ana Paiva, and Marilyn Walker, editors, *Intelligent Virtual Agents*, volume 7502 of *Lecture Notes in Computer Science*, pages 289–302. Springer Berlin Heidelberg.

Y. Wang, L. Wang, M. Rastegar-Mojarad, S. Moon, F. Shen, N. Afzal, S. Liu, Y. Zeng, S. Mehrabi, S. Sohn, and H. Liu. 2018. Clinical information extraction applications: A literature review. *Journal of Biomedical Informatics*, 77:34 – 49.

Joseph Weizenbaum. 1966. Eliza, a computer program for the study of natural language communication between man and machine. *ACM*, 9(1):36–45.

Jason D. Williams, Kavosh Asadi, and Geoffrey Zweig. 2017. Hybrid code networks: practical and efficient end-to-end dialog control with supervised and reinforcement learning. In *Proceedings of ACL*.

World Health Organization. 2013. Mental health action plan 2013 - 2020.

Kaisheng Yao, Baolin Peng, Yu Zhang, Dong Yu, G. Zweig, and Yangyang Shi. 2014. Spoken language understanding using long short-term memory neural networks. In *Spoken Language Technology Workshop (SLT), 2014 IEEE*, pages 189–194.

SJ Young. 2002. Talking to Machines (Statistically Speaking). In *Proceedings of ICSLP*.

Investigating the Challenges of Temporal Relation Extraction from Clinical Text

Diana Galvan[1]   **Naoaki Okazaki**[2]   **Koji Matsuda**[1]   **Kentaro Inui**[1,3]

[1] Tohoku University   [2] Tokyo Institute of Technology

[3] RIKEN Center for Advanced Intelligence Project

{dianags,matsuda,inui}@ecei.tohoku.ac.jp

okazaki@dc.titech.ac.jp

Abstract

Temporal reasoning remains as an unsolved task for Natural Language Processing (NLP), particularly demonstrated in the clinical domain. The complexity of temporal representation in language is evident as results of the 2016 Clinical TempEval challenge indicate: the current state-of-the-art systems perform well in solving mention-identification tasks of event and time expressions but poorly in temporal relation extraction, showing a gap of around 0.25 point below human performance. We explore to adapt the tree-based LSTM-RNN model proposed by Miwa and Bansal (2016) to temporal relation extraction from clinical text, obtaining a five point improvement over the best 2016 Clinical TempEval system and two points over the state-of-the-art. We deliver a deep analysis of the results and discuss the next step towards human-like temporal reasoning.

1 Introduction

Temporal Information Extraction (TIE) is an active research area in NLP, where the ultimate goal is to be able to represent the development of a story over time. TIE is a key to text processing tasks including Question Answering and Text Summarization and follows the traditional pipeline of named entity recognition (NER) and relation extraction separately. Research on this area has been led by TempEval shared tasks (Verhagen et al., 2007, 2010; UzZaman et al., 2013) but in recent years, the target domain has been shifted to the clinical domain. The resulting Clinical TempEval challenges (Bethard et al., 2015, 2016, 2017) introduced the adoption of narrative containers to their annotation schema, based on the widely used TIE annotation standard ISO-TimeML (Pustejovsky et al., 2010). Narrative containers were defined by Pustejovsky and Stubbs

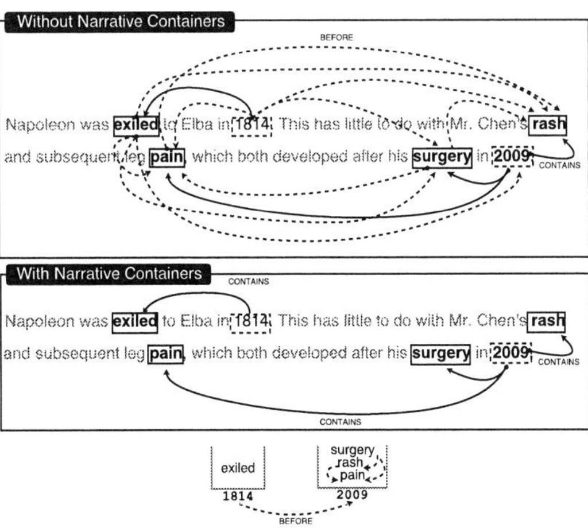

Figure 1: Example temporal relation annotation with and without using narrative containers.

(2011) as an effort to reduce the scope of temporal relations between pairs of events and time expressions. As illustrated in Figure 1, narrative containers can be thought of as temporal buckets in which an event or series of events may fall. They help visualize the temporal relations within a text and facilitate the identification of other temporal relation types. Until now, the only corpus annotated with narrative containers is limited to clinical texts.

Results of the systems participating in Clinical TempEval suggest that they perform well on time-entity identification tasks. Nevertheless, temporal relation extraction has shown to be the most difficult. UTHealth (Lee et al., 2016), the best ranked system in 2016 Clinical TempEval, showed a significant gap of 0.25 when compared to human performance even with gold-standard entity annotations. Recent work by Lin et al. (2016) and Leeuwenberg and Moens (2017) improved UTHealth's results further but the gap with respect to humans is still around 0.21. Regardless of the

Proceedings of the 9th International Workshop on Health Text Mining and Information Analysis (LOUHI 2018), pages 55–64
Brussels, Belgium, October 31, 2018. ©2018 Association for Computational Linguistics

increase in annotation agreement of temporal relations by relying on narrative containers, there is a consensus within the research community regarding TIE difficulty. Still, the reasons of the uneven results between entity and temporal relation predictions remain unclear.

We attribute the complexity of temporal representation in natural language as the main cause of the low performance on temporal relation tasks. Tense and aspect are the two grammatical means to express the notion of time in English but little has been discussed about the latter on clinical text. Furthermore, the focus of previous work on temporal relation extraction is set on narrative containers, which have proved to be useful to locate and relate two events on a timeline. Identification of other temporal relation types has been less frequently tackled. We believe is key to look at the whole set of temporal types to achieve the ultimate goal of developing systems that automatically create a timeline of a patient's health care.

In this paper, we describe the process followed to adapt the neural model proposed by Miwa and Bansal (2016) on TIE, which has already shown competitive results on semantic relation extraction. In our pursuit of understanding the nature of the challenges that characterize the processing of temporal relations, we continue with an error analysis of our system's overall performance and not only on the identification of narrative containers. Our final goal is to shed some light on the difficulties of temporal relation extraction and the necessary efforts to improve further current state-of-the-art systems performance with that of humans on completing the same task.

2   Related Work

Due to the recent shift of TIE to the clinical domain, most related work has been done by Clinical TempEval participating systems. This challenge uses a corpus annotated with five different temporal relation (TLINK) types between events and times ("TIMEX3" in this schema): BEFORE, BEGINS-ON, CONTAINS, ENDS-ON and OVERLAP. However, this challenge only evaluates the identification of a narrative container, marked with the CONTAINS type.

Until 2016 edition of Clinical TempEval, classic machine learning algorithms for classification such as conditional random fields (CRF), support vector machines (SVM) and logistic regression with a variety of features (lexical, syntactic, morphological, and many others) were the predominant approach. In fact, the best performance was achieved by UTHealth team (Lee et al., 2016) using an end-to-end system based on linear and structural Hidden Markov Model (HMM)-SVM. Just a few teams tried a neural based method, including RNN-based models (Fries, 2016) and CNN-based models (Chikka, 2016), (Li and Huang, 2016). Furthermore, among those teams just Chikka (2016) participated in the CONTAINS identification task, being around 0.30 below UTHealth's top performance.

Recent work by Lin et al. (2016), Dligach et al. (2017) and Leeuwenberg and Moens (2017) followed the settings of 2016 Clinical TempEval challenge but they did not participate in the competition. Out of these, our results are only directly comparable to those of Lin et al. (2016) and Leeuwenberg and Moens (2017) since the work of Dligach et al. (2017) was not evaluated using the Clinical TempEval official scorer.

Even though Leeuwenberg and Moens (2017) established a new state-of-the-art in temporal relation extraction, their result is still below human performance. Moreover, none of the aforementioned works provides a detailed discussion of *why* is current performance so low and *how* can we improve further the results on temporal relation extraction, except from Leeuwenberg and Moens (2016), which in their first attempt on tackling this task on 2016 Clinical TempEval identified false negatives as their major problem.

Our contribution is a deep error analysis taking into account the performance of our model on predicting all TLINK types. As a result, we were able to identify important clues on temporal relation extraction and based on these findings, we discuss the next step towards human-like temporal reasoning performance.

3   Method

We adapted the tree-based bidirectional LSTM-RNN end-to-end neural model of Miwa and Bansal (2016) to intra-sentential temporal relation extraction from clinical text. This three-layer model (embedding, sequence and dependency layers) jointly identifies entities and relations between them. For relation classification, the model heavily relies on the dependency structure around the target word pair and the output of the sequence

TLINK	Train	Test
CONTAINS	8653	4554
NONE	43643	20465
Total	52296	25019

Table 1: Label distribution of pre-processed dataset for binary classification.

TLINK	Train	Test
BEFORE	1839	982
BEGINS-ON	717	363
CONTAINS	8653	4554
ENDS-ON	334	138
OVERLAP	2388	1186
NONE	43643	20465
Total	57574	27688

Table 2: Label distribution of pre-processed dataset for multi-class classification.

System	P	R	F1
(Lee et al., 2016)	0.588	0.559	0.573
(Lin et al., 2016)	0.669	0.534	0.594
(Leeuwenberg and Moens, 2017)	-	-	0.608
Our model	0.983	0.462	0.629
Human performance	-	-	0.817

Table 3: Performance of systems and humans on identifying CONTAINS relations.

layer. When tested on nominal relation classification (Hendrickx et al., 2009), it showed competitive results against the state-of-the-art.

We followed the official 2016 Clinical TempEval settings for phase 2 of evaluation, where given the raw text and manual event and time annotations, the task is to identify the temporal relation between a directed pair (e_1, e_2), if any. e_1 and e_2 are entities of either EVENT or TIMEX3 type. For relation classification, Miwa and Bansal (2016) model takes as an input a sentence and a annotation file with a word pair. The output contains the predicted relation type and the directionality of the entities:(e_1, e_2) when e_1 is the source and e_2 the target and (e_2, e_1) otherwise.

4 Experimental settings

4.1 Dataset

Similar to 2016 Clinical TempEval, we used the THYME corpus (Styler IV et al., 2014) for evaluation, a dataset of 600 clinical notes and pathology reports from colon cancer patients at the Mayo Clinic. The corpus is annotated at the document level and identified entities are given a set of attributes depending on their type: *DocTimeRel*, *Type*, *Polarity*, *Degree*, *Contextual Modality* and *Contextual Aspect* for EVENTs and *Class* for TIMEX3. Temporal relation annotations specify source and target entities along with one of the following TLINK types: BEFORE, BEGINS-ON, CONTAINS, ENDS-ON and OVERLAP.

Sentence-level annotations are necessary to meet Miwa and Bansal (2016)'s input requirements. Therefore, we used the Clinical Language Annotation, Modeling and Processing (CLAMP) toolkit[1] for tokenization and sentence boundary detection. We matched all entities spans from the gold standard with the sentence offsets on the CLAMP output to identify those within the same sentence. As a result, the new annotations contain a pair of words, their offsets in the sentence, the temporal relation between them marked on the gold standard and the directionality of the arguments. Example 1 shows an example annotation of the TLINK CONTAINS(*lifelong, nonsmoker*) in the sentence *He is a lifelong nonsmoker.*

$$(1) \quad \begin{array}{lll} \text{T1} & \text{Term 8 16} & \text{lifelong} \\ \text{T2} & \text{Term 17 26} & \text{nonsmoker} \\ \text{R1} & \text{ContainsSource-ContainsTarget} & \text{Arg1:T1} \\ & & \text{Arg2:T2} \end{array}$$

Since any two EVENT/TIMEX3 can be a candidate pair, we took all entities in a sentence to generate all pair combinations as candidates. Pairs that do not have any temporal relation were labeled as NONE. Due to the large number of negative instances produced by this procedure, it was applied only to CONTAINS. No negative instances were generated for the remaining TLINK types and we did not extend the set of TLINKs to its transitive closure (i.e. A CONTAINS B ∧ B CONTAINS C → A CONTAINS C). Table 1 and Table 2 detail the resulting datasets.

---

[1] http://clinicalnlptool.com/index.php

| | Binary classification | | | Multi-class classification | | | | | | | | |
| | *Wikipedia word emb* | | | *Wikipedia word emb* | | | *PubMed word emb* | | | *PubMed word emb + FNE* | | |
TLINK	P	R	F1	P	R	F1	P	R	F1	P	R	F1
BEFORE	-	-	-	0.698	0.185	0.292	0.708	0.198	0.310	0.683	0.202	0.312
BEGINS-ON	-	-	-	0.585	0.062	0.112	0.615	0.103	0.177	0.608	0.116	0.195
CONTAINS	0.983	0.462	**0.629**	0.905	0.472	**0.621**	0.908	0.471	**0.620**	0.889	0.479	**0.623**
ENDS-ON	-	-	-	0.520	0.086	0.148	0.704	0.126	0.213	0.760	0.126	0.216
OVERLAP	-	-	-	0.504	0.134	0.211	0.504	0.134	0.211	0.497	0.140	0.218

Table 4: Results of our four experiments on the THYME test set. FNE refers to filtered negative examples.

4.2 Experiments

We followed the same experimental settings described in Miwa and Bansal (2016). Additional to the model's default Wikipedia word embeddings, we trained word vectors of 200 dimensions using word2vec (Mikolov et al., 2013) on a subset of journal abstracts in Oncology and Gastroenterology from PubMed2014[2]. PubMed data can be easily downloaded without application approval that clinical corpus like MIMIC II (Saeed et al., 2011) require.

We conducted four experiments at the intra-sentential level. The first experiment follows 2016 Clinical TempEval, focusing only on the identification of the CONTAINS type. The remaining experiments include the five annotated TLINKs. Further detail of each experiment is given below:

1. TLINK:CONTAINS binary classification: In order to obtain results comparable to Lee et al. (2016), the best ranked system in 2016 Clinical TempEval, we only considered TLINK:CONTAINS instances. The model chooses between CONTAINS and NONE relations.

2. Multi-class classification: To test the model in a real-world setting, we added to train and test sets the remaining pairs in the gold standard that have any of the other TLINK types. No further negative examples were created for the additional types.

3. Multi-class classification with PubMed word embeddings: In addition to the previous setting (2), we used word embeddings trained on the subset of PubMed instead of the default word vectors trained on Wikipedia.

4. Multi-class classification with PubMed word embeddings and filtered negative examples: In addition to the previous setting (3), we filtered from the dataset NONE pairs that according to the THYME guidelines[3] should never be TLINKed. Thus, we removed a candidate pair whenever e_1 contextual modality value[4] was ACTUAL or HEDGED and the e_2 had HYPOTHETICAL or GENERIC modality, and vice versa.

5 Results

5.1 TLINK:CONTAINS binary classification

Table 3 presents the results of previous approaches compared to human performance. The first row shows the top performance in 2016 Clinical TempEval using binary classification. The second and third rows are the latests results outside the competition. Following the steps of the Clinical TempEval narrative container identification task, we only tried to predict TLINKs of CONTAINS type. In doing so we obtained an F1 score of 0.629, outperforming UTHealth's system. The model shows a high precision but lower recall than UTHealth; this is probably because of NONE relations prevailing in the dataset. By handling the task as binary classification, given a pair of entities we are already assuming there is some kind of temporal relation and the classifier's task is to decide whether it is CONTAINS or not. We performed this experiment in order to have results comparable with those of UTHealth. However, we cannot compare this re-

[2]https://www.nlm.nih.gov/databases/download/pubmed_medline.html

[3]http://savethevowels.org/files/THYMEGuidelines.pdf, Section 6.2.5

[4]Entity attributes introduced in Section 4.1 were not used as features in our model. EVENTs marked with HYPOTHETICAL or GENERIC modality are non-real events. Therefore, they cannot be related to real events marked as ACTUAL or HEDGED.

sult to the state-of-the-art since Leeuwenberg and Moens (2017) was a multi-class classification approach.

5.2 Multi-class classification

Table 4 reports our experimental results of a single run with the four different settings[5]. Switching from binary classification to multi-class classification we observe a significant drop in precision and a lower F1 score. This is expected since the classifier now has more TLINK as options from where to decide. Despite of this change, the model keeps outperforming both UTHealth and the state-of-the-art.

5.3 Multi-class classification with PubMed word embeddings

Once we confirmed the adapted model gives competitive results on the narrative container identification task, we focused on increasing the system's recall. Therefore, we changed the word representations for in-domain word embeddings in comparison with the previous experiment, which uses word vectors trained on Wikipedia. Word representation depends on the words in context and because the clinical domain is a very specific field with a different vocabulary of that used in the general domain, we expected the model to benefit from a resource like PubMed. However, our results suggest this does not have a significant impact on most TLINKs (OVERLAP did not change at all). Only BEGINS-ON and ENDS-ON recall considerably improved.

5.4 Multi-class classification with PubMed word embeddings and filtered negative examples

While we increased recall by using in-domain word embeddings, we can still witness an imbalance between precision and recall. Moreover, we are still below UTHealth recall score (highest on CONTAINS identification task). To improve further the model's recall, around 10% of NONE:EVENT-EVENT pairs were removed from the dataset based on a rule of the annotation guidelines that prevents non-real events (i.e. events that do not actually appear on the patient's timeline) to be linked with real events. Recall was further improved for most TLINKs while it remained the same for ENDS-ON. Under this setting, our model reached its best F1

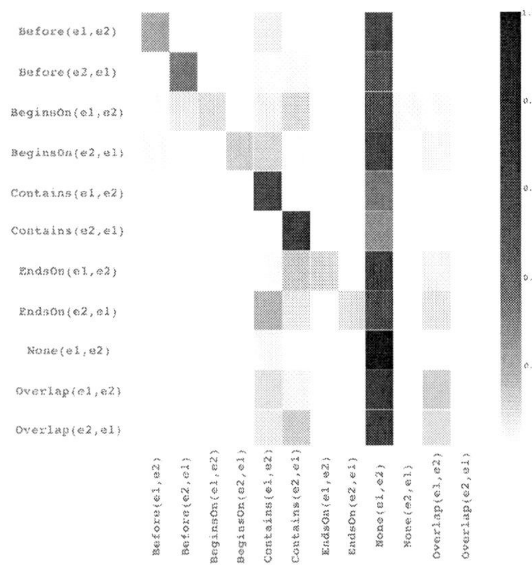

Figure 2: Confusion matrix of our multi-class classification model with PubMed word embedding on the dev set.

scores for all TLINKs, outperforming the state-of-the-art on CONTAINS.

6 Error Analysis

We focused our error analysis on the fourth of our experiments. Systems participating in the Clinical TempEval narrative container identification task only received credit if for a pair of entities, they correctly identified the source, target and the CONTAINS relation between them. Given this setting, we understand that even when using manual event and time annotations the challenge is not only to predict the TLINK type but also the correct directionality of the entities. Part of our analysis is to determine whether type classification or directionality identification is the most difficult task or if they are both equally problematic for the model. Confusion matrix on Figure 2 shows the results on the development set. Overall, due to the high number of negative instances, most of the false positives fall into the $None(e_1, e_2)$ category. At the same time, we can observe that this type of relation is the reason why the system shows high precision. Apart from this, we can identify the performance on OVERLAP as our system's main problem. Accuracy in both $Overlap(e_1, e_2)$ and $Overlap(e_2, e_1)$ is considerably low, with the latter being the lowest among all types with 0.024. Not even the performance on $Before(e_2, e_1)$ with 0.34 is as low, even though

---

[5]We experimented a couple of additional runs but the results were always the same.

True relation	Predicted relation	Sentence
$Overlap(e_1, e_2)$	$Contains(e_1, e_2)$	1. **Tumor** *invades* into the muscularis propria.
$Overlap(e_1, e_2)$	$Contains(e_1, e_2)$	2. Recurrent rectal **adenocarcinoma**, previously *resected* node-negative
$Overlap(e_1, e_2)$	$Contains(e_1, e_2)$	3. June 30, 2009: Due to change in stool, patient underwent **colonoscopy** *noting* mass in the right colon.
$Overlap(e_1, e_2)$	$Contains(e_1, e_2)$	4. A **biopsy** obtained was positive for *adenocarcinoma*, consistent with colorectal primary and confirmed by LCC.
$Overlap(e_1, e_2)$	$Contains(e_1, e_2)$	5. **Pathology** from the extended right hemicolectomy was positive for invasive moderately differentiated *adenocarcinoma* in the ascending colon.
$Overlap(e_1, e_2)$	$Contains(e_1, e_2)$	6. Exploratory **surgery** with *appendicitis* many years ago.
$Overlap(e_1, e_2)$	$Contains(e_1, e_2)$	7. She was seen by a cardiologist in Idyllwild back in **April** when she was *hospitalized* and had an adenosine sestamibi scan after that hospitalization, but if surgery is contemplated I would wish her to be seen by cardiology.
$Overlap(e_2, e_1)$	$Contains(e_2, e_1)$	8. Does have some constipation with her iron supplementations but denies nausea, vomiting, abdominal distention, or worsening constipation, as she does have bowel **movements** *once every several days*.
$Overlap(e_2, e_1)$	$Contains(e_2, e_1)$	9. She is still **moving** her bowels *multiple times a day*.
$Overlap(e_2, e_1)$	$Contains(e_2, e_1)$	10. The patient **smokes** cigars *about once-a-month*.

Table 5: Sample of the analyzed misclassified sentences by our system. e_1 and e_2 are shown in bold and italics, respectively.

they have similar number of instances (290 and 353, respectively). $Overlap(e_1, e_2)$ with 0.14 is comparable to $BeginsOn(e_2, e_1)$, despite of having 7 times more instances (1291 vs. 176). For this reason, we focused our error analysis on OVERLAP.

From Figure 2 we can observe that $Overlap(e_1, e_2)$ is usually predicted as $Contains(e_1, e_2)$ and $Overlap(e_2, e_1)$ is predicted as $Contains(e_2, e_1)$. In both cases the directionality of the entities was correct but the system failed to identify the appropriate temporal relation. For $Overlap(e_1, e_2)$ there were 126 misclassified sentences while in $Overlap(e_2, e_1)$ there were 37. EVENT-EVENT pairs were the predominant type of pair in the former while TIMEX3-EVENT were for the latter, with 116 and 29 instances, respectively. We took all of the aforementioned misclassified sentences for supplementary examination and discuss the reason(s) of this errors in the following section.

6.1 Temporal relations and Aspectual Classes

Before proceeding further, it is important to understand the definition of OVERLAP and CONTAINS. Both temporal relations are closely related since they encompass the notion of two things happening at the same time. However, CONTAINS relations imply that the contained event (i.e. the target) occurs entirely within the temporal bounds of the event it is contained within (i.e. the source) while

OVERLAP relations are those where containment is not entirely sure. Also, OVERLAP is the only symmetrical TLINK type since e_1 OVERLAP e_2 means the same as e_2 OVERLAP e_1.

Strictly speaking, every entity occupies time. An entity's time interval is crucial for understanding its temporal relation with respect to another entity, specially in the case of CONTAINS and OVERLAP relations where the end point of the target is key to determine whether there is complete containment or not. The temporal relations used by the THYME project rely on Allen (1990) interval algebra, a precise way to express time periods using clear start and end points. By comparing those, we can easily indicate the position of two events on the timeline. However, the concept of time is widely discussed across disciplines and Allen's representation is just one among many others. In Linguistics, the expression of time is understood thanks to two important grammatical systems: tense and aspect. It is particularly to our interest the definition of aspect, the means with which speakers discuss a single situation, for example, as beginning, continuation, or completion (Li and Shirai, 2000). One of the best known and widely accepted aspect classifications is that of Vendler, who distinguished four categories for verb and verb phrases: *activities*, *accomplishments*, *achievements* and *states*.

Figure 3 presents Vendler's classification using (Andersen, 1990) schematization. Arrows are used to represent an indefinite time interval, solid lines indicate a homogeneous duration and dashed lines indicate a dynamic duration. An X is used to represent a situation's natural end point.

Category	C-Start	C-End	NC-Start	NC-End
Activity	+			+
Accomplishment	+	+		
Achievement	+*	+		
State			+	+

** Start and end are so close to each other that this category considers <u>no duration</u>*

Activity	- ->
Accomplishment	- X
Achievement	X
State	⎯⎯⎯⎯⎯⎯⎯⎯⎯⎯⎯⎯⎯⎯⎯⎯>

Figure 3: Vendler's four-way classification. Abbreviations: C, Clear; NC, Not Clear

Categorizing the source and target entities of a relation as one of Vendler's types simplifies the TLINK classification task. For example, categories with no clear end points like *activities* and *states* are more likely to overlap with *accomplishments* and *achievements*, which have clear end points. Figure 4 illustrates an OVERLAP and CONTAINS relations using Allen's and Vendler's representation of time periods. Leveraging on aspectual type for temporal relation extraction is a promising approach that has already been explored by Costa and Branco (2012) on TempEval data. However, this approach is limited since aspect is a property of verbs.

When analyzing OVERLAP relations that were mistaken for CONTAINS, we realized that just a few events are verbs. Events in sentences 1, 3 and 9 in Table 5 are some examples of this ("invades", "noting" and "moving"). This pointed out the necessity of discriminating between verbal and non-verbal events to understand how they are temporally related. Our observations suggest that rather than recognizing an entity semantic type (e.g. sign or symptoms, diseases, procedures) it is imperative to take into account the action associated to it. Thus, procedures like colonoscopy, biopsy, pathology and surgery have to be *performed*, a dynamic verb with a natural end point: an *accomplishment*. Diseases like adenocarcinoma and appendicitis are *present*, they exist, and consequently they fall in the *state* category. Following this line of reasoning, it is easier to differentiate an OVERLAP relation from CONTAINS in sentence 5 since we understand the adenocarcinoma was found during the performance of the pathology but there is not enough information to tell whether the adenocarcinoma is still present or not. In other words, its end point is unclear.

In the case of TIMEX3-EVENT pairs like those in sentences 8 to 10 in Table 5, the nature of the OVERLAP relation between the entities is due to the ambiguity of the time expressions combined with actions that we perceive as ongoing. For example, in sentence 9 the action of moving is an *activity*, done indeterminably throughout the day as *multiple times a day* imply. In sentence 7, on the other hand, there is a time expression with a definite time interval overlapping the patient's state of being hospitalized.

Temporally locating two events on a timeline requires a high level of reasoning that even for humans can turn into a complicated task. All of the aforementioned inferences were done heavily relying on the internal constituency of an event, implying Costa and Branco (2012) claim that temporal information processing can profit from information about aspectual type is valid in the clinical domain. Due to the high similarity of CONTAINS and OVERLAP relations it does not come as a surprise that these two types are easily confused by our system, which performed reasonably well on identifying other TLINK types with similar number of instances. This suggests than the main problem is not the amount of data available but how temporal properties are encoded in language.

Aspectual information proved useful for differentiating between two of the most frequent and most similar TLINK types: CONTAINS and OVERLAP. As previously mentioned in Section 4.1, there is a contextual aspect attribute available for EVENT entities with three possible values: *N/A* (default), *NOVEL* and *INTERMITTENT*. The latter could be useful to identify an *activity* or an *accomplishment* but just a small portion of EVENTs were annotated with a value different from the default one. Moreover, aspect is a property of verbs and our analysis insinuates it is more common to find nouns as events. We discuss this finding in more detail in the following section.

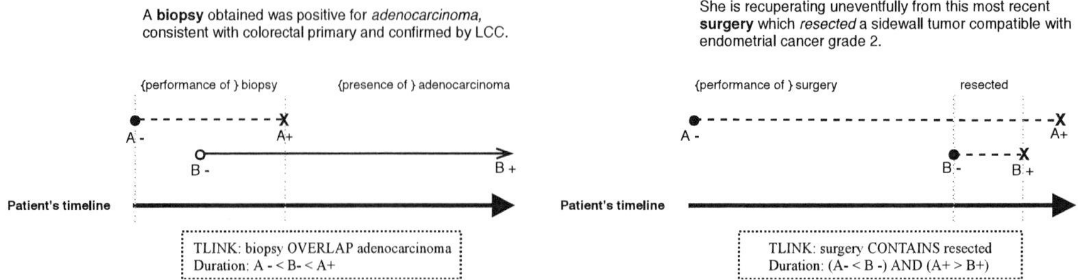

Figure 4: Allen's and Vendler's interval representation of OVERLAP and CONTAINS relations. A- / B- and A+ / B+ represent the start and end of an event, respectively. Filled-dots represent clear start points while an empty-dot represent a not-clear start point.

7 Temporality of nominal events

To deepen our understanding on the complexity of the temporal relation extraction task, we divided all OVERLAP and CONTAINS false negatives into the four possible pair types: EVENT-EVENT, TIMEX3-TIMEX3, EVENT-TIMEX3 and TIMEX3-EVENT. A significant amount of OVERLAP links were EVENT-EVENT relations and they also made around half of CONTAINS links. We looked further into these type of pairs, discriminating between verb and non-verbal events. Table 6 shows the results in more detail.

Dev set: Event-Event pairs				
TLINK	**V-V**	**V-NV**	**NV-V**	**NV-NV**
CONTAINS	6	47	24	103
OVERLAP	6	55	27	193
Total	12	102	51	296

Table 6: Distribution of misclassified CONTAINS and OVERLAP Event-Event pairs by type of EVENT. Abbreviations: V, Verb; NV, Non-Verb

As mentioned by Pustejovsky and Stubbs (2011) and further discussed in Styler IV et al. (2014), EVENT-EVENT pairings are a complex and vital component, particularly in clinical narratives where doctors rely on shared domain knowledge and it is essential to read "between the lines". The distribution of verb/non-verb entities in Table 6 indicates that most of EVENT-EVENT missclasified pairings were either of NV-NV type or include a NV entity. Time intervals of NV entities like "pain" or "resection" are more difficult to understand, while V entities like "removed" or "improving" have their time properties morphologically encoded. Thus, regardless of the low number of V-V relations, temporal information from verb predicates usually have more explicit hints. NV entities are more challenging and require more careful examination.

The high frequency of NV entities is likely to be one of the reasons why not only our system but also previous works in temporal relation extraction are behind human performance. In the previous section we introduced Vendler's aspectual classification and discussed how it helps separate two extremely similar TLINKs. Unfortunately, this is not compatible with nominal predicates. Verb/Non-Verb entities distinction of EVENTs is a first step that could alleviate this problem and positively influence the temporal relation extraction task.

8 Conclusion and Future work

Clinical language processing represents a special challenge to NLP systems. The structure of clinical texts range from telegraphic constructions to long utterances describing a patient's condition or a suggested diagnosis. The high use of domain knowledge to infer temporal relations between events does not make this task any easier. A doctor naturally interprets adenocarcinoma (a type of cancer) as an abnormal, uncontrolled and *progressive* growth of tissue which temporally speaking it is and should be thought as an ongoing process unless explicitly qualified (*"We resected the adenocarcinoma, and since margins were clear, we can say it is gone"*). This is a non-trivial task for a computer even when relying on context information.

Up to now, there have been several attempts on tackling temporal relation extraction from clinical text mostly led by the Clinical TempEval challenges. However, the results are still far from human performance and there is little information of the reasons behind. This encouraged our work to adapt a state-of-the-art system and do a detailed error analysis, which pointed out that one of the major challenges is how to handle the eventive properties of nominals, the predominant type of events on the most frequent type of pairs: EVENT-EVENT.

Existing knowledge bases like the Unified Medical Language System (UMLS) Metathesaurus help to classify entities into semantic types like *Therapeutic or Preventive procedure*, *Sign or Symptom* or *Disease or Syndrome*. Still, the associated events and actions cannot be found in this or any other knowledge base. We hypothesize that a resource containing aspectual information of the actions associated to common nominals like procedures or diseases can further improve temporal relation extraction in the clinical domain. With that in mind, we plan to analyze further EVENT-EVENT relations differentiating events as verbal and non-verbal events.

Acknowledgments

This work was supported by JST CREST Grant Number JPMJCR1513, Japan. We thank the anonymous reviewers for their insightful comments and suggestions.

References

James F Allen. 1990. Maintaining knowledge about temporal intervals. In *Readings in qualitative reasoning about physical systems*, pages 361–372. Elsevier.

Roger W. Andersen. 1990. Unpublished lecture in the seminar on the acquisition of tense and aspect.

Steven Bethard, Leon Derczynski, Guergana Savova, James Pustejovsky, and Marc Verhagen. 2015. Semeval-2015 task 6: Clinical tempeval. In *Proceedings of the 9th International Workshop on Semantic Evaluation (SemEval 2015)*, pages 806–814.

Steven Bethard, Guergana Savova, Wei-Te Chen, Leon Derczynski, James Pustejovsky, and Marc Verhagen. 2016. Semeval-2016 task 12: Clinical tempeval. In *Proceedings of the 10th International Workshop on Semantic Evaluation (SemEval-2016)*, pages 1052–1062.

Steven Bethard, Guergana Savova, Martha Palmer, and James Pustejovsky. 2017. Semeval-2017 task 12: Clinical tempeval. In *Proceedings of the 11th International Workshop on Semantic Evaluation (SemEval-2017)*, pages 565–572.

Veera Raghavendra Chikka. 2016. Cde-iiith at semeval-2016 task 12: Extraction of temporal information from clinical documents using machine learning techniques. In *Proceedings of the 10th International Workshop on Semantic Evaluation (SemEval-2016)*, pages 1237–1240.

Francisco Costa and António Branco. 2012. Aspectual type and temporal relation classification. In *Proceedings of the 13th Conference of the European Chapter of the Association for Computational Linguistics*, pages 266–275. Association for Computational Linguistics.

Dmitriy Dligach, Timothy Miller, Chen Lin, Steven Bethard, and Guergana Savova. 2017. Neural temporal relation extraction. In *Proceedings of the 15th Conference of the European Chapter of the Association for Computational Linguistics: Volume 2, Short Papers*, volume 2, pages 746–751.

Jason Alan Fries. 2016. Brundlefly at semeval-2016 task 12: Recurrent neural networks vs. joint inference for clinical temporal information extraction.

Iris Hendrickx, Su Nam Kim, Zornitsa Kozareva, Preslav Nakov, Diarmuid Ó Séaghdha, Sebastian Padó, Marco Pennacchiotti, Lorenza Romano, and Stan Szpakowicz. 2009. Semeval-2010 task 8: Multi-way classification of semantic relations between pairs of nominals. In *Proceedings of the Workshop on Semantic Evaluations: Recent Achievements and Future Directions*, pages 94–99.

Hee-Jin Lee, Hua Xu, Jingqi Wang, Yaoyun Zhang, Sungrim Moon, Jun Xu, and Yonghui Wu. 2016. Uthealth at semeval-2016 task 12: an end-to-end system for temporal information extraction from clinical notes. In *Proceedings of the 10th International Workshop on Semantic Evaluation (SemEval-2016)*, pages 1292–1297.

Artuur Leeuwenberg and Marie-Francine Moens. 2016. Kuleuven-liir at semeval 2016 task 12: Detecting narrative containment in clinical records. In *Proceedings of the 10th International Workshop on Semantic Evaluation (SemEval-2016)*, pages 1280–1285.

Artuur Leeuwenberg and Marie-Francine Moens. 2017. Structured learning for temporal relation extraction from clinical records. In *Proceedings of the 15th Conference of the European Chapter of the Association for Computational Linguistics*.

Peng Li and Heng Huang. 2016. Uta dlnlp at semeval-2016 task 12: deep learning based natural language processing system for clinical information identification from clinical notes and pathology reports. In *Proceedings of the 10th International Workshop on Semantic Evaluation (SemEval-2016)*, pages 1268–1273.

Ping Li and Yasuhiro Shirai. 2000. *The acquisition of lexical and grammatical aspect*, volume 16. Walter de Gruyter.

Chen Lin, Timothy Miller, Dmitriy Dligach, Steven Bethard, and Guergana Savova. 2016. Improving temporal relation extraction with training instance augmentation. In *Proceedings of the 15th Workshop on Biomedical Natural Language Processing*, pages 108–113.

Tomas Mikolov, Kai Chen, Greg Corrado, and Jeffrey Dean. 2013. Efficient estimation of word representations in vector space. *CoRR abs/1301.3781*.

Makoto Miwa and Mohit Bansal. 2016. End-to-end relation extraction using lstms on sequences and tree structures. In *Proceedings of the 54th Annual Meeting of the Association for Computational Linguistics (Volume 1: Long Papers)*, pages 1105–1116. Association for Computational Linguistics.

James Pustejovsky, Kiyong Lee, Harry Bunt, and Laurent Romary. 2010. Iso-timeml: An international standard for semantic annotation. In *LREC*, volume 10, pages 394–397.

James Pustejovsky and Amber Stubbs. 2011. Increasing informativeness in temporal annotation. In *Proceedings of the 5th Linguistic Annotation Workshop*, pages 152–160.

Mohammed Saeed, Mauricio Villarroel, Andrew T Reisner, Gari Clifford, Li-Wei Lehman, George Moody, Thomas Heldt, Tin H Kyaw, Benjamin Moody, and Roger G Mark. 2011. Multiparameter intelligent monitoring in intensive care ii (mimic-ii): a public-access intensive care unit database. *Critical care medicine*, 39(5):952.

William F Styler IV, Steven Bethard, Sean Finan, Martha Palmer, Sameer Pradhan, Piet C de Groen, Brad Erickson, Timothy Miller, Chen Lin, Guergana Savova, et al. 2014. Temporal annotation in the clinical domain. *Transactions of the Association for Computational Linguistics*, 2:143.

Naushad UzZaman, Hector Llorens, Leon Derczynski, James Allen, Marc Verhagen, and James Pustejovsky. 2013. Semeval-2013 task 1: Tempeval-3: Evaluating time expressions, events, and temporal relations. In *Second Joint Conference on Lexical and Computational Semantics (* SEM), Volume 2: Proceedings of the Seventh International Workshop on Semantic Evaluation (SemEval 2013)*, volume 2, pages 1–9.

Marc Verhagen, Robert Gaizauskas, Frank Schilder, Mark Hepple, Graham Katz, and James Pustejovsky. 2007. Semeval-2007 task 15: Tempeval temporal relation identification. In *Proceedings of the 4th international workshop on semantic evaluations*, pages 75–80.

Marc Verhagen, Roser Sauri, Tommaso Caselli, and James Pustejovsky. 2010. Semeval-2010 task 13: Tempeval-2. In *Proceedings of the 5th international workshop on semantic evaluation*, pages 57–62.

De-identifying Free Text of Japanese Electronic Health Records

**Kohei Kajiyama[1], Hiromasa Horiguchi[2], Takashi Okumura[3],
Mizuki Morita[4] and Yoshinobu Kano[5]**

[1,5]Shizuoka University
[2]National Hospital Organization
[3]Kitami Institute of Technology
[4]Okayama University
[1]kajiyama.kohei.14@shizuoka.ac.jp
[2]horiguchi-hiromasa@hosp.go.jp
[3]tokumura@mail.kitami-it.ac.jp
[4]mizuki@okayama-u.ac.jp
[5]kano@inf.shizuoka.ac.jp

Abstract

A new law was established in Japan to promote utilization of EHRs for research and developments, while de-identification is required to use EHRs. However, studies of automatic de-identification in the healthcare domain is not active for Japanese language, no de-identification tool available in practical performance for Japanese medical domains, as far as we know. Previous work shows that rule-based methods are still effective, while deep learning methods are reported to be better recently. In order to implement and evaluate a de-identification tool in a practical level, we implemented three methods, rule-based, CRF, and LSTM. We prepared three datasets of pseudo EHRs with de-identification tags manually annotated. These datasets are derived from shared task data to compare with previous work, and our new data to increase training data. Our result shows that our LSTM-based method is better and robust, which leads to our future work that plans to apply our system to actual de-identification tasks in hospitals.

1 Introduction

Recently, healthcare data is getting increased both in companies and government. Especially, utilization of Electronic Health Records (EHRs) is one of the most important task in the healthcare domain. While it is required to de-identify EHRs to protect personal information, automatic de-identification of EHRs has not been studied sufficiently for the Japanese language.

Like other countries, there are new laws for medical data treatments established in Japan. "Act Regarding Anonymized Medical Data to Contribute to Research and Development in the Medical Field" was established in 2018. This law allows specific third party institute to handle EHRs. As commercial and non-commercial health data is already increasing in recent years [1], this law promotes more health data to be utilized. At the same time, developers are required to de-identify personal information. "Personal Information Protection Act" was established in 2017, which requires EHRs to be handled more strictly than other personal information. This law defines personal identification codes including individual numbers (e.g. health insurance card, driver license card, and personal number), biometric information (e.g. finger print, DNA, voice, and appearance), and information of disabilities.

---

[1] (Ministry of Internal Affairs and Communication International Strategy Bureau, Information and Communication Economy Office, 2018)

Proceedings of the 9th International Workshop on Health Text Mining and Information Analysis (LOUHI 2018), pages 65–70
Brussels, Belgium, October 31, 2018. ©2018 Association for Computational Linguistics

De-identification of structured data in EHRs is easier than that of unstructured data, because it is straightforward to apply de-identification methods e.g. k-anonymization (Latanya, 2002).

In the i2b2 task, automatic de-identification of clinical records was challenged to clear a hurdle of the Health Insurance Portability and Accountability Act (HIPAA) states (Özlem, Yuan, & Peter, 2007). There have been attempts to make k-anonymization for Japanese plain texts (Maeda, Suzuki, Yoshino, & Satoshi, 2016). Shared tasks of de-identification for Japanese EHRs were also held as MedNLP-1 (Mizuki, Yoshinobu, Tomoko, Mai, & Eiji, 2013) and MedNLP-2 (Aramaki, Morita, Kano, & Ohkuma, 2014).

While rule-based, SVM (Corinna & Vlandimir, 1995) and CRF (Lafferty, McCallum, & Pereira, 2001) were often used in these previous NER tasks, deep neural network model has shown better results recently. However, rule-based methods are still often better than machine learning methods, especially when there is not enough data, e.g. the best system in MedNLPDoc (Aramaki, Morita, Kano, & Ohkuma, Overview of the NTCIR-12 MedNLPDoc Task, 2016). The aim of the MedNLPDoc task was to infer ICD Codes of diagnosis from Japanese EHRs.

In this paper, we focus on de-identification of free text of EHRs written in the Japanese language. We compare three methods, rule, CRF and LSTM based, using three datasets that are derived from EHRs and discharge summaries.

We follow the MedNLP-1's standard of person information which require to de-identify "age", "hospital", "sex" and "time".

Methods

We used the Japanese morphological analyzer kuromoji[2] with our customized dictionary, as same as the best result team (Sakishita & Kano, 2016) in the MedNLPDoc task.

We implemented three methods as described below: rule-based, CRF-based, and LSTM-based.

1.1 Rule-based Method

Unfortunately, details and implementation of the best method of the MedNLP1 de-identification task (Imaichi, Yanase, & Niwa, 2013) are not publicly available. We implemented our own rule-based program based on their descriptions in their

<hr>

[2] https://www.atilika.com/en/kuromoji/

option1	main rule		option2
翌 (next)	一昨年	two yeas ago	より (from)
前 (before)	昨年	last year	まで (until)
入院前 (before hospitalizetion)	先月	last month	代 ('s)
入院後 (after hospitalizetion)	先週	last week	前半 (ealry)
来院から (after visit)	昨日	yesterday	後半 (last)
午前 (a.m.)	今年	this year	〜 (from)
午後 (p.m.)	今月	this month	~ (from)
発症から (after onset)	今週	this week	以上 (over)
発症してから (after onset)	今日	today	以下 (under)
治療してから (after care)	本日	today	から (from)
	来年	next year	時 (when)
	来月	next month	頃 (about)
	来週	next week	ごろ (about)
	翌日	tomorrow	ころ (about)
	再来週	the week after next	上旬 (early)
	明後日	day after tommorow	中旬 (mid)
	同年	same year	下旬 (late)
	同月	same month	春 (spring)
	同日	same day	夏 (summer)
	翌年	following year	秋 (fall)
	翌日	the next day	冬 (winter)
	翌朝	the next morning	朝 (morning)
	前日	the previous day	昼 (Noon)
	未明	early morning	夕 (evening)
	その後	after that	晩 (night)
	xx年	xx(year)	早朝 (early morning)
	xx月	xx(month)	明朝 (early morning)
	xx週間	xx(week)	以前 (before)
	xx日	xx(day)	以降 (after)
	xx時	xx(o'clock)	夕刻 (evening)
	xx分	xx(minutes)	ほど (about)

Table 1: our extraction rules for "age"

paper. Our rules are shown below. For a target word x,

age (subject's years of age with its suffix)

- If the detailed POS is *number*, apply rules in Table 1

hospital (hospital name)

- If one of following keywords appeared, then mark as *hospital*: 近医 (a near clinic or hospital), 当院 (this clinic or hospital), 同院 (same clinic or hospital)

- If POS is *noun* and detailed-POS is not *non-autonomous word*, or x is either "●", "○", "▲" or "■" (these symbols are used for manual de-identification due to the datasets are pseudo EHRs), then if suffix of x is one of following keywords, mark as *hospital*: 病院 (hospital or clinic), クリニック (clinic), 医院 (clinic)

sex

- If either 男性 (man), 女性 (woman), men, women, man, woman, then mark as *sex*

time (subject's time with its suffix)

- If detailed-POS is *number* and x is concatenation of four or two, or one digit number, slash and two-digit number (e.g. yyyy/mm or mm/dd) then mark as *time*

- If detailed-POS of x is *number* and followed with either 歳 (old), 才 (old), 代 ('s), mark as *time*

  - If it is further followed with either "より","まで","前半","後半","以上","以下","時","頃","ごろ","ころ","から","前半から","後半から","頃から","ごろから","ころから" and so on include these words in the marked *time*

1.2   CRF-based Method

As a classic machine learning baseline method of series labelling, we employed CRF. Many teams of the MedNLP1 de-identification task used CRF, including the second best team and the baseline system. We used the mallet library[3] for our CRF implementation. We defined five training features for each token as follows: part-of-speech (POS), detailed POS, character type (Hiragana, Katakana, Kanji, Number,), whether the token is included in our user dictionary or not, and a binary feature whether the token is beginning of sentence or not.

1.3   LSTM-based Method

We used a machine learning method that combines bi-LSTM and CRF using character-based and word-based embedding, originally suggested by other group (Misawa, Taniguchi, Yasuhiro, & Ohkuma, 2017). In this method, both characters and words are embedded into feature vectors. Then a bi-LSTM is trained using these feature vectors. Finally, a CRF is trained using the output of the bi-LSTM, using character level tags.

The original method uses a skip-gram model to embed words and characters by seven years of Mainichi newspaper articles of almost 500 million words. However, we did not use skip-gram model but GloVe[4], because GloVe is more effective than skip-gram (Pennington, Socher, & Manning, 2014). We used existing word vectors[5] instead of the pre-training in the original method. Our training and prediction is word based while the original method is character based. Our implementation is based on an open source API[6].

2   Experiment

2.1   Data

Our dataset is derived from two different sources. We used the MedNLP-1 de-identification task data to compare with previous work. This data includes pseudo EHRs of 50 patients. Although there were training data and test data provided, the test data is not publicly available now, which makes direct comparison with previous work impossible. However, both training and test data are written by the same writer and was originally one piece of data. Therefore, we assume that the training data can be regarded as almost same as the test data in their characteristics.

Another source is our dummy EHRs. We built our own dummy EHRs of 32 patients, assuming that the patients are hospitalized. Documents of our dummy EHRs were written by medical professionals (doctors). We added manual annotations for de-identification following a guideline of the MedNLP-1 task. These annotations were assigned by ourselves.

---

[3] http://mallet.cs.umass.edu/sequences.php
[4] https://nlp.stanford.edu/projects/glove/
[5] http://www.cl.ecei.tohoku.ac.jp/~m-suzuki/jawiki_vector/
[6] https://github.com/guillaumegenthial/sequence_tagging

All of these data are assigned five types of de-identification tag; *age*, *hospital*, *sex*, *time* and *person*. MedNLP-1 data includes 2244 sentences and our dummy EHRs include 8327 sentences. Writers hold doctor's licenses in both sources, assuming fake patients to describe pseudo medical records. However, descriptions are not similar between the two sources, probably because of the difference of the writers.

	Rule based	CRF	CRF Mix	LSTM	LSTM Mix
ALL	**84.23**	82.62	26.40	80.61	66.25
age	**93.43**	71.12	32.55	88.49	91.68
hospital	84.73	87.09	26.02	**92.90**	84.82
person	N/A	N/A	N/A	N/A	N/A
sex	**50.00**	16.67	14.65	0.00	**50.00**
time	82.61	83.88	26.12	**94.32**	87.53

Table 2: F1 value testing of MedNLP1's dataset. There were no "person" annotations in this dataset..

	Rule based	CRF	CRF Mix	LSTM	LSTM Mix
ALL	43.74	66.97	67.13	77.20	**77.66**
age	51.13	48.46	38.87	75.69	**79.16**
hospital	15.98	47.85	48.62	67.57	**68.70**
person	N/A	26.96	28.36	**65.60**	65.06
sex	93.75	35.92	90.08	45.51	**98.08**
time	49.48	71.28	70.60	89.17	**90.92**

Table 3: F1 value testing of dummy-EHR dataset. We did not implement rules for "person".

2.2 Evaluation method

Our evaluation method followed MedNLP-1, using the IOB2 tagging (Tjong & Jorn, 1999). We applied four hold cross validation, while the rule-based method does not require training data. From the two sources described above, we derived three datasets: MedNLP-1, dummy EHRs, and both of MedNLP1 and dummy EHRs (mixture). We trained CRF and LSTM by this mixture data. We divided each data source for our cross-fold validation to hold the same balance of these two sources. Our evaluation metrics is strict match of named entities.

3 Result and Discussion

3.1 Result of MedNLP-1 dataset

Table 2 shows the evaluation results. The best F1 score is by the rule-based method. This is because the rules were tuned for the MedNLP-1 data. In both of datasets, CRF and LSTM are not significantly different from the rule-based one. LSTM performed best for the *hospital* tag and the

	MedNLP1	dummy	Mix
ALL	26.40	67.13	47.10
age	32.55	38.87	36.28
hospital	26.02	48.62	32.27
person	N/A	28.36	18.04
sex	14.65	90.08	53.83
time	26.12	70.60	51.01

Table 4: F1 value of trained Mix dataset by CRF

	MedNLP1	dummy	Mix
ALL	66.25	77.66	76.21
age	91.68	79.16	86.35
hospital	84.82	68.70	72.18
person	N/A	65.06	65.06
sex	50.00	98.08	98.08
time	87.53	90.92	90.55

Table 5: F1 value of trained Mix dataset by LSTM

time tag, probably because they might have typical patterns of less variations. Total occurrence of *sex* is very small, *person* is zero, in the MedNLP-1 dataset.

3.2 Result of Dummy-EHR dataset

The result is shown at Table 3. The best score is performed by LSTM trained by the mixture dataset. Despite the data size is four times larger than that of MedNLP-1, the result is a little better. Regarding CRF, training with mixture dataset is worse than the dummy her dataset only. This is not true for LSTM, which shows better results when trained by mixture dataset.

3.3 Overall

We trained CRF and LSTM by the mixture dataset and evaluated on MedNLP-1, dummy-EHR and mixture dataset individually. These results are shown in Table 4 and Table. Regarding CRF, there is 26 point difference in average between evaluations with MedNLP-1 and dummy-EHR datasets. On the other hand, LSTM shows 7 point difference in average. These results suggest that the datasets are quite different, but LSTM absorbed these differences well.

4 Conclusion and Future Work

We implemented three different de-identification methods for Japanese EHRs. We applied these methods to three datasets derived from two different pseudo EHR sources with de-identification tags manually annotated. Our results show that LSTM is better than other methods also shows robustness between different sources compared with CRF. Machine learning methods could extract named entities of de-identification comparable to the rule based method that is manually tuned to specific target data. However, machine learning method is still weak for expressions with low occurrences. Combination of LSTM and rule-based method could be a future work.

Because the current performance is enough high among publicly available Japanese de-identification tools, we plan to apply our system to actual de-identification tasks in hospitals. Although it is still difficult to make real EHRs publicly available, we could use our large amount of EHRs inside our hospitals. Increasing the annotated dataset for such internal usage would be another future work.

5 Acknowledgement

This work was partially supported by Japanese Health Labour Sciences Research Grant and JST CREST.

References

Aramaki, Eiji, Mizuki Morita, Yoshinobu Kano, and Tomoko Ohkuma. "Overview of the NTCIR-11 MedNLP-2 Task." *Proceedings of the 11th NTCIR conference*, 2014: 147-154.

Aramaki, Eiji, Mizuki Morita, Yoshinobu Kano, and Tomoko Ohkuma. "Overview of the NTCIR-12 MedNLPDoc Task." *Proceedings of the 12th NTCIR Conference on Evaluation of Information Access Technologies* (Proceedings of the 12th NTCIR Conference on Evaluation of Information Access Technologies), 2016: 147-154.

Corinna, Cortes, and Vapnink Vlandimir. "Support-Vector Networks." *Machine Learning*, 1995: 20:273-297.

Hochreiter, Sepp, and Jürgen Schmidhunber. "LONG SHORT-TERM MEMORY." *NEURAL COMPUTATION 9(8)*, 1997: 1735-1780.

Imaichi, Osamu, Toshihiko Yanase, and Yoshiki Niwa. "A Comparison of Aramaki, E., Morita, M., Kano, Y., & Ohkuma, T. (2014). Overview of the NTCIR-11 MedNLP-2 Task. *Proceedings of the 11th NTCIR conference*, 147-154.

Aramaki, E., Morita, M., Kano, Y., & Ohkuma, T. (2016). Overview of the NTCIR-12 MedNLPDoc Task. *Proceedings of the 12th NTCIR Conference on Evaluation of Information Access Technologies*, 147-154.

Corinna, C., & Vlandimir, V. (1995). Support-Vector Networks. *Machine Learning*, 20:273-297.

Hochreiter, S., & Schmidhunber, J. (1997). LONG SHORT-TERM MEMORY. *NEURAL COMPUTATION 9(8)*, 1735-1780.

Imaichi, O., Yanase, T., & Niwa, Y. (2013). A Comparison of Rule-Based and Machine Learning Methods for Medical Information Extraction. *International Joint Conference on Natural Language Processing Workshop on Natural Language Processing for Medical and Healthcare Fields*, 38-42.

Lafferty, J. D., McCallum, A., & Pereira, F. C. (2001). Conditional Random Fields: Probabilistic Models for Segmenting and Labeling Sequence Data. *ICML2001*, 282-289.

Latanya, S. (2002). k-ANONYMITY: A MODEL FOR PROTECTING PRIVACY. *Int J Uncerainty, Fuzziness Knowledge-Based Systems*, 10:557-570.

Ma, X., & Hovy, E. (2016). End-to-end Sequence Labeling via Bi-directional LSTM-CNNs-CRF. *Proceedings of the 54th Annual Meeting of the Association for Computational Linguistics*, 1064-1074.

Maeda, W., Suzuki, Y., Yoshino, K., & Satoshi, N. (2016). Anonymization technique for Unstructured text data considering inference from context. *Forum on Information Tecnology Vol.2*, 47-48.

Ministry of Internal Affairs and Communication International Strategy Bureau, Information and Communication Economy Office. (2018, 7 8). *Survey research report on the weekly report on information distribution / accumulation volume.* Japan, Tokyo. Retrieved from http://www.soumu.go.jp/johotsusintokei/linkdata/h2 5_03_houkoku.pdf

Misawa, S., Taniguchi, M., Yasuhiro, M., & Ohkuma, T. (2017). Character-based Bidirectional LSTM-CRF with words and characters for Japanese Named Entity Recognition. *Proceedings of the First Workshop on Subword and Character Level Models in NLP*, 97-102.

Mizuki, M., Yoshinobu, K., Tomoko, O., Mai, M., & Eiji, A. (2013). *Overview of the NTCIR-10 MedNLP task.* Tokyo, Japan: In Proceedings of NTCIR-10.

Özlem, U., Yuan, L., & Peter, S. (2007). Evaluating the State-of-the-Art in Automatic De-identification. *J Am Med Inform Assoc, Sep-Oct*(14), 550-563.

Imaichi, Osamu, Toshihiko Yanase, and Yoshiki Niwa. "A Comparison of Aramaki, E., Morita, M., Kano, Y., & Ohkuma, T. (2014). Overview of the NTCIR-11 MedNLP-2 Task. *Proceedings of the 11th NTCIR conference*, 147-154.

Pennington, J., Socher, R., & Manning, C. D. (2014). GloVe: Global Vectors for Word Representation. *Proceedings of the 2014 Conference on Empirical Methods in Natural Language Processing (EMNLP)*, 1532-1543.

Sakishita, M., & Kano, Y. (2016). *Inference of ICD Codes from Japanese Medical Records by Searching Disease Names.* Clinical Natural Language Processing Workshop at the 26th International Conference on Computational Linguistics (COLING 2016).

Tjong, K. S., & Jorn, V. (1999). *Representing Text Chunks.* Proceedings of EACL '99.

Unsupervised Identification of Study Descriptors in Toxicology Research: An Experimental Study

Drahomira Herrmannova, Steven R. Young, Robert M. Patton, Christopher G. Stahl
Oak Ridge National Laboratory, TN, USA
{herrmannovad, youngsr, pattonrm, stahlcg}@ornl.gov

Nicole C. Kleinstreuer
NICEATM, NTP, NIEHS, NIH
Research Triangle Park, NC, USA
nicole.kleinstreuer@nih.gov

Mary S. Wolfe
NTP, NIEHS, NIH
Research Triangle Park, NC, USA
wolfe@niehs.nih.gov

Abstract

Identifying and extracting data elements such as study descriptors in publication full texts is a critical yet manual and labor-intensive step required in a number of tasks. In this paper we address the question of identifying data elements in an unsupervised manner. Specifically, provided a set of criteria describing specific study parameters, such as species, route of administration, and dosing regimen, we develop an unsupervised approach to identify text segments (sentences) relevant to the criteria. A binary classifier trained to identify publications that met the criteria performs better when trained on the candidate sentences than when trained on sentences randomly picked from the text, supporting the intuition that our method is able to accurately identify study descriptors.

Acknowledgments

Support for this research was provided by a grant from the National Institute of Environmental Health Sciences (AES 16002-001), National Institutes of Health to Oak Ridge National Laboratory.

This research was supported in part by an appointment to the Oak Ridge National Laboratory ASTRO Program, sponsored by the U.S. Department of Energy and administered by the Oak Ridge Institute for Science and Education.

This manuscript has been authored by UT-Battelle, LLC under Contract No. DE-AC05-00OR22725 with the U.S. Department of Energy. The United States Government retains and the publisher, by accepting the article for publication, acknowledges that the United States Government retains a non-exclusive, paid-up, irrevocable, worldwide license to publish or reproduce the published form of this manuscript, or allow others to do so, for United States Government purposes. The Department of Energy will provide public access to these results of federally sponsored research in accordance with the DOE Public Access Plan[1].

1 Introduction

Extracting data elements such as study descriptors from publication full texts is an essential step in a number of tasks including systematic review preparation (Jonnalagadda et al., 2015), construction of reference databases (Kleinstreuer et al., 2016), and knowledge discovery (Smalheiser, 2012). These tasks typically involve domain experts identifying relevant literature pertaining to a specific research question or a topic being investigated, identifying passages in the retrieved articles that discuss the sought after information, and extracting structured data from these passages. The extracted data is then analyzed, for example to assess adherence to existing guidelines (Kleinstreuer et al., 2016). Figure 1 shows an example text excerpt with information relevant to a specific task (assessment of adherence to existing guidelines (Kleinstreuer et al., 2016)) highlighted.

---

[1]http://energy.gov/downloads/
doe-public-access-plan

71

The intact female weanling version in the Organization for Economic Cooperation and Development (OECD) uterotrophic assay Test Guideline (TG) 440 is proposed as an alternative to the adult ovariectomized female version, because it does not involve surgical intervention (vs the ovariectomized version) and detects direct/indirect-acting estrogenic/anti-estrogenic substances (vs the ovariectomized version which detects only direct-acting estrogenic/antiestrogenic substances binding to the estrogen receptor). This validation study followed OECD TG 440, with six female weanling rats (postnatal day 21) per dose group and six treatment groups. Females were weighed and dosed once daily by oral gavage for three consecutive days, with one of six doses of 17α-ethinyl estradiol in corn oil at 5 ml kg^{-1} at 0 and 0.1–10 μg kg^{-1} per day. On postnatal day 24, the juvenile females were euthanized by CO_2 asphyxiation, weighed, livers weighed and uteri weighed wet and blotted. The presence or absence of vaginal patency was recorded. Absolute and relative (to terminal body weight) uterine wet and blotted weights and uterine luminal fluid weights were significantly increased at 3.0 and 10.0 (both P $<$ 0.01) μg kg^{-1} per day, and increased to ~140% of control values at 1.0 μg kg^{-1} per day (not statistically significantly). In vivo body weights, weight changes, feed consumption, liver weights and terminal body weights were unaffected. Vaginal patency was not acquired in any female at any dose, although vaginal puckering was observed in one female at 10.0 μg kg^{-1} per day. Therefore, this intact weanling uterotrophic assay is validated in our laboratory for use under US and European endocrine toxicity testing programs/legislation.

Figure 1: Text excerpt from a reference database of rodent uterotrophic bioassay publications (Kleinstreuer et al., 2016). The text in this example was manually annotated by one of the authors to highlight information relevant to guidelines for performing uterotrophic bioassays set forth by (OECD, 2007).

Extracting the data elements needed in these tasks is a time-consuming and at present a largely manual process which requires domain expertise. For example, in systematic review preparation, information extraction generally constitutes the most time consuming task (Tsafnat et al., 2014). This situation is made worse by the rapidly expanding body of potentially relevant literature with more than one million papers added into PubMed each year (Landhuis, 2016). Therefore, data annotation and extraction presents an important challenge for automation.

A typical approach to automated identification of relevant information in biomedical texts is to infer a prediction model from labeled training data – such a model can then be used to assign predicted labels to new data instances. However, obtaining training data for creating such prediction models can be very costly as it involves the step which these models are trying to automate – manual data extraction. Furthermore, depending on the task at hand, the types of information being extracted may vary significantly. For example, in systematic reviews of randomized controlled trials this information generally includes the *patient* group, the *intervention* being tested, the *comparison*, and the *outcomes* of the study (PICO elements) (Tsafnat et al., 2014). In toxicology research the extraction may focus on routes of exposure, dose, and necropsy timing (Kleinstreuer et al., 2016). Previous work has largely focused on identifying specific pieces of information such as biomedical events (Gonzalez et al., 2015) or PICO elements (Jonnalagadda et al., 2015). However, depending on the domain and the end goal of the extraction, these may be insufficient to comprehensively describe a given study.

Therefore, in this paper we focus on *unsupervised methods* for identifying text segments (such as sentences or fixed length sequences of words) relevant to the information being extracted. We develop a model that can be used to identify text segments from text documents without labeled data and that only requires the current document itself, rather than an entire training corpus linked to the target document. More specifically, we utilize representation learning methods (Mikolov et al., 2013a), where words or phrases are embedded into the same vector space. This allows us to compute semantic relatedness among text fragments, in particular sentences or text segments in a given document and a short description of the type of information being extracted from the document, by using similarity measures in the feature space. The model has the potential to speed up identification of relevant segments in text and therefore to expedite annotation of domain specific information without reliance on costly labeled data.

We have developed and tested our approach on a reference database of rodent uterotropic bioassays[2] (Kleinstreuer et al., 2016) which are labeled according to their adherence to test guidelines set forth in (OECD, 2007). Each study in the database is assigned a label determining whether or not it met each of six main criteria defined by the

[2] https://ntp.niehs.nih.gov/pubhealth/ evalatm/test-method-evaluations/ endocrine-disruptors/ref-data/edhts.html

guidelines; however, the database does not contain sentence-level annotations or any information about where the criteria was mentioned in each publication. Due to the lack of fine-grained annotations, supervised learning methods cannot be easily applied to aid annotating new publications or to annotate related but distinct types of studies. This database therefore presents an ideal use-case for unsupervised approaches.

While our approach doesn't require any labeled data to work, we use the labels available in the dataset to evaluate the approach. We train a binary classification model for identifying publications which satisfied given criteria and show the model performs better when trained on relevant sentences identified by our method than when trained on sentences randomly picked from the text. Furthermore, for three out of the six criteria, a model trained solely on the relevant sentences outperforms a model which utilizes full text. The results of our evaluation support the intuition that semantic relatedness to criteria descriptions can help in identifying text sequences discussing sought after information.

There are two main contributions of this work. We present an unsupervised method that employs representation learning to identify text segments from publication full text which are relevant to/contain specific sought after information (such as number of dose groups). In addition, we explore a new dataset which hasn't been previously used in the field of information extraction.

The remainder of this paper is organized as follows. In the following section we provide more details of the task and the dataset used in this study. In Section 3 we describe our approach. In Section 4 we evaluate our model and discuss our results. In Section 5 we compare our work to existing approaches. Finally, in Section 6 we provide ideas for further study.

2 The Task and the Data

This section provides more details about the specific task and the dataset used in our study which motivated the development of our model.

2.1 Task Description

Significant efforts in toxicology research are being devoted towards developing new *in vitro* methods for testing chemicals due to the large number of untested chemicals in use ($>$75,000-80,000

(Judson et al., 2009; Kleinstreuer et al., 2016)) and the cost and time required by existing *in vivo* methods (2-3 years and millions of dollars per chemical (Judson et al., 2009)). To facilitate the development of novel *in vitro* methods and assess the adherence to existing study guidelines, a curated database of high-quality *in vivo* rodent uterotrophic bioassay data extracted from research publications has recently been developed and published (Kleinstreuer et al., 2016).

The creation of the database followed the study protocol design set forth in (OECD, 2007), which is composed of six minimum criteria (MC, Table 1). An example of information pertaining to the criteria is shown in Figure 1. Only studies which met all six minimum criteria were considered guideline-like (GL) and were included in a follow-up detailed study and the final database (Kleinstreuer et al., 2016). However, of the 670 publications initially considered for inclusion, only 93 ($\sim$14%) were found to contain studies which met all six MC and could therefore be included in the final database; the remaining 577 publications could not be used in the final reference set. Therefore, significant time and resources could be saved by automating the identification and extraction of the MC.

While each study present in the database is assigned a label for each MC determining whether a given MC was met and the pertinent protocol information was manually extracted, there exist no fine-grained text annotations showing the exact location within each publication's full text where a given criteria was met. Therefore, our goal was to develop a model not requiring detailed text annotations that could be used to expedite the annotation of new publications being added into the database and potentially support the development of new reference databases focusing on different domains and sets of guidelines. Due to the lack of detailed annotations, our focus was on identification of potentially relevant text segments.

2.2 The Dataset

The version of the database which contains both GL and non-GL studies consists of 670 publications (spanning the years 1938 through 2014) with results from 2,615 uterotrophic bioassays. Specifically, each entry in the database describes one study, and studies are linked to publications using PubMed reference numbers (PMIDs). Each study

Criteria name	Description
MC 1: Animal model	Immature rats, ovariectomized (OVX) adult rats, or OVX adult mice are acceptable (immature mice are not acceptable). OVX animals: OVX should be performed between six and eight weeks of age (allowing at least 14 days post-surgery before dosing for rats and seven days post-surgery for mice). Immature rats: dosing should begin between postnatal day (PND) 18 and PND 21, and be completed by PND 25.
MC 2: Group size	Each control group should have a minimum of three animals and each test group should have a minimum of five animals.
MC 3: Route of administration	Acceptable routes of administration: oral gavage (p.o.), subcutaneous (s.c.) injection, or intraperitoneal (i.p.) injection.
MC 4: Number of dose groups	Minimum of two dose level groups. Must have positive control and negative control.
MC 5: Dosing interval	Dosing for a minimum of three consecutive days. Complete by PND 25 in immature animals.
MC 6: Necropsy timing	Should be carried out 18-36 hours after the last dose.

Table 1: Minimum criteria for guideline-like studies. The descriptions are reprinted here from (Kleinstreuer et al., 2016).

is assigned seven 0/1 labels – one for each of the minimum criteria and one for the overall GL/non-GL label. The database also contains more detailed subcategories for each label (for example "species" label for MC 1) which were not used in this study. The publication PDFs were provided to us by the database creators. We have used the Grobid[3] library to convert the PDF files into structured text. After removing documents with missing PDF files and documents which were not converted successfully, we were left with 624 full text documents.

Each publication contains on average 3.7 studies (separate bioassays), 194 publications contain a single study, while the rest contain two or more studies (with 82 being the most bioassays per publication). The following excerpt shows an example sentence mentioning multiple bioassays (with different study protocols):

> *With the exception of the first study (experiment 1), which had group sizes of 12, all other studies had group sizes of 8.*

For this experiment we did not distinguish between publications describing a single or multiple studies. Instead, our focus was on retrieving all text segments (which may be related to multiple studies) relevant to each of the criteria. For

Criteria	0	1	Total	% of 1
MC 1	414	175	589	29.71
MC 2	35	577	612	94.28
MC 3	70	536	606	88.45
MC 4	309	206	515	40.00
MC 5	96	490	586	83.62
MC 6	228	340	568	59.86
GL	522	72	594	12.12

Table 2: Label statistics. Column *0* shows number of publications per MC which did not meet the criteria and column *1* shows number of publications which met the criteria. The last column in the table shows proportion of positive (i.e. criteria met) labels.

each MC, if a document contained multiple studies with different labels, we discarded that document from our analysis of that criteria; if a document contained multiple studies with the same label, we simply combine all those labels into a single label. Table 2 shows the final size of the dataset.

3 Approach

In this section we describe the method we have used for retrieving text segments related to the criteria described in the previous section. The intuition is based off question answering systems. We treat the criteria descriptions (Table 1) as the question and the text segments within the publication that discusses the criteria as the answer. Given a full text publication, the goal is to find the text segments most likely to contain the answer.

We represent the criteria descriptions and text segments extracted from the documents as vectors of features, and utilize relatedness measures to retrieve text segments most similar to the descriptions. A similar step is typically performed by most question answering (QA) systems – in QA systems both the input documents and the question are represented as a sequence of embedding vectors and a retrieval system then compares the document and question representations to retrieve text segments most likely containing the answer (Mishra and Jain, 2016).

To account for the variations in language that can be used to describe the criteria, we represent words as vectors generated using Word2Vec (Mikolov et al., 2013a). The following two excerpts show two different ways MC 6 was described in text:

> *Animals were killed 24 h after being injected and their uteri were removed and weighed.*

> *All animals were euthanized by exposure to ethyl ether 24 h after the final treatment.*

We hypothesize that the use of word embedding features will allow us to detect relevant words which are not present in the criteria descriptions. (Mikolov et al., 2013b) have shown that an important feature of Word2Vec embeddings is that similar words will have similar vectors because they appear in similar contexts. We utilize this feature to calculate similarity between the criteria descriptions and text segments (such as sentences) extracted from each document. A high-level overview of our approach is shown in Figure 2.

We use the following method to retrieve the most relevant text segments:

Segment extraction: First, we break each document down into shorter sequences such as sentences or word sequences of fixed length. While the first option (sentences) results in text which is easier to process, it has the disadvantage of resulting in sequences of varying length which may affect the resulting similarity value. However, for simplicity, in this study we utilize the sentence version.

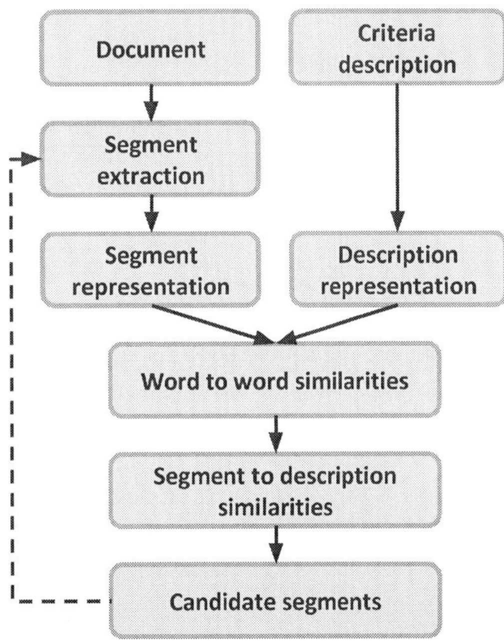

Figure 2: High level overview of our approach. The dotted line represents an optional step of finding smaller sub-segments within the candidate segments. For example, in our case, we first retrieve the most similar sentences and in the second step find most similar continuous 5-grams found withing those sentences.

Segment/description representation: We represent each sequence and the input description as a sequence of vector representations. For this study we have utilized Word2Vec embeddings (Mikolov et al., 2013a) trained using the Gensim library on our corpus of 624 full text publications.

Word to word similarities: Next we calculate similarity between each word vector from each sequence s_i and each word vector from the input description d using *cosine similarity*. The output of this step is a similarity matrix $\mathbf{S}_i \in \mathbb{R}^{N_i \times M_d}$ for each sequence s_i, where N_i is the number of unique words in the sequence and M_d is the number of unique words in the description d.

Segment to description similarities: To obtain a similarity value representing the relatedness of each sequence to the input description we first convert each input matrix $\mathbf{S}_i$ into a vector $v_i \in \mathbb{R}^{N_i}$ by choosing the maximum similarity value for each word in the sequence, that is $v_i = \max_{rows}(\mathbf{S}_i)$. Each sequence is then assigned a similarity value $r_i \in \mathbb{R}$ which is calculated as $r_i = \text{avg}(v_i)$. In the future we are planning to experiment with different ways of calculating relatedness of the sequences to the descriptions, such as with computing similarity of embeddings created from the text fragments using approaches like Doc2Vec (Le and Mikolov, 2014). In this study, after finding the top sentences, we further break each sentence down into continuous n-grams to find the specific part of the sentence discussing the MC. We repeat the same process described above to calculate the relatedness of each n-gram to the description.

Candidate segments: For each document we select the top k text segments (sentences in the first step and 5-grams in the second step) most similar to the description.

3.1 Example Results

Figures 3, 4, and 5 show example annotations generated using our method for the first three criteria. For this example we ran our method on the abstract of the target document rather than the full text and highlighted only the single most similar sentence. The abstract used to produce these figures is the same as the abstract shown in Figure 1. In all three figures, the lighter yellow color highlights the sentence which was found to be the most similar to a given MC description, the darker red color shows the top 5-gram found within the top sentence, and the bold underlined text is the text we are looking for (the correct answer). Annotations generated for the remaining three criteria are shown in Appendix A.

Due to space limitations, Figures 3, 4, and 5 show results generated on abstracts rather than on full text; however, we have observed similarly accurate results when we applied our method to full text. The only difference between the abstracts and the full text version is how many top sentences we retrieved. When working with abstracts only, we observed that if the criteria was discussed in the abstract, it was generally sufficient to retrieve the single most similar sentence. However, as the criteria may be mentioned in multiple places within the document, when working with full text documents we have retrieved and analyzed the top k sentences instead of just a single sentence. In this case we have typically found the correct sentence/sentences among the top 5 sentences. We have also observed that the similar sentences which don't discuss the criteria directly (i.e. the "incorrect" sentences) typically discuss related topics. For example, consider the following three sentences:

After weaning on pnd 21, the dams were euthanized by CO2 asphyxiation and the juvenile females were individually housed.

Six CD(SD) rat dams, each with reconstituted litters of six female pups, were received from Charles River Laboratories (Raleigh, NC, USA) on offspring postnatal day (pnd) 16.

This validation study followed OECD TG 440, with six female <u>weanling rats (postnatal day 21)</u> per dose group and six treatment groups.

These three sentences were extracted from the abstract and the full text of a single document (document 20981862, the abstract of which is shown in Figures 1 and 3-8). These three sentences were retrieved as the most similar to MC 1, with similarity scores of 70.61, 65.31, and 63.69, respectively. The third sentence contains the "answer" to MC 1 (underlined). However, it can be seen the top two sentences also discuss the animals used in the study (more specifically, the sentences discuss the animals' housing and their origin).

4 Evaluation

The goal of this experiment was to explore empirically whether our approach truly identifies mentions of the minimum criteria in text. As we did not have any fine-grained annotations that could be used to directly evaluate whether our model identifies the correct sequences, we have used a different methodology. We have utilized the existing 0/1 labels which were available in the database (these were discussed in Section 2) to train one binary classifier for each MC. The task of each of the classifiers is to determine whether a publication met the given criteria or not. We have then compared a baseline classifier trained on all full text with three other models:

- A model which, instead of all full text, utilized only the top k sentences most similar to the given MC. The top k sentences were identified using our model introduced in the previous section.

- A model which utilized only the k least similar sentences.

The intact female weanling version in the Organization for Economic Cooperation and Development (OECD) uterotrophic assay Test Guideline (TG) 440 is proposed as an alternative to the adult ovariectomized female version , because it does not involve surgical intervention (vs the ovariectomized version) and detects direct/indirect-acting estrogenic/anti-estrogenic substances (vs the ovariectomized version which detects only direct-acting estrogenic/anti-estrogenic substances binding to the estrogen receptor) . This validation study followed OECD TG 440 , with six female weanling rats (postnatal day 21) per dose group and six treatment groups . Females were weighed and dosed once daily by oral gavage for three consecutive days , with one of six doses of 17alpha-ethinyl estradiol in corn oil at 5 ml kg (-) (1) at 0 and 0.1-10 microg kg (-) (1) per day . On postnatal day 24 , the juvenile females were euthanized by CO (2) asphyxiation , weighed , livers weighed and uteri weighed wet and blotted . The presence or absence of vaginal patency was recorded . Absolute and relative (to terminal body weight) uterine wet and blotted weights and uterine luminal fluid weights were significantly increased at 3.0 and 10.0 (both P < 0.01) microg kg (-) (1) per day , and increased to ~140 % of control values at 1.0 microg kg (-) (1) per day (not statistically significantly) . In vivo body weights , weight changes , feed consumption , liver weights and terminal body weights were unaffected . Vaginal patency was not acquired in any female at any dose , although vaginal puckering was observed in one female at 10.0 microg kg (-) (1) per day . Therefore , this intact weanling uterotrophic assay is validated in our laboratory for use under US and European endocrine toxicity testing programs/legislation .

Figure 3: Annotations generated using our method for the abstract from Figure 1. The sentence which was found to be the most similar to the description for "MC 1: Animal model" is highlighted in yellow and the most similar sequence of words within that sentence is highlighted in red. The text we are looking for is highlighted with bold underlined text. For this example we ran our method on the abstract of the target document rather than the full text and highlighted only the single most similar sentence.

The intact female weanling version in the Organization for Economic Cooperation and Development (OECD) uterotrophic assay Test Guideline (TG) 440 is proposed as an alternative to the adult ovariectomized female version , because it does not involve surgical intervention (vs the ovariectomized version) and detects direct/indirect-acting estrogenic/anti-estrogenic substances (vs the ovariectomized version which detects only direct-acting estrogenic/anti-estrogenic substances binding to the estrogen receptor) . This validation study followed OECD TG 440 , with six female weanling rats (postnatal day 21) per dose group and six treatment groups . Females were weighed and dosed once daily by oral gavage for three consecutive days , with one of six doses of 17alpha-ethinyl estradiol in corn oil at 5 ml kg (-) (1) at 0 and 0.1-10 microg kg (-) (1) per day . On postnatal day 24 , the juvenile females were euthanized by CO (2) asphyxiation , weighed , livers weighed and uteri weighed wet and blotted . The presence or absence of vaginal patency was recorded . Absolute and relative (to terminal body weight) uterine wet and blotted weights and uterine luminal fluid weights were significantly increased at 3.0 and 10.0 (both P < 0.01) microg kg (-) (1) per day , and increased to ~140 % of control values at 1.0 microg kg (-) (1) per day (not statistically significantly) . In vivo body weights , weight changes , feed consumption , liver weights and terminal body weights were unaffected . Vaginal patency was not acquired in any female at any dose , although vaginal puckering was observed in one female at 10.0 microg kg (-) (1) per day . Therefore , this intact weanling uterotrophic assay is validated in our laboratory for use under US and European endocrine toxicity testing programs/legislation .

Figure 4: Annotations generated using our method for "MC 2: Group size". The highlighting used is the same as in Figure 3.

- A model which utilized only k random sentences (but none of the top or bottom k sentences – the sentences were chosen at random from the interval $(k, n - k)$ where n is the number of sentences in the document and where sentences are sorted from the most similar to the least similar).

The only difference between the four models is which sentences from each document are passed to the classifier for training and testing. The intuition is that a classifier utilizing the correct sentences should outperform both other models.

To avoid selecting the same sentences across the three models we removed documents which contained less than $3 * k$ sentences (Table 3, row *Number of documents* shows how many documents satisfied this condition). In all of the experiments presented in this section, the publication full text was tokenized, lower-cased, stemmed, and stop words were removed. All models used a Bernoulli Naïve Bayes classifier (scikit-learn implementation which used a uniform class prior) trained on binary occurrence matrices created using 1-3-grams extracted from the publications, with n-grams appearing in only one document removed. The complete results obtained from leave-one-out cross validation are shown in Table 3. In all cases we report classification accuracy. In the case of the *random-k sentences* model the accuracy was averaged over 10 runs of the model.

We compare the results to two baselines: (1) a baseline obtained by classifying all documents as belonging to the majority class (*baseline 1* in Table 3) and (2) a baseline obtained using the same setup (features and classification algorithm) as in the case of the *top-/random-/bottom-k sentences* models but which utilized all full text instead of selected sentences extracted from the text only (*baseline 2* in Table 3).

4.1 Results analysis

Table 3 shows that for four out of the six criteria (MC 1, MC 4, MC 5, and MC 6) the *top-k sentences* model outperforms *baseline 1* as well the *bottom-k* and the *random-k sentences* models by a significant margin. Furthermore, for three of the

The intact female weanling version in the Organization for Economic Cooperation and Development (OECD) uterotrophic assay Test Guideline (TG) 440 is proposed as an alternative to the adult ovariectomized female version , because it does not involve surgical intervention (vs the ovariectomized version) and detects direct/indirect-acting estrogenic/anti-estrogenic substances (vs the ovariectomized version which detects only direct-acting estrogenic/anti-estrogenic substances binding to the estrogen receptor) . This validation study followed OECD TG 440 , with six female weanling rats (postnatal day 21) per dose group and six treatment groups . Females were weighed and dosed once daily by oral gavage for three consecutive days , with one of six doses of 17alpha-ethinyl estradiol in corn oil at 5 ml kg (-) (1) at 0 and 0.1-10 microg kg (-) (1) per day . On postnatal day 24 , the juvenile females were euthanized by CO (2) asphyxiation , weighed , livers weighed and uteri weighed wet and blotted . The presence or absence of vaginal patency was recorded . Absolute and relative (to terminal body weight) uterine wet and blotted weights and uterine luminal fluid weights were significantly increased at 3.0 and 10.0 (both P < 0.01) microg kg (-) (1) per day , and increased to ~140 % of control values at 1.0 microg kg (-) (1) per day (not statistically significantly) . In vivo body weights , weight changes , feed consumption , liver weights and terminal body weights were unaffected . Vaginal patency was not acquired in any female at any dose , although vaginal puckering was observed in one female at 10.0 microg kg (-) (1) per day . Therefore , this intact weanling uterotrophic assay is validated in our laboratory for use under US and European endocrine toxicity testing programs/legislation .

Figure 5: Annotations generated using our method for "MC 3: Route of administration". The highlighting used is the same as in Figure 3.

Approach	MC1	MC2	MC3	MC4	MC5	MC6
Baseline 1: Most frequent label	70.35	94.43	88.74	59.48	84.30	60.44
Baseline 2: All full text	78.25	92.06	89.59	67.94	84.83	74.05
Top-k sentence	76.84	91.55	87.71	68.35	88.54	74.23
Bottom-k sentences	70.00	91.39	88.23	63.10	80.60	63.70
Random-k sentences	73.26	93.72	88.43	65.65	85.29	68.28
Number of documents	570	592	586	496	567	551
Number of pos. labels	169	559	520	201	478	333

Table 3: Evaluation results.

six criteria (MC 4, MC 5, and MC 6) the *top-k sentences* model also outperforms the *baseline 2* model (model which utilized all full text). This seems to confirm our hypothesis that semantic relatedness of sentences to the criteria descriptions helps in identifying sentences discussing the criteria. These seems to be the case especially given that for three of the six criteria the *top-k sentences* model outperforms the model which utilizes all full text (*baseline 2*) despite being given less information to learn from (selected sentences only in the case of the *top-k sentences* model vs. all full text in the case of the *baseline 2* model).

For two of the criteria (MC 2 and MC 3) this is not the case and the *top-k sentences* model performs worse than both other models in the case of MC 3 and worse than the *random-k* model in the case of MC 2. One possible explanation for this is class imbalance. In the case of MC 2, only 33 out of 592 publications (5.57%) represent negative examples (Table 3). As the *top-k sentences* model picks only sentences closely related to MC 2, it is possible that due to the class imbalance the top sentences don't contain enough negative examples to learn from. On the other hand, the *bottom-k* and *random-k sentences* models may select text not necessarily related to the criteria but

potentially containing linguistic patterns which the model learns to associate with the criteria; for example, certain chemicals may require the use of a certain study protocol which may not be aligned with the MC and the model may key in on the appearance of these chemicals in text rather than the appearance of MC indicators. The situation is similar in the case of MC 3. We would like to emphasize that the goal of this experiment was not to achieve state-of-the-art results but to investigate empirically the viability of utilizing semantic relatedness of text segments to criteria descriptions for identifying relevant segments.

5   Related Work

In this section we present studies most similar to our work. We focus on unsupervised methods for information extraction from biomedical texts.

Many methods for biomedical data annotation and extraction exist which utilize labeled data and supervised learning approaches ((Liu et al., 2016) and (Gonzalez et al., 2015) provided a good overview of a number of these methods); however, unsupervised approaches in this area are much scarcer. One such approach has been introduced by (Zhang and Elhadad, 2013), who have proposed a model for unsupervised Named Entity Recognition. Similar to our approach, their model is based on calculating the similarity between vector representations of candidate phrases and existing entities. However, their vector representations are created using a combination of TF-IDF weights and word context information, and their method relies on a terminology. More recently, (Chen and Sokolova, 2018) have utilized Word2Vec and Doc2Vec embeddings for unsupervised sentiment classification in medical discharge summaries.

A number of previous studies have focused on unsupervised extraction of relations such as protein-protein interactions (PPI) from biomedical texts. For example, (Quan et al., 2014) have utilized several techniques, namely kernel-based pattern clustering and dependency parsing, to extract PPI from biomedical texts. (Alicante et al., 2016) have introduced a system for unsupervised extraction of entities and relations between these entities from clinical texts written in Italian, which utilized a thesaurus for extraction of entities and clustering methods for relation extraction. (Rink and Harabagiu, 2011) also used clinical texts and proposed a generative model for unsupervised relation extraction. Another approach focusing on relation extraction has been proposed by (Madkour et al., 2007). Their approach is based on constructing a graph which is used to construct domain-independent patterns for extracting protein-protein interactions.

A similar but distinct approach to unsupervised extraction is distant supervision. Similarly as unsupervised extraction methods, distant supervision methods don't require any labeled data, but make use of weakly labeled data, such as data extracted from a knowledge base. Distant supervision has been applied to relation extraction (Liu et al., 2014), extraction of gene interactions (Mallory et al., 2015), PPI extraction (Thomas et al., 2012; Bobić et al., 2012), and identification of PICO elements (Wallace et al., 2016). The advantage of our approach compared to the distantly supervised methods is that it does not require any underlying knowledge base or a similar source of data.

6 Conclusions and Future Work

In this paper we presented a method for unsupervised identification of text segments relevant to specific sought after information being extracted from scientific documents. Our method is entirely unsupervised and only requires the current document itself and the input descriptions instead of corpus linked to this document. The method utilizes short descriptions of the information being extracted from the documents and the ability of word embeddings to capture word context. Consequently, it is domain independent and can potentially be applied to another set of documents and criteria with minimal effort. We have used the method on a corpus of toxicology documents and a set of guideline protocol criteria needed to be extracted from the documents. We have shown the identified text segments are very accurate. Furthermore, a binary classifier trained to identify publications that met the criteria performed better when trained on the candidate sentences than when trained on sentences randomly picked from the text, supporting our intuition that our method is able to accurately identify relevant text segments from full text documents.

There are a number of things we plan on investigating next. In our initial experiment we have utilized criteria descriptions which were not designed to be used by our model. One possible improvement of our method could be replacing the current descriptions with example sentences taken from the documents containing the sought after information. We also plan on testing our method on an annotated dataset, for example using existing annotated PICO element datasets (Boudin et al., 2010).

References

Anita Alicante, Anna Corazza, Francesco Isgrò, and Stefano Silvestri. 2016. Unsupervised entity and relation extraction from clinical records in italian. *Computers in biology and medicine*, 72:263–275.

Tamara Bobić, Roman Klinger, Philippe Thomas, and Martin Hofmann-Apitius. 2012. Improving distantly supervised extraction of drug-drug and protein-protein interactions. In *Proceedings of the Joint Workshop on Unsupervised and Semi-Supervised Learning in NLP*, pages 35–43. Association for Computational Linguistics.

Florian Boudin, Jian-Yun Nie, Joan C Bartlett, Roland Grad, Pierre Pluye, and Martin Dawes. 2010. Combining classifiers for robust pico element detection. *BMC medical informatics and decision making*, 10(1):29.

Qufei Chen and Marina Sokolova. 2018. Word2vec and doc2vec in unsupervised sentiment analysis of clinical discharge summaries. *arXiv preprint arXiv:1805.00352*.

Graciela H Gonzalez, Tasnia Tahsin, Britton C Goodale, Anna C Greene, and Casey S Greene. 2015. Recent advances and emerging applications in text and data mining for biomedical discovery. *Briefings in bioinformatics*, 17(1):33–42.

Siddhartha R. Jonnalagadda, Pawan Goyal, and Mark D. Huffman. 2015. Automating data extraction in systematic reviews: a systematic review. *Systematic Reviews*, 4(1).

Richard Judson, Ann Richard, David J Dix, Keith Houck, Matthew Martin, Robert Kavlock, Vicki Dellarco, Tala Henry, Todd Holderman, Philip Sayre, et al. 2009. The toxicity data landscape for environmental chemicals. *Environmental health perspectives*, 117(5):685.

Nicole C. Kleinstreuer, Patricia C. Ceger, David G. Allen, Judy Strickland, Xiaoqing Chang, Jonathan T. Hamm, and Warren M. Casey. 2016. A curated database of rodent uterotrophic bioactivity. *Environmental Health Perspectives*, 124(5).

Esther Landhuis. 2016. Scientific literature: Information overload. *Nature*, 535(7612):457–458.

Quoc Le and Tomas Mikolov. 2014. Distributed representations of sentences and documents. In *International Conference on Machine Learning*, pages 1188–1196.

Feifan Liu, Jinying Chen, Abhyuday Jagannatha, and Hong Yu. 2016. Learning for biomedical information extraction: Methodological review of recent advances. *arXiv preprint arXiv:1606.07993*.

Mengwen Liu, Yuan Ling, Yuan An, and Xiaohua Hu. 2014. Relation extraction from biomedical literature with minimal supervision and grouping strategy. In *Bioinformatics and Biomedicine (BIBM), 2014 IEEE International Conference on*, pages 444–449. IEEE.

Amgad Madkour, Kareem Darwish, Hany Hassan, Ahmed Hassan, and Ossama Emam. 2007. Bionoculars: extracting protein-protein interactions from biomedical text. In *Proceedings of the Workshop on BioNLP 2007: Biological, Translational, and Clinical Language Processing*, pages 89–96. Association for Computational Linguistics.

Emily K Mallory, Ce Zhang, Christopher Ré, and Russ B Altman. 2015. Large-scale extraction of gene interactions from full-text literature using deepdive. *Bioinformatics*, 32(1):106–113.

Tomas Mikolov, Kai Chen, Greg Corrado, and Jeffrey Dean. 2013a. Efficient estimation of word representations in vector space. In *ICLR Workshop*.

Tomas Mikolov, Wen-tau Yih, and Geoffrey Zweig. 2013b. Linguistic regularities in continuous space word representations. In *Proceedings of the 2013 Conference of the North American Chapter of the Association for Computational Linguistics: Human Language Technologies*, pages 746–751.

Amit Mishra and Sanjay Kumar Jain. 2016. A survey on question answering systems with classification. *Journal of King Saud University-Computer and Information Sciences*, 28(3):345–361.

OECD. 2007. Test No. 440: Uterotrophic Bioassay in Rodents. In *OECD Guidelines for the Testing of Chemicals, Section 4*. OECD Publishing, Paris.

Changqin Quan, Meng Wang, and Fuji Ren. 2014. An unsupervised text mining method for relation extraction from biomedical literature. *PloS one*, 9(7):e102039.

Bryan Rink and Sanda Harabagiu. 2011. A generative model for unsupervised discovery of relations and argument classes from clinical texts. In *Proceedings of the Conference on Empirical Methods in Natural Language Processing*, pages 519–528. Association for Computational Linguistics.

Neil R. Smalheiser. 2012. Literature-based discovery: Beyond the ABCs. *Journal of the American Society for Information Science and Technology*, 63(2):218–224.

Philippe Thomas, Tamara Bobić, Ulf Leser, Martin Hofmann-Apitius, and Roman Klinger. 2012. Weakly labeled corpora as silver standard for drug-drug and protein-protein interaction. In *Proceedings of the Workshop on Building and Evaluating Resources for Biomedical Text Mining (BioTxtM) on Language Resources and Evaluation Conference (LREC)*.

Guy Tsafnat, Paul Glasziou, Miew Keen Choong, Adam Dunn, Filippo Galgani, and Enrico Coiera. 2014. Systematic review automation technologies. *Systematic reviews*, 3(1):74.

Byron C Wallace, Joel Kuiper, Aakash Sharma, Mingxi Zhu, and Iain J Marshall. 2016. Extracting pico sentences from clinical trial reports using supervised distant supervision. *The Journal of Machine Learning Research*, 17(1):4572–4596.

Shaodian Zhang and Noémie Elhadad. 2013. Unsupervised biomedical named entity recognition: Experiments with clinical and biological texts. *Journal of biomedical informatics*, 46(6):1088–1098.

A Supplemental Material

This section provides additional details and results. Figures 6, 7, and 8 show example annotations generated for criteria MC 4, MC 5, and MC 6.

The intact female weanling version in the Organization for Economic Cooperation and Development (OECD) uterotrophic assay Test Guideline (TG) 440 is proposed as an alternative to the adult ovariectomized female version , because it does not involve surgical intervention (vs the ovariectomized version) and detects direct/indirect-acting estrogenic/anti-estrogenic substances (vs the ovariectomized version which detects only direct-acting estrogenic/anti-estrogenic substances binding to the estrogen receptor) . This validation study followed OECD TG 440 , with six female weanling rats (postnatal day 21) per dose group and six treatment groups . Females were weighed and dosed once daily by oral gavage for three consecutive days , with one of six doses of 17alpha-ethinyl estradiol in corn oil at 5 ml kg (-) (1) at 0 and 0.1-10 microg kg (-) (1) per day . On postnatal day 24 , the juvenile females were euthanized by CO (2) asphyxiation , weighed , livers weighed and uteri weighed wet and blotted . The presence or absence of vaginal patency was recorded . Absolute and relative (to terminal body weight) uterine wet and blotted weights and uterine luminal fluid weights were significantly increased at 3.0 and 10.0 (both P < 0.01) microg kg (-) (1) per day , and increased to ~140 % of control values at 1.0 microg kg (-) (1) per day (not statistically significantly) . In vivo body weights , weight changes , feed consumption , liver weights and terminal body weights were unaffected . Vaginal patency was not acquired in any female at any dose , although vaginal puckering was observed in one female at 10.0 microg kg (-) (1) per day . Therefore , this intact weanling uterotrophic assay is validated in our laboratory for use under US and European endocrine toxicity testing programs/legislation .

Figure 6: Annotations generated using our method for abstract from Figure 1. The sentence which was found to be the most similar to the description for "MC 4: Number of dose groups" is highlighted in yellow and the most similar sequence of words within that sentence is highlighted in red. The text we are looking for is highlighted with bold underlined text. For this example we ran our method on the abstract of the target document rather than the full text and highlighted only the single most similar sentence.

The intact female weanling version in the Organization for Economic Cooperation and Development (OECD) uterotrophic assay Test Guideline (TG) 440 is proposed as an alternative to the adult ovariectomized female version , because it does not involve surgical intervention (vs the ovariectomized version) and detects direct/indirect-acting estrogenic/anti-estrogenic substances (vs the ovariectomized version which detects only direct-acting estrogenic/anti-estrogenic substances binding to the estrogen receptor) . This validation study followed OECD TG 440 , with six female weanling rats (postnatal day 21) per dose group and six treatment groups . Females were weighed and dosed once daily by oral gavage for three consecutive days , with one of six doses of 17alpha-ethinyl estradiol in corn oil at 5 ml kg (-) (1) at 0 and 0.1-10 microg kg (-) (1) per day . On postnatal day 24 , the juvenile females were euthanized by CO (2) asphyxiation , weighed , livers weighed and uteri weighed wet and blotted . The presence or absence of vaginal patency was recorded . Absolute and relative (to terminal body weight) uterine wet and blotted weights and uterine luminal fluid weights were significantly increased at 3.0 and 10.0 (both P < 0.01) microg kg (-) (1) per day , and increased to ~140 % of control values at 1.0 microg kg (-) (1) per day (not statistically significantly) . In vivo body weights , weight changes , feed consumption , liver weights and terminal body weights were unaffected . Vaginal patency was not acquired in any female at any dose , although vaginal puckering was observed in one female at 10.0 microg kg (-) (1) per day . Therefore , this intact weanling uterotrophic assay is validated in our laboratory for use under US and European endocrine toxicity testing programs/legislation .

Figure 7: Annotations generated using our method for "MC 5: Dosing interval".

The intact female weanling version in the Organization for Economic Cooperation and Development (OECD) uterotrophic assay Test Guideline (TG) 440 is proposed as an alternative to the adult ovariectomized female version , because it does not involve surgical intervention (vs the ovariectomized version) and detects direct/indirect-acting estrogenic/anti-estrogenic substances (vs the ovariectomized version which detects only direct-acting estrogenic/anti-estrogenic substances binding to the estrogen receptor) . This validation study followed OECD TG 440 , with six female weanling rats (postnatal day 21) per dose group and six treatment groups . Females were weighed and dosed once daily by oral gavage for three consecutive days , with one of six doses of 17alpha-ethinyl estradiol in corn oil at 5 ml kg (-) (1) at 0 and 0.1-10 microg kg (-) (1) per day . On postnatal day 24 , the juvenile females were euthanized by CO (2) asphyxiation , weighed , livers weighed and uteri weighed wet and blotted . The presence or absence of vaginal patency was recorded . Absolute and relative (to terminal body weight) uterine wet and blotted weights and uterine luminal fluid weights were significantly increased at 3.0 and 10.0 (both P < 0.01) microg kg (-) (1) per day , and increased to ~140 % of control values at 1.0 microg kg (-) (1) per day (not statistically significantly) . In vivo body weights , weight changes , feed consumption , liver weights and terminal body weights were unaffected . Vaginal patency was not acquired in any female at any dose , although vaginal puckering was observed in one female at 10.0 microg kg (-) (1) per day . Therefore , this intact weanling uterotrophic assay is validated in our laboratory for use under US and European endocrine toxicity testing programs/legislation .

Figure 8: Annotations generated using our method for "MC 6: Necropsy timing".

Identification of Parallel Sentences in Comparable Monolingual Corpora from Different Registers

Rémi Cardon
UMR CNRS 8163 – STL
F-59000 Lille, France
remi.cardon@univ-lille.fr

Natalia Grabar
UMR CNRS 8163 – STL
F-59000 Lille, France
natalia.grabar@univ-lille.fr

Abstract

Parallel aligned sentences provide useful information for different NLP applications. Yet, this kind of data is seldom available, especially for languages other than English. We propose to exploit comparable corpora in French which are distinguished by their registers (specialized and simplified versions) to detect and align parallel sentences. These corpora are related to the biomedical area. Our purpose is to state whether a given pair of specialized and simplified sentences is to be aligned or not. Manually created reference data show 0.76 inter-annotator agreement. We exploit a set of features and several automatic classifiers. The automatic alignment reaches up to 0.93 Precision, Recall and F-measure. In order to better evaluate the method, it is applied to data in English from the *SemEval* STS competitions. The same features and models are applied in monolingual and cross-lingual contexts, in which they show up to 0.90 and 0.73 F-measure, respectively.

1 Introduction

The purpose of text simplification is to provide simplified versions of texts, in order to remove or replace difficult words or information. Simplification can be concerned with different linguistic aspects, such as lexicon, syntax, semantics, pragmatics and even document structure. Simplification can address needs of people or NLP applications (Brunato et al., 2014). In the first case, simplified documents are typically created for children (Son et al., 1008; De Belder and Moens, 2010; Vu et al., 2014), people with low literacy or foreigners (Paetzold and Specia, 2016), people with mental or neurodegenerative disorders (Chen et al., 2016), or laypeople who face specialized documents (Arya et al., 2011; Leroy et al., 2013). In the second case, the purpose of simplification is to transform documents in order to make them easier to process within other NLP tasks, such as syntactic analysis (Chandrasekar and Srinivas, 1997; Jonnalagadda et al., 2009), semantic annotation (Vickrey and Koller, 2008), summarization (Blake et al., 2007), machine translation (Stymne et al., 2013; Štajner and Popović, 2016), indexing (Wei et al., 2014), or information retrieval and extraction (Beigman Klebanov et al., 2004). Hence, parallel sentences, which align difficult and simple information, provide crucial indicators for the text simplification. Indeed such pairs of sentences contain cues on transformations which are suitable for the simplification, such as lexical substitutes and syntactic modifications. Yet, this kind of resources is seldom available, especially in languages other than English. The purpose of our work is to detect and align parallel sentences from comparable monolingual corpora, that are differentiated by their registers. Besides, comparable corpora are easier to obtain. More precisely, we work with texts written for specialists and their simplified versions. We work with corpora in French.

2 Existing Work

In parallel corpora, sentence alignment can rely on empirical information, such as relative length of the sentences in each language (Gale and Church, 1993), or lexical information (Chen, 1993). In comparable corpora, both monolingual and bilingual, sentences present relatively loose common semantics and do not necessarily occur in the same order. It should also be noted that (1) the degree of parallelism can vary from nearly parallel corpora, with a lot of parallel sentences, to *very-non-parallel corpora* (Fung and Cheung, 2004); and that (2) such corpora can contain parallel information at various degrees of granularity, such as documents, sentences or sub-phrastic segments (Hewavitharana and Vogel, 2011). Detection of

Proceedings of the 9th International Workshop on Health Text Mining and Information Analysis (LOUHI 2018), pages 83–93
Brussels, Belgium, October 31, 2018. ©2018 Association for Computational Linguistics

parallel sentences in comparable corpora is thus a substantial challenge and requires specific methods.

Several existing works are related to machine translation: bilingual comparable corpora are exploited for creation of parallel and aligned corpora. Usually, these methods rely on three steps:

1. detection of comparable documents using for instance generative models (Zhao and Vogel, 2002) or similarity scores (Utiyama and Isahara, 2003; Fung and Cheung, 2004);

2. detection of candidate sentences, or sub-phrastic segments, for the alignment using for instance cross-lingual information retrieval (Utiyama and Isahara, 2003; Munteanu and Marcu, 2006), sequence alignment trees (Munteanu and Marcu, 2002), mutual translations (Munteanu and Marcu, 2005; Kumano et al., 2007; Abdul-Rauf and Schwenk, 2009), or dynamic programming (Yang and Li, 2003);

3. filtering and selection of correct extractions using classification (Munteanu and Marcu, 2005; Tillmann and Xu, 2009; Hewavitharana and Vogel, 2011; Ștefănescu et al., 2012), similarity measure of translations (Fung and Cheung, 2004; Hewavitharana and Vogel, 2011), error rate (Abdul-Rauf and Schwenk, 2009), generative models (Zhao and Vogel, 2002; Quirk et al., 2007), or specific rules (Munteanu and Marcu, 2002; Yang and Li, 2003).

In relation with monolingual comparable corpora, the main difficulty is that sentences may show low lexical overlap but be nevertheless parallel. Recently, this task gained in popularity thanks to the semantic text similarity (STS) initiative. Dedicated *SemEval* competitions have been proposed for several years (Agirre et al., 2013, 2015, 2016). The objective, for a given pair of sentences, is to predict if they are semantically similar and to assign similarity score going from 0 (independent semantics) to 5 (semantic equivalence). This task is usually explored in general-language corpora. Among the exploited methods, we can notice:

- lexicon-based methods which rely on similarity of subwords or words from the processed texts or on machine translation (Madnani et al., 2012). The features exploited can be: lexical overlap, sentence length, string edition distance, numbers, named entities, the longest common substring (Clough et al., 2002; Zhang and Patrick, 2005; Qiu et al., 2006; Zhao et al., 2014; Nelken and Shieber, 2006; Zhu et al., 2010);

- knowledge-based methods which exploit external resources, such as WordNet (Miller et al., 1993) or PPDB (Ganitkevitch et al., 2013). The features exploited can be: overlap with external resources, distance between the synsets, intersection of synsets, semantic similarity of resource graphs, presence of synonyms, hyperonyms or antonyms (Mihalcea et al., 2006; Fernando and Stevenson, 2008; Lai and Hockenmaier, 2014);

- syntax-based methods which exploit the syntactic modelling of sentences. The features often exploited are: syntactic categories, syntactic overlap, syntactic dependencies and constituents, predicat-argument relations, edition distance between syntactic trees (Wan et al., 2006; Severyn et al., 2013; Tai et al., 2015; Tsubaki et al., 2016);

- corpus-based methods which exploit distributional methods, latent semantic analysis (LSA), topics modelling, word embeddings, etc. (Barzilay and Elhadad, 2003; Guo and Diab, 2012; Zhao et al., 2014; Kiros et al., 2015; He et al., 2015; Mueller and Thyagarajan, 2016).

These methods and types of features can of course be combined for optimizing the results (Bjerva et al., 2014; Lai and Hockenmaier, 2014; Zhao et al., 2014; Rychalska et al., 2016; Severyn et al., 2013; Kiros et al., 2015; He et al., 2015; Tsubaki et al., 2016; Mueller and Thyagarajan, 2016).

Our objective is close to the second type of works: we want to detect and align parallel sentences from monologual comparable corpora. Yet, there are some differences: (1) we work with corpora related to the biomedical area and not to the general language, (2) we have to state if two sentences have to be aligned (binary statement) and not to compute their similarity score, and (3) we work with data in French which were not exploited for this kind of task yet. To our knowledge, the only work which exploited articles from French encyclopedia performed manual alignment of sentences (Brouwers et al., 2014).

In what follows, we first present the linguistic material used, and the methods proposed. We then present and discuss the results obtained, and conclude with directions of future work.

3 Linguistic Material

We use three comparable corpora in French. They are related to the biomedical domain and are contrasted by the technicity of information they contain with typically specialized and simplified versions of a given text. These corpora cover three genres: drug information, summaries of scientific articles, and encyclopedia articles (Sec. 3.1). We also exploit a set of stopwords (Sec. 3.2), and the reference data with sentences manually aligned by two annotators (Sec. 3.3).

3.1 Comparable Corpora

Table 1 indicates the size of the source corpora (number of documents, number of words in specialized and simplified versions). The three corpora are built with French data.

The *Drug* corpus contains drug information such as provided to health professionals and patients. Indeed, two distinct sets of documents exist, each of which contains common and specific information. This corpus is built from the public drug database[1] of the French Health ministry. These data have been downloaded in June 2017. We can see that the specialized versions of documents provide more word occurrences.

The *Scientific* corpus contains summaries of meta-reviews of high evidence health-related articles, such as proposed by the Cochrane collaboration (Sackett et al., 1996). These reviews have been first intended for health professionals but recently the collaborators started to create simplified versions of the reviews (*Plain language summary*) so that they can be read and understood by the whole population. This corpus has been built from the online library of the Cochrane collaboration[2]. The data have been downloaded in November 2017. We can see that specialized version of summaries is also larger than the simplified version, although the difference is not very important.

The *Encyclopedia* corpus contains encyclopedia articles from Wikipedia[3] and Vikidia[4].

Wikipedia articles are considered as technical texts while Vikidia articles are considered as their simplified versions (they are created for children 8 to 13 year old). Similarly to the works done in English, we associate Vikidia with Simple Wikipedia[5]. Only articles related to the medical portal are exploited in this work. These *encyclopedia* articles have been downloaded in August and September 2017. From Table 1, we can see that specialized versions (from Wikipedia) are also longer than simplified versions.

These three corpora are more or less parallel: Wikipedia and Vikidia articles are written independently from each other, drug information documents are related to the same drugs but the types of information presented for experts and laypeople vary a lot, while simplified summaries from the *scientific* corpus are created starting from the expert summaries.

3.2 Stopwords

We use a set of 83 stopwords in French, which are mostly grammatical words, like prepositions (*de, et, à, ou (of, and, in, or)*), auxiliary verbs (*est, a (is, has)*) or adverbs (*tout, plusieurs (all, several)*).

3.3 Reference Data

In this section we describe the data that are used for training and evaluation of the automatic sentence alignments.

The reference data are created manually. We have randomly selected 2*14 *encyclopedia* articles, 2*12 *drug* documents, and 2*13 *scientific* summaries. The sentence alignment is done by two annotators following these guidelines:

1. exclude identical sentences or sentences with only punctuation and stopword difference ;

2. include sentence pairs with morphological variations (e.g. *Ne pas dépasser la posologie recommandée.* and *Ne dépassez pas la posologie recommandée.* – both examples can be translated by *Do not take more than the recommended dose.*);

3. exclude sentence pairs with overlapping semantics, when each sentence brings own information, in addition to the common semantics;

85

corpus	# docs	# occ_{sp}	# occ_{simpl}	# $lemmas_{sp}$	# $lemmas_{simpl}$
Drugs	11,800*2	52,313,126	33,682,889	43,515	25,725
Scient.	3,815*2	2,840,003	1,515,051	11,558	7,567
Encyc.	575*2	2,293,078	197,672	19,287	3,117

Table 1: Size of the three source corpora. (column headers : number of documents, total of occurrences (specialized and simple), total of unique words (specialized and simple))

corpus	# doc.	Specialized				Simplified				Alignment rate (%)	
		source		aligned		source		aligned			
		# pairs.	# occ.	# pairs.	# occ.	# pairs	# occ.	# pairs.	# occ.	sp.	simp.
Drugs	12*2	4,416	44,709	502	5,751	2,736	27,820	502	10,398	18	11
Scient.	13*2	553	8,854	112	3,166	263	4,688	112	3,306	20	43
Encyc.	14*2	2,494	36,002	49	1,100	238	2,659	49	853	2	21

Table 2: Size of the reference data with consensual alignment of sentences. (number of sentence pairs and word occurrences for each subset)

4. include sentence pairs in which one sentence is included in the other, which enables many-to-one matching (e.g. *C'est un organe fait de tissus membraneux et musculaires, d'environ 10 à 15 mm de long, qui pend à la partie moyenne du voile du palais.* and *Elle est constituée d' un tissu membraneux et musculaire. – It is an organ made of membranous and muscular tissues, approximately 10 to 15 mm long, that hangs from the medium part of the soft palate.* and *It is made of a membranous and muscular tissue.*);

5. include sentence pairs with equivalent semantics – other than semantic intersection and inclusion (e.g. *Les médicaments inhibant le péristaltisme sont contre-indiqués dans cette situation.* and *Dans ce cas, ne prenez pas de médicaments destinés à bloquer ou ralentir le transit intestinal. – Drugs that inhibit the peristalsis are contraindicated in that situation.* and *In that case, do not take drugs intended for blocking or slowing down the intestinal transit.*)

The judgement on semantic closeness may vary according to the annotators. For this reason, the alignments provided by each annotator undergo consensus discussions. This alignment process provides a set of 663 aligned sentence pairs. The inter-annotator agreement is 0.76 (Cohen, 1960). It is computed within the two sets of sentences proposed for alignment by the two annotators.

Table 2 indicates the size of the reference data before (*source* columns) and after (*aligned* columns) the alignment. In the two last columns (*Alignment rate*), we indicate the percentage of sentences aligned in each register and corpus. We can observe that *scientific* corpus is the most parallel with the highest alignment rate of sentences from specialized and simplified documents, while the two other corpora (*drugs* and *encylopedia*) contain proportionnally less parallel sentences. Another interesting observation is that sentences from simplified documents in the *scientific* and *drugs* corpora are longer than sentences from specialized documents because they often add explanations for technical notions, like in this example: *We considered studies involving bulking agents (a fibre supplement), antispasmodics (smooth muscle relaxants) or antidepressants (drugs used to treat depression that can also change pain perceptions) that used outcome measures including improvement of abdominal pain, global assessment (overall relief of IBS symptoms) or symptom score.* In the *encyclopedia* corpus such notions are replaced by simpler words, or removed. Finally, in all corpora, we observe frequent substitutions by synonyms, like in these pairs: {*nutrition*; *food*}, {*enteral*; *directly in the stomach*}, {*hypersensitivity*; *allergy*}, {*incidence*; *possible complications*}. Notice that with such substitutions, lexical similarity between sentences is reduced.

4 Automatic Alignment of Parallel Sentences

As already indicated, our objective is to detect and align parallel sentences within monologual comparable corpora in French. We already have the information on which documents are comparable. So, the task is really dedicated to the alignment of sentences from specialized and simplified versions of documents. The method is composed of several steps: pre-processing of data (Sec. 4.1), generation of features (Sec. 4.2), automatic alignment of sentences (Sec. 4.3), and evaluation (Sec. 4.4).

4.1 Pre-processing of Data

The documents are first pre-processed: they are POS-tagged with TreeTagger (Schmid, 1994), which permits to obtain their lemmatized versions. Then, the documents are segmented into sentences using strong punctuation (*i.e.* *.?!;:*). The same pre-processing and segmentation have been applied when creating the reference data.

4.2 Feature Generation

Our goal is to propose features that can work on textual data in different languages. We use several features which are mainly lexicon-based and corpus-based, so that they can be easily applied to textual data in other languages or transposed to data in other languages. The features are computed on word forms and on lemmas:

1. Number of common non-stopwords. This feature permits to compute the basic lexical overlap between specialized and simplified versions of sentences (Barzilay and Elhadad, 2003). This feature exploits external knowledge (set of stopwords), which are nevertheless very common linguistic data;

2. Number of common stopwords. This feature also exploits external knowledge (set of stopwords). It concentrates on non-lexical content of sentences;

3. Percentage of words from one sentence included in the other sentence, computed in both directions. This features represents possible lexical and semantic inclusion relations between the sentences;

4. Sentence length difference between specialized and simplified sentences. This feature assumes that simplification may imply stable association with the sentence length;

5. Average length difference in words between specialized and simplified sentences. This feature is similar to the previous one but takes into account average difference in sentence length;

6. Total number of common bigrams and trigrams. This feature is computed on character ngrams. The assumption is that, at the sub-word level, some sequences of characters may be meaningful for the alignment of sentences if they are shared by them;

7. Word-based similarity measure exploits three scores (cosine, Dice and Jaccard). This feature provides a more sophisticated indication on word overlap between the two compared sentences. Weight assigned to each word is set to 1;

8. Word-based similarity measure with the tf*idf weighting of words (Nelken and Shieber, 2006). This feature is similar to the previous one but it also exploits information on context by incorporating the tf*idf weighting (Salton and Buckley, 1988) of words. For this, sentences are considered as documents and documents as corpora. This feature permits to weigh words in a sentence with respect to their occurrences in other sentences of the document;

9. Character-based minimal edit distance (Levenshtein, 1966). This is a classical acception of edit distance. It takes into account basic edit operations (insertion, deletion and substitution) at the level of characters. The cost of each operation is set to 1;

10. Word-based minimal edit distance (Levenshtein, 1966). This feature is computed with words as units within sentence. It takes into account the same three edit operations with the same cost set to 1. This feature permits to compute the cost of lexical transformation of one sentence into another.

4.3 Automatic Alignment of Sentences

The task is to find parallel sentences within the whole set of sentences we described in section

3.3. Hence, we have to categorize the pairs of sentences in one of the two categories:

- alignment: the sentences are parallel and can be aligned;

- non-alignment: the sentences are non-parallel and cannot be aligned.

The reference data provide positive examples (663 parallel sentences), while negative examples are obtained by randomly pairing some of the remaining sentences (800 non-parallel sentences) from the same documents.

We use several linear classifiers with their default parameters if not indicated otherwise: Perceptron (Rosenblatt, 1958), Multilayer Perceptron (MLP) (Rosenblatt, 1961), Linear discriminant analysis (LDA) (Fisher, 1936) with the LSQR solver, Quadratic discriminant analysis (QDA) (Cover, 1965), Logistic regression (Berkson, 1944), Stochastic gradient descent (SGD) (Ferguson, 1982) with the log loss, Linear SVM (Vapnik and Lerner, 1963). We also tested hinge and modified huber as loss functions with the SGD, and Eigen and SVD solvers with the LDA, but the results were either lower or very close to the best parameters and we abandoned the idea to use them.

4.4 Evaluation

The training of the system is performed on two thirds of the sentence pairs, and the test is performed on the remaining third. Several classifiers and several combinations of features are tested. Classical evaluation measures are computed: Precision, Recall, F-measure, Mean Square Errors, and True Positives. Our baseline is the combination of length measures with the common words (features 1, 2, 4 and 5). These features are indeed traditionnally exploited in the existing work.

We also evaluate the system on data in English that were released for STS competitions[6]: we use 750 sentence pairs from *SemEval 2012*, 1,500 sentence pairs from *SemEval 2013*, 3,750 sentence pairs from *SemEval 2014*. Each pair of sentences is associated with the similarity score [0;5]. We apply our system to these data in two ways: (1) the system is trained and tested on the STS dataset, and (2) the system is trained on our dataset in French and tested on the STS dataset in English.

We assume indeed that the features used and even the models generated can be transposed to data in other languages. For the experiments with the English data, we use the same evaluation measures (Precision, Recall, F-measure, Mean Square Errors, and True Positives). The set of stopwords in English contains 150 entities.

5 Results and Discussion

Classifier	R	P	F1	MSE	TP
Perceptron	0.87	0.84	0.84	0.63	142
MLP	0.87	0.87	0.86	0.53	167
LDA	0.90	0.90	0.90	0.40	175
QDA	0.89	0.89	0.89	0.45	197
LogReg	0.93	0.93	0.93	0.30	191
SGD	0.87	0.84	0.84	0.84	210
LinSVM	0.81	0.81	0.81	0.74	166

Table 3: Alignment results obtained with different classifiers on French data, test set, whole featureset without tf*idf similarity scores, and non-lemmatized text.

In Table 3, we present the results obtained on French data using the whole set of features (but without the tf*idf similarity scores) on test set, and non-lemmatized texts. The results are indicated in terms of Recall R, Precision P, F-measure F, Mean Square Errors MSE and True positives TP (out of the 221 positive sentence pairs in the test set). We can see that all the classifiers are competitive with F-measure above 0.80. Overall, several classifiers (LDA, QDA, LogReg, LinSVM) provide stable results, for which we indicate the evaluation scores obtained in one iteration. Other classifiers (Perceptron, MLP, SGD) provide fluctuating results, and we indicate then the average scores obtained after 20 iterations. Another positive observation is that Precision and Recall values are well balanced. Logistic regression seems to be the best classifier for this task, with Precision, Recall and F-measure at 0.93. This classifier is used for the experiments described in the next sections.

We first present and discuss the exploitation of various featuresets on French data (Sec. 5.1), and then the exploitation of the features and models on the STS data in English in monolingual (Sec. 5.2) and cross-lingual (Sec. 5.3) contexts. As our final objective (text simplification in French) and the data we work on (French texts from the biomedical domain) are different from the STS context, we believe it should be noted that there are intrin-

---

[6]http://ixa2.si.ehu.es/stswiki/index.php/Main_Page

sic limitations as to the comparison we can make.

5.1 Different Featuresets

Feature set	R	P	F1	MSE	TP
BL	0.87	0.87	0.86	0.54	173
S	0.84	0.84	0.84	0.64	174
L	0.79	0.78	0.78	0.86	146
N	0.89	0.88	0.88	0.48	168
L+S	0.88	0.88	0.88	0.48	170
L+N	0.91	0.91	0.91	0.37	187
S+N	0.91	0.91	0.91	0.37	183
BL+L	0.90	0.90	0.90	0.40	184
BL+S	0.89	0.89	0.89	0.46	180
BL+N	0.91	0.91	0.91	0.35	187
BL+L+S	0.90	0.90	0.90	0.40	184
BL+L+N	0.93	0.93	0.93	0.29	191
BL+S+N	0.91	0.91	0.91	0.36	189
L+S+N	0.91	0.91	0.91	0.36	189
BL+L+S+N	0.93	0.93	0.93	0.29	191

Table 4: Alignment results obtained with various featuresets, logistic regression, non-lemmatized text.

The purpose of these experiments is to detect the most suitable combinations of features. We present the results obtained on our data. We distinguish four sets of features, which are used in isolation and in various combinations. We indicate the corresponding numbers from section 4.2 between brackets :

1. BL: baseline (1, 2, 3, 4 5);

2. L: Levenshtein-based features (9, 10);

3. S: similarity-based features (7, 8);

4. N: ngram-based features (6).

Contrary to the previous work (Nelken and Shieber, 2006; Zhu et al., 2010), the tf*idf weighting of words is not efficient on our data. For this reason, this set of features was not used in the experiments.

The results are presented in Table 4. The lowest results are obtained with the Levenshtein-based features (F-measure 0.78), they are followed by the similarity-based features (F-measure 0.84). We obtain 0.86 F-measure with the baseline. Other combinations indicate that each set of features exploited is useful to gain efficiency for this task. Hence, the best results are obtained with the combination BL+L+N and with the whole set of features (BL+L+S+N), which shows 0.93 F-measure. We use the whole set of features for the experiments with the STS dataset.

5.2 Classification of the STS Sentence Pairs

STSset score	R	P	F1	MSE	TP
STS2012 2.5	0.82	0.82	0.82	0.71	477
STS2012 3.5	0.74	0.74	0.74	1.04	277
STS2012 4.5	0.79	0.81	0.78	0.74	37
STS2013 2.5	0.73	0.73	0.73	1.09	176
STS2013 3.5	0.78	0.78	0.78	0.87	96
STS2013 4.5	0.89	0.93	0.90	0.29	2
STS2014 2.5	0.75	0.76	0.75	0.97	653
STS2014 3.5	0.70	0.71	0.71	1.17	306
STS2014 4.5	0.89	0.93	0.90	0.29	2

Table 5: Alignment results obtained on the STS data in English, test set, whole featureset, logistic regression, non-lemmatized text and training on the STS data.

In this set of experiments, the classification model is trained and tested on the STS reference data in English. Our assumption is that the features exploited are transferable from one language to another. The reference data and categories in English and in French differ. One difference is that the STS pairs of sentences are scored from 0 to 5 according to their similarity, while in French we do binary classification (a given pair of sentences should be aligned or not). To make the two datasets comparable, we propose to transform the STS scoring in binary categories. We test similarity thresholds within the interval [2.5;4.5] by step of 0.5, which permits not to consider identical sentences (scores close to 5) and very distant sentences (scores lower than 2.5). As indicated in Table 5, we obtain up to 0.90 F-measure with the similarity threshold 4.5 on data from 2013 and 2014, while in 2012 the best F-measure (0.82) is obtained with the similarity score 2.5. It is difficult to compare our results with those of the participating teams and already published results because our categories and evaluation differ from the STS protocols – we rate sentence pairs as either aligned or not aligned, while STS offers a scale from 0 to 5. Yet, the MSE rate (0.308) published by one of the top participants in 2014 (Bjerva et al., 2014) indicates that our MSE rate is improved, as it is at 0.29 on the 2014 data.

5.3 Cross-lingual Classification of the STS Sentence Pairs

STSset score	R	P	F1	MSE	TP
STS2012 2.5	0.83	0.81	0.82	0.19	1378
STS2012 3.5	0.74	0.72	0.71	0.28	1035
STS2012 4.5	0.81	0.49	0.52	0.51	413
STS2013 2.5	0.74	0.74	0.74	0.26	523
STS2013 3.5	0.78	0.73	0.74	0.27	396
STS2013 4.5	0.92	0.57	0.67	0.43	88
STS2014 2.5	0.74	0.72	0.73	0.28	1688
STS2014 3.5	0.72	0.69	0.69	0.31	1216
STS2014 4.5	0.88	0.54	0.61	0.46	384

Table 6: Alignment results obtained on the STS data in English, test set, whole featureset, Logistic regression, non-lemmatized text and training on the French data.

In this set of experiments, the classification model is trained on French data and tested on the STS data in English. Here, our assumption is that the models generated on one language can be transferable to another language in order to detect parallel sentences. Here as well, we test several similarity thresholds. As we can see in Table 6, in this cross-lingual experiment, the best F-measures are obtained with the score 2.5 in 2012 (0.82) and in 2014 (0.73), and with scores 2.5 and 3.0 in 2013 (0.74). These thresholds indicate that the models generated on our French data can be exploited on the STS data in English quite efficiently and that the features that are used show cross-lingual relevance for the French-English language pair. These results also indicate that, for the targeted task of text simplification, we need quite a strong similarity between sentences.

6 Conclusion and Future Work

In this work, we proposed to address the task of detection and alignment of parallel sentences from monolingual comparable corpora in French. The comparable dimension is due to the technicality of documents, which contrast specialized and simplified versions of documents and sentences. We use three corpora which are related to the biomedical area. Several features and classifiers and exploited. Our results reach up to 0.93 F-measure on the French data, with a very good balance between Precision and Recall. Linear regression appears to be the best classifier for this task. Our approach is then tested on the STS data in English, such as proposed by several *SemEval* competitions between 2012 and 2014. We first test the features, with training and testing done on the STS data. This gives up to 0.90 F-measure with the 4.5 similarity threshold. Then, we test the models: they are generated on the French data and tested on the STS data. This gives 0.82 F-measure. We assume that the proposed approach (features and classifiers) show a good transferability to another language. This is a good point because it validates our approach on data from another language.

In future, we plan to exploit the best models generated in French for enriching the set of parallel sentences. This will permit to prepare data necessary for the developement of simplification methods for French. Parallel sentences may also be helpful for othe NLP applications. Other directions for future work are concerned with the exploitation of other features for the alignment of sentences, such as use of word embeddings to smooth lexical variation or exploitation of external knowledge. Besides, our appoach will be further evaluated on data from other languages.

7 Acknowledgements

This work was funded by the French National Agency for Research (ANR) as part of the *CLEAR* project (*Communication, Literacy, Education, Accessibility, Readability*), ANR-17-CE19-0016-01.

The authors would like to thank the reviewers for their helpful comments.

References

Sadaf Abdul-Rauf and Holger Schwenk. 2009. On the use of comparable corpora to improve smt performance. In *European Chapter of the ACL*, pages 16–23.

Eneko Agirre, Carmen Banea, Claire Cardie, Daniel Cer, Mona Diab, Aitor Gonzalez-Agirre, Weiwei Guo, Inigo Lopez-Gazpio, Montse Maritxalar, Rada Mihalcea, German Rigau, Larraitz Uria, and Janyce Wiebe. 2015. SemEval-2015 task 2: Semantic textual similarity, english, spanish and pilot on interpretability. In *SemEval 2015*, pages 252–263.

Eneko Agirre, Carmen Banea, Daniel Cer, Mona Diab, Aitor Gonzalez-Agirre, Rada Mihalcea, German Rigau, and Janyce Wiebe. 2016. SemEval-2016 task 1: Semantic textual similarity, monolingual and cross-lingual evaluation. In *SemEval 2016*, pages 497–511.

Eneko Agirre, Daniel Cer, Mona Diab, Aitor Gonzalez-Agirre, and WeiWei Guo. 2013. *sem 2013 shared task: Semantic textual similarity. In **SEM*, pages 32–43.

Diana J. Arya, Elfrieda H. Hiebert, and P. David Pearson. 2011. The effects of syntactic and lexical complexity on the comprehension of elementary science texts. *Int Electronic Journal of Elementary Education*, 4(1):107–125.

Regina Barzilay and Noemie Elhadad. 2003. Sentence alignment for monolingual comparable corpora. In *EMNLP*, pages 25–32.

Beata Beigman Klebanov, Kevin Knight, and Daniel Marcu. 2004. Text simplification for information-seeking applications. In Robert Meersman and Zahir Tari, editors, *On the Move to Meaningful Internet Systems 2004: CoopIS, DOA, and ODBASE*. Springer, LNCS vol 3290, Berlin, Heidelberg.

Joseph Berkson. 1944. Application of the logistic function to bio-essay. *Journal of the American Statistical Association*, 39:357–365.

Johannes Bjerva, Johan Bos, Rob van der Goot, and Malvina Nissim. 2014. The meaning factory: Formal semantics for recognizing textual entailment and determining semantic similarity. In *Workshop on Semantic Evaluation (SemEval 2014)*, pages 642–646, Dublin, Ireland.

Catherine Blake, Julia Kampov, Andreas Orphanides, David West, and Cory Lown. 2007. Query expansion, lexical simplification, and sentence selection strategies for multi-document summarization. In *DUC*.

Laetitia Brouwers, Delphine Bernhard, Anne-Laure Ligozat, and Thomas Francois. 2014. Syntactic sentence simplification for French. In *PITR workshop*, pages 47–56.

Dominique Brunato, Felice Dell'Orletta, Giulia Venturi, and Simonetta Montemagni. 2014. Defining an annotation scheme with a view to automatic text simplification. In *CLICIT*, pages 87–92.

Raman Chandrasekar and Bangalore Srinivas. 1997. Automatic induction of rules for text simplification. *Knowledge Based Systems*, 10(3):183–190.

Ping Chen, John Rochford, David N. Kennedy, Soussan Djamasbi, Peter Fay, and Will Scott. 2016. Automatic text simplification for people with intellectual disabilities. In *AIST*, pages 1–9.

Stanley F Chen. 1993. Aligning sentences in bilingual corpora using lexical information. In *Annual Meeting of the Association for Computational Linguistics*, pages 9–16.

Paul Clough, Robert Gaizauskas, Scott S.L. Piao, and Yorick Wilks. 2002. METER: Measuring text reuse. In *ACL*, pages 152–159.

Jacob Cohen. 1960. A coefficient of agreement for nominal scales. *Educational and Psychological Measurement*, 20(1):37–46.

Thomas M. Cover. 1965. Geometrical and statistical properties of systems of linear inequalities with applications in pattern recognition. *IEEE Transactions on Electronic Computers*, 14(3):326–334.

Jan De Belder and Marie-Francine Moens. 2010. Text simplification for children. In *Workshop on Accessible Search Systems of SIGIR*, pages 1–8.

Thomas S. Ferguson. 1982. An inconsistent maximum likelihood estimate. *Journal of the American Statistical Association*, 77(380):831–834.

Samuel Fernando and Mark Stevenson. 2008. A semantic similarity approach to paraphrase detection. In *Comp Ling UK*, pages 1–7.

Ronald A. Fisher. 1936. The use of multiple measurements in taxonomic problems. *Annals of Eugenics*, 7(2):179–188.

Pascale Fung and Percy Cheung. 2004. Mining very non-parallel corpora: Parallel sentence and lexicon extraction vie bootstrapping and em. In *Conference on Empirical Methods in Natural Language Processing*, pages 57–63.

William A. Gale and Kenneth W. Church. 1993. A program for aligning sentences in bilingual corpora. *Comp Linguistics*, 19(1):75–102.

Juri Ganitkevitch, Benjamin Van Durme, and Chris Callison-Burch. 2013. PPDB: The paraphrase database. In *NAACL-HLT*, pages 758–764.

Weiwei Guo and Mona Diab. 2012. Modeling sentences in the latent space. In *ACL*, pages 864–872.

Hua He, Kevin Gimpel, and Jimmy Lin. 2015. Multi-perspective sentence similarity modeling with convolutional neural networks. In *EMNLP*, pages 1576–1586, Lisbon, Portugal.

Sanjika Hewavitharana and Stephan Vogel. 2011. Extracting parallel phrases from comparable data. In *4th Workshop on Building and Using Comparable Corpora*, pages 61–68.

Siddhartha Jonnalagadda, Luis Tari, Jörg Hakenberg, Chitta Baral, and Graciela Gonzalez. 2009. Towards effective sentence simplification for automatic processing of biomedical text. In *NAACL HLT 2009*, pages 177–180.

Ryan Kiros, Yukun Zhu, Ruslan Salakhutdinov, Richard S. Zemel, Antonio Torralba, Raquel Urtasun, and Sanja Fidler. 2015. Skip-thought vectors. In *Neural Information Processing Systems (NIPS)*, pages 3294–3302.

Tadashi Kumano, Hideki Tanaka, and Takenobu Tokunaga. 2007. Extracting phrasal alignments from comparable corpora by using joint probability SMT model. In *Int Conf on Theoretical and Methodological Issues in Machine Translation*.

Alice Lai and Julia Hockenmaier. 2014. Illinois-LH: A denotational and distributional approach to semantics. In *Workshop on Semantic Evaluation (SemEval 2014)*, pages 239–334, Dublin, Ireland.

Gondy Leroy, David Kauchak, and Obay Mouradi. 2013. A user-study measuring the effects of lexical simplification and coherence enhancement on perceived and actual text difficulty. *Int J Med Inform*, 82(8):717–730.

Vladimir I. Levenshtein. 1966. Binary codes capable of correcting deletions, insertions and reversals. *Soviet physics. Doklady*, 707(10).

Nitin Madnani, Joel Tetreault, and Martin Chodorow. 2012. Re-examining machine translation metrics for paraphrase identification. In *NAACL-HLT*, pages 182–190.

Rada Mihalcea, Courtney Corley, and Carlo Strapparava. 2006. Corpus-based and knowledge-based measures of text semantic similarity. In *AAAI*, pages 1–6.

George A. Miller, Richard Beckwith, Christiane Fellbaum, Derek Gross, and Katherine Miller. 1993. Introduction to wordnet: An on-line lexical database. Technical report, WordNet.

Jonas Mueller and Aditya Thyagarajan. 2016. Siamese recurrent architectures for learning sentence similarity. In *AAAI Conference on Artificial Intelligence*, pages 2786–2792.

Dragos Stefan Munteanu and Daniel Marcu. 2002. Processing comparable corpora with bilingual suffix trees. In *EMNLP*, pages 289–295.

Dragos Stefan Munteanu and Daniel Marcu. 2005. Improving machine translation performance by exploiting non-parallel corpora. *Computational Linguistics*, 31(4):477–504.

Dragos Stefan Munteanu and Daniel Marcu. 2006. Extracting parallel sub-sentential fragments from non-parallel corpora. In *COLING-ACL*, pages 81–88.

Rani Nelken and Stuart M. Shieber. 2006. Towards robust context-sensitive sentence alignment for monolingual corpora. In *EACL*, pages 161–168.

Gustavo H. Paetzold and Lucia Specia. 2016. Benchmarking lexical simplification systems. In *LREC*, pages 3074–3080.

Long Qiu, Min-Yen Kan, and Tat-Seng Chua. 2006. Paraphrase recognition via dissimilarity significance classification. In *Empirical Methods in Natural Language Processing*, pages 18–26, Sydney, Australia.

Chris Quirk, Raghavendra U. Udupa, and Arul Menezes. 2007. Generative models of noisy translations with applications to parallel fragment extraction. In *Machine Translation Summit XI*, pages 377–384.

Frank Rosenblatt. 1958. The perceptron: a probabilistic model for information storage and organization in the brain. *Psychological Review*, 65(6):386–408.

Frank Rosenblatt. 1961. *Principles of Neurodynamics: Perceptrons and the Theory of Brain Mechanisms*. Spartan Books, Washington DC.

Barbara Rychalska, Katarzyna Pakulska, Krystyna Chodorowska, WojciechWalczak, and Piotr Andruszkiewicz. 2016. Samsung Poland NLP team at SemEval-2016 task 1: Necessity for diversity; combining recursive autoencoders, wordnet and ensemble methods to measure semantic similarity. In *SemEval-2016*, pages 614–620.

David L. Sackett, William M. C. Rosenberg, Jeffrey A. MuirGray, R. Brian Haynes, and W. Scott Richardson. 1996. Evidence based medicine: what it is and what it isn't. *BMJ*, 312(7023):71–2.

Gerard Salton and Chris Buckley. 1988. Term weighting approaches in automatic text retrieval. *Information Processing & Management*, 24(5):513–523.

Helmut Schmid. 1994. Probabilistic part-of-speech tagging using decision trees. In *Int Conf on New Methods in Language Processing*, pages 44–49.

Aliaksei Severyn, Massimo Nicosia, and Alessandro Moschitti. 2013. Learning semantic textual similarity with structural representations. In *Annual Meeting of the Association for Computational Linguistics*, pages 714–718.

Ji Y. Son, Linda B. Smith, and Robert L. Goldstone. 1008. Simplicity and generalization: Shortcutting abstraction in children's object categorizations. *Cognition*, 108:626–638.

Sara Stymne, Jörg Tiedemann, Christian Hardmeier, and Joakim Nivre. 2013. Statistical machine translation with readability constraints. In *NODALIDA*, pages 1–12.

Kai Sheng Tai, Richard Socher, and Christopher D. Manning. 2015. Improved semantic representations from tree-structured long short-term memory networks. In *Annual Meeting of the Association for Computational Linguistics*, pages 1556–1566, Beijing, China.

Christoph Tillmann and Jian-Ming Xu. 2009. A simple sentence-level extraction algorithm for comparable data. In *Companion Vol. of NAACL HLT*.

Masashi Tsubaki, Kevin Duh, Masashi Shimbo, and Yuji Matsumoto. 2016. Non-linear similarity learning for compositionality. In *AAAI Conference on Artificial Intelligence*, pages 2828–2834.

Masao Utiyama and Hitoshi Isahara. 2003. Reliable measures for aligning Japanese-English news articles and sentences. In *Annual Meeting of the Association for Computational Linguistics*, pages 72–79.

Vladimir Vapnik and A. Lerner. 1963. Pattern recognition using generalized portrait method. *Automation and Remote Control*, 24:709–715.

David Vickrey and Daphne Koller. 2008. Sentence simplification for semantic role labeling. In *Annual Meeting of the Association for Computational Linguistics-HLT*, pages 344–352.

Sanja Štajner and Maja Popović. 2016. Can text simplification help machine translation? *Baltic J. Modern Computing*, 4(2):230–242.

Tu Thanh Vu, Giang Binh Tran, and Son Bao Pham. 2014. Learning to simplify children stories with limited data. In *Intelligent Information and Database Systems*, pages 31–41.

Stephen Wan, Mark Dras, Robert Dale, and Cecile Paris. 2006. Using dependency-based features to take the "para-farce" out of paraphrase. In *Australasian Language Technology Workshop*, pages 131–138.

Chih-Hsuan Wei, Robert Leaman, and Zhiyong Lu. 2014. Simconcept: A hybrid approach for simplifying composite named entities in biomedicine. In *BCB '14*, pages 138–146.

Christopher C. Yang and Kar Wing Li. 2003. Automatic construction of English/Chinese parallel corpora. *J. Am. Soc. Inf. Sci. Technol.*, 54(8):730–742.

Yitao Zhang and Jon Patrick. 2005. Paraphrase identification by text canonicalization. In *Australasian Language Technology Workshop*, pages 160–166.

Bing Zhao and Stephan Vogel. 2002. Adaptive parallel sentences mining from web bilingual news collection. In *IEEE Int Conf on Data Mining*, pages 745–748.

Jiang Zhao, Tian Tian Zhu, and Man Lan. 2014. ECNU: One stone two birds: Ensemble of heterogenous measures for semantic relatedness and textual entailment. In *Workshop on Semantic Evaluation (SemEval 2014)*, page 271–277.

Zhemin Zhu, Delphine Bernhard, and Iryna Gurevych. 2010. A monolingual tree-based translation model for sentence simplification. In *COLING 2010*, pages 1353–1361.

Dan Ștefănescu, Radu Ion, and Sabine Hunsicker. 2012. Hybrid parallel sentence mining from comparable corpora. In *16th EAMT Conference*, pages 137–144.

Evaluation of a Prototype System that Automatically Assigns Subject Headings to Nursing Narratives Using Recurrent Neural Network

Hans Moen[1], Kai Hakala[1], Laura-Maria Peltonen[2,3], Henry Suhonen[2,3],
Petri Loukasmäki[1], Tapio Salakoski[1], Filip Ginter[1] and Sanna Salanterä[2,3]
[1]Turku NLP Group, Department of Future Technologies, University of Turku, Finland
[2]Department of Nursing Science, University of Turku, Finland
[3]Turku University Hospital, Finland
{hanmoe,kahaka,lmemur,hajsuh,
peerlo,figint,tapio.salakoski,sansala}@utu.fi

Abstract

We present our initial evaluation of a prototype system designed to assist nurses in assigning subject headings to nursing narratives – written in the context of documenting patient care in hospitals. Currently nurses may need to memorize several hundred subject headings from standardized nursing terminologies when structuring and assigning the right section/subject headings to their text. Our aim is to allow nurses to write in a narrative manner without having to plan and structure the text with respect to sections and subject headings, instead the system should assist with the assignment of subject headings and restructuring afterwards. We hypothesize that this could reduce the time and effort needed for nursing documentation in hospitals. A central component of the system is a text classification model based on a long short-term memory (LSTM) recurrent neural network architecture, trained on a large data set of nursing notes. A simple Web-based interface has been implemented for user interaction. To evaluate the system, three nurses write a set of artificial nursing shift notes in a fully unstructured narrative manner, without planning for or consider the use of sections and subject headings. These are then fed to the system which assigns subject headings to each sentence and then groups them into paragraphs. Manual evaluation is conducted by a group of nurses. The results show that about 70% of the sentences are assigned to correct subject headings. The nurses believe that such a system can be of great help in making nursing documentation in hospitals easier and less time consuming. Finally, various measures and approaches for improving the system are discussed.

1 Introduction

An important task for hospital nurses is to document the administrated patient care in order to ensure care continuity. These nursing (shift) notes are typically stored in patients' electronic health records. However, documentation constitutes a relatively large portion of nurses time, up to 35%, and an average of 19% (Yee et al., 2012). Reducing the time spent on documentation will free up more time for direct patient care. As a means to make the documented text easier to navigate and process, e.g., for the purpose of planning and extracting statistics, nurses in many countries are required to perform some sort of structuring of the text they write (Saranto et al., 2014). Such structuring approaches include the use of documentation standards, classifications and standardized terminologies (Hyppönen et al., 2014). Compared to using fully unstructured free (narrative) text, certain restrictions and requirements to the documentation process are added. As an example, in Finland nurses are nowadays expected to structure the text they write by using subject headings from the Finnish Care Classification (FinCC) standard (Hoffrén et al., 2008). FinCC consist primarily of two taxonomy resources, the Finnish Classification of Nursing Diagnoses (FiCND) and the Finnish Classification of Nursing Interventions (FiCNI), and both of these have a three-level hierarchy. For example, one branch in FiCND is: "Tissue integrity" (level 1), "Chronic wound" (level 2) and "Infected wound" (level 3). Another example, a branch from FiCNI is: "Medication" (level 1), "Pharmacotherapy" (level 2) and "Pharmaceutical treatment, oral instructions" (level 3). In sum, FinCC consist of more than 500 subject headings, making it challenging and time consuming for nurses to use since they are required to memorize, use and structure the text they write according to such a large number of subject headings (Häyrinen et al., 2010).

Our goal is to assist nursing documentation by developing a system that is able to automatically, or semi-automatically, assign subject headings to

Proceedings of the 9th International Workshop on Health Text Mining and Information Analysis (LOUHI 2018), pages 94–100
Brussels, Belgium, October 31, 2018. ©2018 Association for Computational Linguistics

nursing narratives according to the current care classification standard. A central component is a text classification model based on a long short-term memory (LSTM) recurrent neural network architecture (Hochreiter and Schmidhuber, 1997; Gers et al., 2000). We hypothesize that such a system has the potential to reduce the time and effort needed for documentation. It could also increase the consistency in the use of subject headings, and potentially improve the documentation quality. We see two use-cases for such a system: One is where the system assists nurses in selecting appropriate headings when they write, in a suggestive manner, e.g., per sentence or paragraph; A second use-case is where nurses are allowed to write in an unstructured narrative manner, without having to take into consideration the use of subject headings. Instead the system should assign subject headings afterwards and restructure the text under the various subject headings when such a representation is needed. In the presented experiment we focus on the second use-case, where we evaluate the performance of a prototype system developed for this purpose.

2 Related Work

Natural language text is among the most complex data types commonly used for storing and managing information. Thanks to continuous advancements in the field of natural language processing (NLP), computers are becoming capable of performing increasingly complex tasks on this type of data.

Denny et al. (2009) present an algorithm called "SecTag" for detecting section headers in clinical notes based on the free text. More precisely, they focus on history and physical examination documents where the goal is to identify and normalize section headers as well as to detect section boundaries, evaluated with 29 section headers to choose from. For this they use various NLP techniques including word recognition, terminology-based rules, and naive Bayesian classifier. Li et al. (2010) present a system that categorizes sections in clinical notes into one of 15 pre-defined section labels. They use a Hidden Markov model which expects as input clinical notes that have already been split into sections. In Haug et al. (2014) the goal is to develop a "Clinical Section Labeler" which assigns standardized topics to the sections found in clinical notes. These topics, 28 in total,

are here seen as separate from the section headings used by the clinicians when writing, thus the section headings are considered as input to the classifier along with the free text. As classifiers they use two variations of Bayesian networks.

Deep learning methods based on artificial neural networks (ANNs) are currently representing state of the art in many NLP tasks (Zhang et al., 2015; Tang et al., 2015), including text classification, relation extraction and translation. In the presented experiment/prototype system we use the popular long short-term memory (LSTM) recurrent neural network architecture (Hochreiter and Schmidhuber, 1997; Gers et al., 2000) for conducting the text classification. In the data set used here there are 676 unique headings to choose from by the classifier.

3 Methods

3.1 User Interface

The prototype system is implemented in Python with a simple Web interface using the Flask framework (Grinberg, 2018). The interface allows users to upload a text document (i.e. nursing narrative), as shown in Figure 1. When pressing the *create headings button*, the system first splits the text into sentences and then performs word-level tokenization. Each sentence is then fed to the pre-trained text classification model (described below) which assigns subject headings the one subject heading with the highest confidence score according to the classifier. Based on their assigned subject headings, sentences are grouped into paragraphs – one paragraph per unique subject heading. Figure 2 shows a translated example of how a nursing note without subject headings (upper) is converted into paragraphs with assigned subject headings (lower) using the system. Although not utilized in the experiment presented here, the interface also allows the user to move sentences between paragraphs, edit existing subject headings and add new subject headings/paragraphs. In addition, when holding the mouse cursor over a sentence the system shows its top 10 subject heading suggestions according to the classifier. These features provide the user with ways to quickly correct the initial subject heading assignments conducted by the system.

3.2 Text Classification

A central component of the system is the text classification model. The classification task is ap-

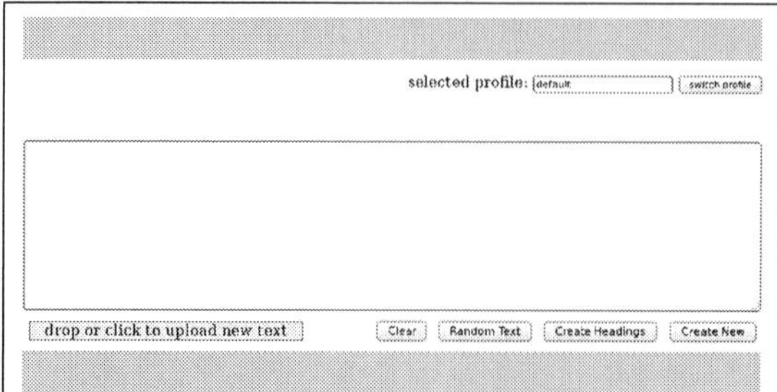

Figure 1: Prototype Web interface.

Oxynorm 10 mg p.o. for abdominal pain when needed to relieve pain. Eaten breakfast. NaCl 0.9 1———-1 cannula removed. Reads news and watches TV in recreation room after breakfast. Feeling well and pain free at the time, the oxynorm administered in the morning helped. CRP decreased now 63, leuc 7.4, also in decline. Eaten lunch. Sister visits after lunch. No need for a sickness certificate. Wound treatment instructions and pain prescriptions given. Has permission to go home in the evening, sister comes to pick up at some point. Left for home at 18.30.

PAIN
Oxynorm 10mg p.o. for abdominal pain when needed to relieve pain.

NUTRITION
Eaten breakfast.
Eaten lunch.

FLUID THERAPY
NaCl 0,9 1———-1 cannula removed.

CURRENT HEALTH AND FUNCTIONALITY
Reads news and watches TV in recreation room after breakfast.
Feeling well and pain free at the time, the oxynorm administered in the morning helped.
Sister visits after lunch.
Left for home at 18.30.

DOCTORS VISIT
CRP decreased now 63, leuc 7.4, also in decline.
No need for a sickness certificate.
Has permission to go home in the evening, sister comes to pick up at some point.

EDUCATION OF RELATIVES
Wound treatment instructions and pain prescriptions given.

Figure 2: An example showing how a nursing note written in a purely narrative manner (upper) is assigned headings and structured using the system (lower). This has been translated from Finnish to English.

proached as a multiclass classification task, where each sentence is assumed to have one correct sub-

ject heading (i.e. class/label). There exist a number of different methods and tools that are suitable for this type of text classification, including the already mentioned LSTM networks (Hochreiter and Schmidhuber, 1997; Gers et al., 2000), convolutional neural networks (CNNs) (LeCun et al., 1998), Random Forest classifiers (Liaw et al., 2002) and support vector machine classifiers (SVM) (Joachims, 1999). However, the focus of this study is not to find the optimal text classification method and parameter settings for this task. This has been the focus of a previous study (under review), where a range of different state-of-the-art and baseline text classification methods are tested and compared. The mentioned study indicated that a bidirectional version of LSTM networks performs best when compared to other classification methods/models, including CNN, SVM and Random Forest. A LSTM network is designed to process sequential data in that it makes its final classification decision after having iteratively observed each element in a sequence, where the order of the elements matters. In our case, a sequence is a list of words belonging to a sentence. This ability to utilize word ordering and to detect long distance word relations in the input sentences is a strength of LSTM networks compared to other text classification approaches relying on bag of word features. In the bidirectional version of LSTM that we use, a sentence is read from both left to right and right to left. This network has been trained on the training set described below. We use the Python-based Keras deep learning library (Chollet et al., 2015) with Theano tensor manipulation library (Bastien et al., 2012) as backend engine.

3.3 Training Data

The data set used for training the classifier is a collection of approximately 0.5 million patients' nursing notes extracted from a hospital in Finland. Ethical approval for using the data was obtained from the hospital district's ethics committee and research approval was obtained from the medical director of the hospital district. The selection criteria were patients with any type of heart-related problem in the period 2005 to 2009. This includes nursing notes from all units in the hospital visited during their hospital stay. The data is collected during a transition period between an older care classification standard and the mentioned FinCC standard, thus only a subset of the headings found

CANNULA CARE
Taken care of the cannula himself. Bandage contains stringy colourful mucus. NaCl cleaning + change of bandages.

taken care of the cannula himself .	*cannula_care*
bandage contains stringy colourful mucus .	*cannula_care*
nacl cleaning + change of bandages .	*cannula_care*

Figure 3: An example showing how a paragraph (upper) is converted into a set of sentence-level training examples (lower). This has been translated from Finnish to English.

there are from FinCC. We only use sentences occurring in a paragraph with a subject heading, which amounts to approximately 5.5 million sentences, 133,890 unique tokens and approximately 38.5 million tokens in total. The average sentence length is 7 tokens and the average number of sentences per paragraph is 2.1. To reduce the number of unique subject headings and to ensure that each included subject heading has a fair number of training examples, we apply a lower frequency threshold of 100. This result in 676 unique subject headings, where their frequency count range from 100 to 222,984, with an average of 4,896. We convert the data into training examples by splitting each paragraph into sentences, each representing a training example with input (X) being the sentence and the output (y) being the associated subject heading of the paragraph. See Table 3 for an example. This enables classification on sentence level, which further allows restructuring and grouping of sentences that are classified as having the same or similar headings. The data set was split into training (60%), development (20%) and test (20%) sets.

Although not the focus of this paper, we report the performance of the bidirectional LSTM classifier when used to predict subject headings for the test set, as a comparison to the experiment presented below. Performance is calculated as recall at N (R@N), which is the average of how many times the correct subject heading is found among the top N suggested subject headings by the system. R@1 is here equal to the classifier's accuracy score on the test set. These results are presented in Table 1. We refer to this evaluation as an automatic evaluation since no (additional) manual evaluation is required.

Measure	Score
R@1 / Accuracy	54.35%
R@10	89.54%

Table 1: The classifiers performance on the test set. R@N is recall at N, reflecting the average of how many times the correct subject heading is found among the top N retrieved ones, over all sentences, in the test set. R@1 is equal to accuracy.

4 Experiment

The main objective of the experiment is to assess how well the described system is able to assign relevant subject headings to nursing notes that are written in a narrative manner, without using or considering subject headings. A secondary objective is to report on feedback from nurses concerning the potential use of such a system in a clinical setting.

The nursing notes that we have in the existing data set are all planned, written and structured according to the ruling documentation standard – where the text is split into sections labeled with subject headings. Thus, to acquire relevant nursing notes for the evaluation – nursing notes written in a way where the authors does not plan for or consider the use of sections and subject headings – we asked three domain experts with nursing background to write a couple of notes each in this way based on made up artificial patients. This resulted in a total of 20 nursing notes. These were then presented to the system, one by one, which classified and assigned subject headings on sentence level before grouping sentences under each heading. The results were stored in a spreadsheet for evaluation, containing a short description of the patient case, the original nursing note and the version with assigned subject headings on sentence level. See Figure 2 for an example of one of the nursing narratives/notes used in the evaluation, both without and with the assigned headings and restructuring conducted by the system.

Next, two domain experts (hereby referred to as evaluators) were given the task of assessing how well the system performed. For this the evaluators were (a) instructed to use a four class scale when manually assessing each sentence with respect to their assigned headings, and (b) asked to answer the open ended question "what do you think about the current performance and functionality of the system and its potential use in a clinical setting?".

Class	Count	Percentage
1 / Accuracy$_{min}$	311	68.05%
2	93	20.35%
3	48	10.50%
4	5	1.10%
1 + 2 / Accuracy$_{max}$	404	88.40%

Table 2: Average results from the manual evaluation. Class description: 1 - Correct heading. 2 - Maybe correct heading. 3 - Wrong heading. 4 - Unable to assess.

The four classes are as follows:

1 - Correct heading (it correctly describes the content of the sentence)

2 - Maybe correct heading

3 - Wrong heading

4 - Unable to assess

The proportion of sentences assigned to Class 1 is equal to the accuracy$_{min}$ score of the system for this task, while the sum of Class 1 and 2 can be considered as the accuracy$_{max}$ score. So the actual accuracy score would be somewhere between accuracy$_{min}$ and accuracy$_{max}$.

5   Results

Initially the two evaluators disagreed in their assessments of 30.45% of the sentences. To reach a common consensus, the two evaluators discussed these cases together with a third domain expert. The results from the manual evaluation (consensus) are presented as average counts and percentages for each class in Table 2.

The percentage of correctly classified sentences in the manual evaluation experiment is 68.05% (Table 2). However, the actual accuracy score of the system can be assumed to be somewhere between 68.05% (accuracy$_{min}$) and 88.40% (accuracy$_{max}$). This is roughly 13% to 34% points up from the R@1/accuracy score resulting from the automatic evaluation in Table 1. When the system is allowed to suggest 10 headings, R@10, the correct heading is found among these for about 90% of the sentences in the test set. I.e. at least one of the suggested 10 headings for a sentence has been considered correct for about 90% of the test set sentences in the manual evaluation.

The evaluators reported that they were generally satisfied with the performance of the system. They think that such a system/functionality could be very useful to have as an integrated part of a hospital information system/electronic health record system, and could reduce the time and effort required to perform the documentation. They also think that it has the potential to increase the quality of documentation by supporting the correct use of such standardized terminologies. The evaluators reported that the system showed a tendency to assign subject headings with a high level of specificity, and sometimes even too specific than what would be practical. For example, for two or more sentences describing different aspects of pain management in the same nursing note, such as treatment and medication, the system would in some cases assign these to different subject headings, and/or headings of different level of specificity/abstraction. Another observation was that the system had sometimes difficulties in correctly classifying sentences that covers multiple subjects.

6   Discussion

One obvious observation is that there is a relatively large gap between the scores resulting from the conducted manual evaluation (68.05% $\leq$ accuracy $\leq$ 88.40%, Table 2) and the automatic evaluation scores (accuracy = 54.35%, Table 1). We believe that this is caused by primarily two underlying problems: First, the data set spans two different documentation standards (as described in Section 3), which could be somewhat confusing to the classifier. Second, the nurses do not necessarily always use the correct subject headings when they write. Thus it is likely, in particular for this type of automatic evaluation, that higher scores will be achieved when the classifier is trained and evaluated on a data set consisting of only one documentation standard. When looking at the R@10 scores (Table 1), the system suggests the correct heading for about 90% of the sentences in the test set. However, it is likely that the same problem of "classification standard confusion" negatively influences this score too. For a use-case where, let us say, the system suggests 10 headings per sentence to the user when he/she is writing the nursing notes, this would mean that there is a very high probability ($\geq$ 90%) of finding a suitable/correct subject heading among the suggested ones.

Based on their observations, the evaluators found the system to sometimes assign subject headings with an artificial detail level. One way to deal with this would be to allow the users to pre-select the level in the hierarchy of the documentation standard that the system should aim for when assigning subject headings. In addition, since a unit in the hospital would typically not use all the headings in the documentation standard, it should be possible to limit the headings that the system can choose from for different units.

To further improve the performance of the system there are several, possibly complementary, approaches that could be explored. One approach is to allow the user to manually correct the initial classifications done by the system, e.g. by moving sentences to their correct subject headings, and allowing the user to add and remove subject headings at will. Additionally, this type of manual corrections could be used to further improve the system/classifier. A possibly complementary approach could be to apply some form of classification heuristic and/or feedback based on the confidence scores produced by the classifier. For example, when classifying a sentence, if the classifier shows very similar confidence scores for the top suggested subject headings, and if a subject heading used in the same or a previous nursing note, from the same patient and care episode, is among these, one could have the system select this one. Another example, if the classifier does not show a clear preference for a single subject heading when classifying a sentence, this could be communicated to the user. Some type of clustering of subject headings that are very similar (in terms of form and/or meaning) within a single nursing note could also be tried. It would also make sense to exploit the taxonomic hierarchy underlying the nursing documentation standard, e.g. during training and/or prediction as well as in the grouping of sentences and possibly for merging some of the assigned subject headings. Another approach would be to try using a more balanced data set for training the classifier – balanced in terms of label/subject heading frequencies. The use of class weighting when training the classifier could also be tried. With enough training data, it could also be an idea to train a separate classifier per hospital unit. Further performance gains could be achieved by also training a classifier on the level of paragraphs as a supplement to the sentence-level classification.

Although the focus of this work has been on assisting nursing documentation, other professions use subject headings in a similar fashion when they write. One example is physicians and the notes they write in relation to diagnosis and treatment of patients. Thus we assume that the same type of classification-based system could be useful to other professions too.

7 Conclusions and Future Work

The presented prototype system for automated assignment of subject headings to nursing notes is shown to perform well based on the reported experiment. It achieves a classification accuracy somewhere between 68.05% (accuracy$_{min}$) and 88.40% (accuracy$_{max}$). The domain experts evaluating the system reported that they believe such a system could save both time and effort when it comes to writing nursing shift notes in hospitals. We argue that future improvements of the system's classification performance could be gained through user feedback or by applying some heuristic based on its confidence scores. In the presented experiment we have the classification system learn to classify text on the level of sentences. As future work we are also considering exploring paragraph-level classification for this task, primarily as a supplement to sentence-level classification. Since there are other professions who use subject headings in a similar way as nurses when they document, we believe that a similar system could also be useful in other domains, for other professions.

As future work we aim to test this system/classifier on a larger scale, where it will also be evaluated when used in the initial writing of nursing notes, by suggesting N subject headings to the user for each sentence being written. We will also strive to acquire a data set containing only one documentation standard – the one currently being used in the targeted hospital district. Then the following step would be clinical testing and assessment of the impact of such a system (extrinsic evaluation).

References

Frédéric Bastien, Pascal Lamblin, Razvan Pascanu, James Bergstra, Ian Goodfellow, Arnaud Bergeron, Nicolas Bouchard, David Warde-Farley, and Yoshua Bengio. 2012. Theano: New features and speed improvements. *arXiv preprint arXiv:1211.5590 (2012)*.

François Chollet et al. 2015. Keras. `https://keras.io`.

Joshua C. Denny, Anderson Spickard, III, Kevin B. Johnson, Neeraja B. Peterson, Josh F. Peterson, and Randolph A. Miller. 2009. Evaluation of a method to identify and categorize section headers in clinical documents. *Journal of the American Medical Informatics Association*, 16(6):806–815.

Felix A Gers, Jürgen Schmidhuber, and Fred Cummins. 2000. Learning to forget: Continual prediction with LSTM. *Neural Computation*, 12(10):2451–2471.

Miguel Grinberg. 2018. *Flask web development: developing web applications with Python.* " O'Reilly Media, Inc.".

Peter J Haug, Xinzi Wu, Jeffery P Ferraro, Guergana K Savova, Stanley M Huff, and Christopher G Chute. 2014. Developing a section labeler for clinical documents. In *AMIA Annual Symposium Proceedings*, volume 2014, page 636. American Medical Informatics Association.

Kristiina Häyrinen, Johanna Lammintakanen, and Kaija Saranto. 2010. Evaluation of electronic nursing documentation – Nursing process model and standardized terminologies as keys to visible and transparent nursing. *International Journal of Medical Informatics*, 79(8):554–564.

Sepp Hochreiter and Jürgen Schmidhuber. 1997. Long short-term memory. *Neural Computation*, 9(8):1735–1780.

Päivi Hoffrén, Kirsi Leivonen, and Merja Miettinen. 2008. Nursing standardized documentation in kuopio university hospital. *Studies in Health Technology and Informatics*, 146:776–777.

Hannele Hyppönen, Kaija Saranto, Riikka Vuokko, Päivi Mäkelä-Bengs, Persephone Doupi, Minna Lindqvist, and Marjukka Mkelä. 2014. Impacts of structuring the electronic health record: A systematic review protocol and results of previous reviews. *International Journal of Medical Informatics*, 83(3):159–169.

Thorsten Joachims. 1999. Making large-scale SVM learning practical. In B. Schölkopf, C. Burges, and A. Smola, editors, *Advances in Kernel Methods – Support Vector Learning*, chapter 11, pages 169–184. MIT Press, Cambridge, MA.

Yann LeCun, Léon Bottou, Yoshua Bengio, and Patrick Haffner. 1998. Gradient-based learning applied to document recognition. *Proceedings of the IEEE*, 86(11):2278–2324.

Ying Li, Sharon Lipsky Gorman, and Noémie Elhadad. 2010. Section classification in clinical notes using supervised hidden markov model. In *Proceedings of the 1st ACM International Health Informatics Symposium*, IHI '10, pages 744–750, New York, NY, USA. ACM.

Andy Liaw, Matthew Wiener, et al. 2002. Classification and regression by randomForest. *R news*, 2(3):18–22.

Kaija Saranto, Ulla-Mari Kinnunen, Eija Kivekäs, Anna-Mari Lappalainen, Pia Liljamo, Elina Rajalahti, and Hannele Hyppönen. 2014. Impacts of structuring nursing records: a systematic review. *Scandinavian Journal of Caring Sciences*, 28(4):629–647.

Duyu Tang, Bing Qin, and Ting Liu. 2015. Document modeling with gated recurrent neural network for sentiment classification. In *Proceedings of the 2015 Conference on Empirical Methods in Natural Language Processing*, pages 1422–1432.

Tracy Yee, Jack Needleman, Marjorie Pearson, Patricia Parkerton, Melissa Parkerton, and Joelle Wolstein. 2012. The influence of integrated electronic medical records and computerized nursing notes on nurses time spent in documentation. *Computers Informatics Nursing*, 30(6):287–292.

Xiang Zhang, Junbo Zhao, and Yann LeCun. 2015. Character-level convolutional networks for text classification. In C. Cortes, N. D. Lawrence, D. D. Lee, M. Sugiyama, and R. Garnett, editors, *Advances in Neural Information Processing Systems 28*, pages 649–657. Curran Associates, Inc.

Automatically Detecting the Position and Type of Psychiatric Evaluation Report Sections

Deya M. Banisakher, Naphtali Rishe, Mark A. Finlayson
School of Computing and Information Sciences
Florida International University
Miami, FL 33199
{dbani001,rishe,markaf}@fiu.edu

Abstract

Psychiatric evaluation reports represent a rich and still mostly-untapped source of information for developing systems for automatic diagnosis and treatment of mental health problems. These reports contain free-text structured within sections using a convention of headings. We present a model for automatically detecting the position and type of different psychiatric evaluation report sections. We developed this model using a corpus of 150 sample reports that we gathered from the Web, and used sentences as a processing unit while section headings were used as labels of section type. From these labels we generated a unified hierarchy of labels of section types, and then learned n-gram models of the language found in each section. To model conventions for section order, we integrated these n-gram models with a Hierarchical Hidden Markov Model (HHMM) representing the probabilities of observed section orders found in the corpus, and then used this HHMM n-gram model in a decoding framework to infer the most likely section boundaries and section types for documents with their section labels removed. We evaluated our model over two tasks, namely, identifying section boundaries and identifying section types and orders. Our model significantly outperformed baselines for each task with an F_1 of 0.88 for identifying section types, and a 0.26 WindowDiff (W_d) and 0.20 and (P_k) scores, respectively, for identifying section boundaries.

1 Introduction

With the exponential growth of free text in electronic health records (EHRs)—which includes mental health documents—it is ever more important to develop natural language processing (NLP) models that automatically understand and parse such text. When incorporated in other systems, these models may aid (1) clinical decision support, (2) the extraction of key population information and trends, and (3) precision medicine efforts where personalized information and trends are extracted and used in the treatment process (Demner-Fushman et al., 2009; Hripcsak et al., 2003).

The majority of clinical NLP work has focused on semantic parsing of clinical notes found in EHRs. There are several challenges in automatic understanding of unstructured text in EHRs, encompassing many levels of linguistic processing: identifying document layouts, their discourse organization, mapping lexical information to semantic concepts found in biomedical ontologies, as well as understanding inter-concept co-reference and temporal relations (Li et al., 2010). These challenges are also present for mental health NLP applications.

We present an approach to automatically model the discourse structure of psychiatric reports as well as segment these reports into various sections. Our model learns the section types, positions, and sequence and can automatically segment unlabeled text in a psychiatric report into the corresponding sections. We hypothesize that knowledge of the ordering of the sections can improve the performance of a section classifier and a text segmenter. To test this hypothesis, we train a Hierarchical Hidden Markov Model (HHMM) that categorizes sections in psychiatric reports into one of 25 pre-defined section labels.

The remainder of this paper is organized as follows: we first introduce psychiatric reports and their various types and conventions (§2). Next, we discuss the task definition in detail (§3). We then describe our approach including the corpus used, and the two main components of our model (§4). Additionally, we present and discuss the baselines and experiments performed as well as the results obtained from those experiments (§5). We follow this with a review of related work on document

101

Proceedings of the 9th International Workshop on Health Text Mining and Information Analysis (LOUHI 2018), pages 101–110
Brussels, Belgium, October 31, 2018. ©2018 Association for Computational Linguistics

section identification and text segmentation (§6). Finally, we conclude and specify our contributions (§7).

2 Psychiatric Evaluation Reports

A mental health assessment is the process through which a psychiatrist or a psychologist obtains and organizes necessary information about mental health patients. This process usually involves a series of psychological and medical tests (clinical and non-clinical), examinations, and interviews (Reeves and Rosner, 2016). These procedures serve the purpose of making a diagnosis that then guides a treatment or a treatment plan (Association, 2018).

The output of a mental health assessment is a mental health report. Psychiatric reports are simpler subtype of this document type, and mainly consist of long-form unstructured text. They are the end product of psychiatric assessments in which psychiatrists summarize the information they gathered, as well as integrate the patient history, their evaluation, patient diagnosis, and suggested treatments or future steps (Groth-Marnat, 2009; Goldfinger and Pomerantz, 2013). There are several types of psychiatric reports that vary depending on the type and purpose of assessment: Psychiatric evaluation reports, crisis evaluation reports, daily SOAP reports (Subjective, Objective, Assessment, Plan), mental status exam reports, and mini mental status exam reports, to name a few (Association, 2006). Our study focuses on psychiatric evaluation reports. Although there is no one strict format, there are general guidelines that psychiatrists follow when writing psychiatric evaluation reports. Drawing from the general psychiatric evaluation domains, these reports start with the patient's identifying information, followed by the patient's chief complaints, presenting illness and its history, personal and family's medical history, mental status examination, and ending with the psychiatric medical diagnosis and treatment plan. This information is typically structured into an ordered list of headed sections (Association, 2006). Table 1 contains a detailed list of the main sections of a psychiatric evaluation report in general order of appearance. Not all listed sections appear in all psychiatric evaluation reports, and they also do not necessarily appear in the same order, although there is usually a general pattern to the order.

> *Family History: Her mother was depressed and was treated. Her mother is currently age 55 ... There is no family history of bipolar disorder, anxiety ... Medical history in the family is significant for her son, age 4, who is having seizures ... and several paternal great aunts had breast cancer.*

Figure 1: Excerpt from a psychiatric report showing an example of implicitly including two different sections within another (namely, *FAMILY PSYCHIATRIC HISTORY* in the first underlined portion, and *FAMILY MEDICAL HISTORY* in the second underlined portion within *FAMILY HISTORY*).

3 Task Definition

Our goal was to build models that learn the section structure of an evaluation psychiatric report. As discussed earlier, a psychiatric evaluation report consists of several sections, often ordered in a usual way. Therefore the task we tackle here is to segment and classify blocks of unstructured text (at the sentence level) drawn from psychiatric evaluation reports into their appropriate section types. We assume that the reports follow the general guidelines of psychiatric evaluation report writing discussed in (§2).

There are four main challenges in section classification of clinical notes and mental health reports. First, labels that psychiatrists use to designate sections are ambiguous and various (Li et al., 2010), for example, a section titled *IDENTIFICATION OF PATIENT* by one psychiatrist might be named *REFERRAL DATA* or *IDENTIFYING INFORMATION* by another. Second, psychiatrists often omit some sections entirely or include them implicitly within other sections or under other labels, for example, the section *CHILDHOOD EVENTS* can be included in a larger section such as *FAMILY HISTORY* while *STRENGTHS AND SUPPORTS* can be listed within *Mental Status*. Figure 1 shows an example. Third, the sections' order can be different between different psychiatric reports. Fourth, some section labels are omitted or skipped, especially if the information that would be placed in that section is not relevant to the patient being evaluated.

Additionally, With the section labels removed from the reports, our segmentation task was to find the section boundaries using sentences as the processing unit. This task is similar to topic shift detection in meeting minute, newscasts, and doctor-patient counseling conversations (both, written and spoken). Psychiatric reports are highly structured , with specific types of information (e.g., prescribed medications) found in particular sections (e.g., Treatment Plan), and with various general conventions for what information should appear in which sections, and in what order. However, the segmentation task is not trivial as it faces the same aforementioned challenges. Additionally, one must find highly distinctive features to distinguish individual sentences (and thus, boundaries) in various sections as some of these sections can contain similar linguistic and structural features and may even contain similar topic keywords (e.g. language in *FAMILY PSYCHIATRIC HISTORY* and *SOCIAL HISTORY*.

We identify the subtasks of this problem as (1) learning and building a model for the sections' order and presence in a report, (2) learning and building models that describe the distinctive features of the various section types, and (3) applying a combination of these two model to simultaneously identifying section boundaries and label section types.

4   Approach

Given the sequential nature of the reports' sections, we treat this ordering task as a sequence labeling task. That is, given a psychiatric report with n sections $S = (S_1, \ldots, S_n)$, determine the optimal sequence of section labels $O^* = (O_1^*, \ldots, O_n^*)$ among all possible section sequences. Hidden Markov Models (HMMs) have been used successfully for sequence labeling in a wide variety of applications, including specifically natural language processing and medical informatics. In our problem formulation and approach, we follow and combine work presented by Sherman and Liu (2008) and Li et al. (2010). Both of these approaches used HMM-based models coupled with section or topic-specific n-gram models to segment text. Sherman and Liu (2008) focused on segmenting sentences within meeting minutes into a set of predefined topics, while Li et al. (2010) focused on identifying sections within a clinical note documents. We take a supervised learning approach where we learn the HMM parameters using a labeled corpus. Our implementation was generally guided by the work described in Barzilay and Lee (2004) and (Rabiner, 1989).

To overcome the challenges outlined in (§3), we first created a unified hierarchy of standardize section labels types, based on observations in a 150 report corpus that we assembled. Second, while Li et al. (2010) focused on the section level when building their n-gram language models, we focus on the sentence level, similar to Sherman and Liu (2008). Additionally, to model the inclusion of some sections within others as discussed in (§3) we built a two-level Hierarchical HMM (HHMM) (Bui et al., 2004) in which some states contain HMM models for their implicit subsections. This is in contrast to the approach presented by Li et al. (2010), who used a flat HMM, disregarding any hierarchy within the clinical notes' sections. The HHMM model was first proposed by Fine et al. (1998) as a strict tree structure where each state in the HHMM is an HHMM itself. This approach was extended and tailored by researchers for various tasks such as the approach proposed by Bui et al. (2004) who relaxed the original model to fit general HMM structures and implementations.

In summary, to tackle the first subtask from (§3) we built a two-level HHMM that models the positions and order of the reports' sections. To tackle the second subtask, we built language models (namely, n-gram models) per section type that describe distinctive lexical information for each of those sections. We then couple the HHMM with the n-gram models where the HHMM and HMM states represent the known section labels, while the states' observations are the n-grams contained within each of the individual sections. Finally, to tackle the the the third subtask, that is identifying section boundaries, we follow a decoding scheme using the Viterbi algorithm (discussed briefly in §4.4).

In the remainder of this section we describe the corpus we collected and annotated. Next, we present the two components of the HHMM model, that is, the states (modeling the section order) and the observations (modeling the section language). Finally we briefly discuss the process by which we use the model to identify section boundaries.

4.1   Corpus

To the best of our knowledge there is no corpus of psychiatric reports annotated with section labels, so we created our own. We collected 150 publicly available psychiatric evaluation report samples by crawling the web through custom search engines (Google Custom Search Engine for Med-

Parent Label	Section Label	# Words	# Sentences	Avg. Sent. Length	% Present	% Implicit
-	*IDENTIFYING DATA*	12	2	6	100	-
	CHIEF COMPLAINT	27	3	9	100	-
MEDICAL HISTORY	*HISTORY OF PRESENT ILLNESS*	232	29	8	95	10
	PSYCHIATRIC HISTORY	85	8	11	82	36
	SUBSTANCE ABUSE HISTORY	98	10	10	88	44
	REVIEW OF SYMPTOMS	150	19	8	96	51
-	*SURGERIES*	28	3	7	33	-
	ALLERGIES	4	2	2	98	-
	CURRENT MEDICATIONS	40	9	4	100	-
FAMILY HISTORY	*BIRTH AND DEVELOPMENTAL HISTORY*	59	5	10	31	51
	ABUSE HISTORY / TRAUMA	110	9	12	79	34
	FAMILY PSYCHIATRIC HISTORY	44	5	9	73	80
	FAMILY MEDICAL HISTORY	48	7	7	92	38
	SOCIAL HISTORY	80	7	11	76	45
	PREGNANCY	29	3	8	47	64
-	*SPIRITUAL BELIEFS*	12	2	5	24	-
	EDUCATION	32	3	8	68	-
	EMPLOYMENT	31	3	9	79	-
	LEGAL	10	1	5	20	-
MENTAL STATUS	*MENTAL STATUS*	155	18	9	95	11
	STRENGTHS AND SUPPORTS	8	1	8	71	43
-	*FORMULATION*	35	4	8	62	-
	DIAGNOSES	63	12	5	100	-
	PROGNOSIS	8	2	3	74	-
	TREATMENT PLAN	121	12	10	100	-

Table 1: List of possible sections in a psychiatric report used in the corpus.

ical Transcriptionists[1] and GoogleMT[2]) and other sources [3]. The reports we selected were complete and adhere to the general guidelines for psychiatric report writing discussed in the previous sections. Some of the reports were anonymized samples of real reports, while others were mock reports written for educational purposes.

We prepared the corpus in two stages. First, we standardized the labels' names, selecting a single uniform name for each section type and mapping corresponding section labels found in the corpus to those names. For example, some reports contained the section *SCHOOL* while others listed it as *EDUCATION*. Here we selected *EDUCATION* as the uniform section label across all reports.

Second, we created a hierarchy for the section names which reflected implicit embedded sections types that we found in the corpus. There were only three section types that included implicit subsections in our data, namely, *MEDICAL HISTORY*, *FAMILY HISTORY*, and *MENTAL STATUS*. For example, some reports containing the section *MENTAL STATUS* might in turn include information in that section about both *MENTAL STATUS EXAM* and *STRENGTHS AND SUPPORTS*. In this case we identified these implicit subsection boundaries (that is, the boundaries were not identified with a section header) and labeled those subsections with both the parent and child label. Table 1 lists the the parent sections that sometimes included other sections implicitly (first column), the unified list of section types found in the collected reports (second column), word and sentence level statistics (columns 3-5), and percentage of reports containing those sections in the corpus (last two columns). For both of these stages we used all 150 reports.

---

[1]`https://cse.google.com/cse/publicurl?cx=010964806533120826279:kyuedntb2fy`

[2]`https://www.googlemt.com/#gsc.tab=0`

[3]`http://www.medicaltranscriptionsamples.com/`

`http://mtsamples.com/`

`https://medword.com/psychiatry5.html`

`http://www.medicaltranscriptionsamplereport.com/`

`http://onwe.bioinnovate.co/psychological-assessment-example/`

Following standard procedure for supervised machine learning, we split our corpus under a cross-validation paradigm into two sets for training and testing, where 80% of the reports were used in training and 20% for testing. This amounted to 120 and 30 reports for training and testing respectively.

4.2 Modeling the Section Orders

As discussed before, we built an HHMM where each state corresponds to a distinct section label. We introduce the terms *state* and *parent state* when discussing the HHMM. A *state* is simply an HMM state corresponding to a distinct section. A *parent state* is an HHMM state corresponding to a collection of ordered sections. To account for sections listed implicitly, we created a two-level HHMM where *parent states* contained *states* representing the ordered subsections found in the *parent state* section. Thus our model contained 25 *states* and three *parent states* corresponding to information in Table 1. The first HHMM layer contained both *states* and *parent states*, while the second layer contained a total of 12 *states* corresponding to the potential implicit subsections for the three *parent states*. In our HHMM, each *parent state* is simply an HMM itself. Thus our discussion of HMM parameter calculation applies to both *states* and *parent states*.

Our model learned transition probabilities from the labeled corpus. The state transition probabilities capture constraints on section orderings. We estimated the probabilities between each state s using Equation 1. Additionally, to account for sparsity (that is, unseen section orders) we smoothed the probabilities by the total number of section labels t_S following Laplace smoothing.

$$P(s_j|s_i) = \frac{count(s_i, s_j) + 1}{count(s_i) + t_S} \quad (1)$$

The second level HMM models contained within the *parent states* follow the same scheme in probability estimation, but differ in the smoothing parameter (t_S). Here, the total number of section labels t_S depends on the number of subsections in each of the *parent states*. For example, the *parent state MEDICAL HISTORY* contains a total of four subsections or *states*, and thus its HMM model is smoothed by $t_S = 4$. Finally, all of the model's states were linked with empty transitions in addition to self-looping ones to account for missing sections as well as a section continuation, respectively (i.e. indicating a section shift or a continuation).

4.3 Modeling Section Language

To tackle the second subtask identified in (§3), we built n-gram language models (Jain et al., 2015) that captured distinctive lexical information contained within the individual sections. This, in turn, helped classify unknown blocks of text (that is, text unseen previously by the trained models) within a report into their respective sections. We opted to use bigrams as our training corpus because higher n-gram models were extremely sparse, and had poor performance. This is consistent with significant research showing that in most applications bigrams work well and better than others (Reynar, 1998).

We built independent bigram models for each section type in the reports, using only text from that section type. Additionally, for each of the three section types represented by the *parent states* (discussed above) we built bigram models using text found in all of the contained subsections. A common problem that arises with n-gram models is sparsity of phrases or words. This is especially the case when training on a small corpus. Given our relatively small corpus, our models were quite sparse at first, however, we used Laplace Smoothing as a solution.

Similar to transition probabilities, our HHMM learned observation probabilities from the labeled corpus. We trained a bigram model for each state s of the HHMM. Equation 2 shows the computation for the likelihood of a sentence sequence w_0^k (i.e., a long sequence of words) to be generated by a state s. Equation 3 shows the computation for estimating the specific state bigram probability along with Laplace smoothing counts for the corresponding section S (V_S represents the vocabulary size for that section state).

$$P(w_0^k|s) = \prod_0^{k-1} P_s(w_{i+1}|w_i) \quad (2)$$

$$P_s(w_{i+1}|w_i) = \frac{count_S(w_i^{i+1}) + 1}{count_S(w_i) + |V_S|} \quad (3)$$

We used a rule-based approach to detect uniformly structured sections containing only standard medical terms such as medications and additional key terms. The sections mapped with hard-coded rules are the *CURRENT MEDICATIONS* and the standard *DSM-IV* multiaxal assessment contained within the *DIAGNOSIS* section, one of which is illustrated in Figure 2. We recognize that this standard has been dropped with the introduction of *DSM-5* in 2013, however, our dataset follows the older standard as most psychiatric reports in existence do since the new standard is relatively new.

Axis I	296.32	Major depressive disorder, recurrent, moderate
	305.00	Alcohol use disorder, mild
Axis II	V71.09	No diagnosis
Axis III		Hypertension
Axis IV		Problems with primary support group
Axis V	GAF = 48 (Current)	

Figure 2: Example of *DSM-IV* multiaxal diagnosis assessment.

For the *MEDICATIONS* section we used publicly available datasets containing lists of medications (eMedicineHealth, 2018), and the U.S. National Library of Medicine's RxNorm dataset (Liu et al., 2005). String-matching was additionally used to locate the *DIAGNOSIS* sections as our algorithm would search for the key headers "Axis I, II, III, IV, V".

Therefore we generated 26 bigram models, one for each section type (except for the two rule-based types) plus three parent section types.

4.4 Decoding

We integrated the bigram models with the HHMM and then used this bigram-HHMM model in a decoding framework to infer the most likely section boundaries and section types for documents with their section labels removed. We used the Viterbi algorithm and applied the following equation to obtain the most likely labeling of sections O^*, where n is the section index, and k_n is the word index for section n:

$$O^* = \arg\max_{s} P(s)P(w_0^{k_n}|s)$$
$$= \arg\max_{s_1 s_2 \dots s_n} P(s_1)P(w_0^{k_n}|s_1) \times$$
$$\prod_{i=0}^{n} P(s_i|s_{i-1})P(w_1^{k_n}|s_i)$$

5 Results and Discussion

As discussed above, we randomly split the corpus into training and testing sets in a cross-validation setup, using ten folds, resulting in 120 reports for training and 30 for testing in each fold. Our models were trained to learn a total of 25 distinct sections. Here we present our evaluation methods and results, describing our baseline approaches, as well as the performance of both the baselines and our method averaged across the test sets.

5.1 Evaluation Methods

There are two problems that our system solves: 1) the section labeling problem —applying the correct section type to each section—and 2) the section segmentation problem—identifying the correct section boundaries. We evaluate our system's performance on these two problems separately.

For the section ordering, we evaluated the performance of the model on each section using the F_1 measure averaged across all folds. As for the boundary detection problem, we use the WindowDiff (W_d) (Pevzner and Hearst, 2002) and P_k (Beeferman et al., 1999) metrics. These metrics compare the number of segmentation boundaries between a system's output and a gold standard by observing a scrolling window of text in the document, and run from 0 to 1, with scores closer to 0 being better. W_d increases (gets worse) when the boundaries are different. Similarly P_k increases when a section type transition (i.e., a section type for this study) is different. The W_d score represents the probability that the number of boundaries found by the system is different from that in the gold standard, while the P_k score represents the probability that any two sentences are incorrectly listed as being in the same section.

5.2 Baseline Methods

We compared our system's performance in finding the correct labels of sections in a report to two baseline methods. The first method was introduced as a baseline by Li et al. (2010). This method uses bigrams to independently classify each section, disregarding any section order information. For the second baseline, we followed the primary approach proposed by Li et al. (2010) which is a flat HMM model built similarly to our model as described previously (§4), but operates on a section level rather than a sentence level. Li's method ignores hierarchical information where some report sections are implicitly included within other sections. Our implementation of this model included 25 states corresponding to each section within the reports. Both of these methods assume that the section boundaries are given, and as such they only generate a sequence labeling for section types.

We compared our system's performance in identifying section boundaries to two other baseline methods. The first is LCSeg—a popular text segmentation baseline (Galley et al., 2003). LC-

Section	Independent Bigram			Flat HMM			HHMM		
	P	*R*	*F₁*	*P*	*R*	*F₁*	*P*	*R*	*F₁*
IDENTIFYING DATA	0.83	0.81	0.82	0.96	0.94	0.95	0.98	0.95	**0.97**
CHIEF COMPLAINT	0.68	0.65	0.67	0.88	0.74	0.80	0.94	0.89	**0.91**
MEDICAL HISTORY	0.66	0.66	0.65	0.93	0.88	**0.90**	0.93	0.88	0.90
HISTORY OF PRESENT ILLNESS	0.69	0.67	0.68	0.91	0.86	0.88	0.94	0.86	**0.90**
PSYCHIATRIC HISTORY	0.65	0.60	0.62	0.74	0.85	0.79	0.93	0.86	**0.89**
SUBSTANCE ABUSE HISTORY	0.69	0.69	0.69	0.88	0.80	0.84	0.95	0.83	**0.89**
REVIEW OF SYMPTOMS	0.8	0.67	0.73	0.79	0.86	0.82	0.94	0.87	**0.90**
SURGERIES	0.4	0.31	0.35	0.79	0.51	0.62	0.85	0.64	**0.73**
ALLERGIES	0.6	0.80	0.69	0.90	0.86	0.88	0.88	0.91	**0.89**
CURRENT MEDICATIONS	0.87	0.74	0.80	0.90	0.84	0.87	0.91	0.93	**0.92**
FAMILY HISTORY	0.60	0.56	0.58	0.92	0.86	**0.89**	0.92	0.86	0.89
BIRTH AND DEVELOPMENTAL HISTORY	0.68	0.50	0.57	0.71	0.68	0.69	0.89	0.80	**0.84**
ABUSE HISTORY / TRAUMA	0.42	0.33	0.37	0.87	0.77	0.82	0.96	0.81	**0.88**
FAMILY PSYCHIATRIC HISTORY	0.57	0.59	0.58	0.92	0.87	0.89	0.92	0.90	**0.91**
FAMILY MEDICAL HISTORY	0.65	0.60	0.62	0.92	0.89	0.90	0.94	0.89	**0.91**
SOCIAL HISTORY	0.67	0.69	0.68	0.66	0.89	0.76	0.93	0.81	**0.87**
PREGNANCY	0.6	0.67	0.63	0.89	0.51	0.65	0.92	0.80	**0.86**
SPIRITUAL BELIEFS	0.73	0.46	0.56	0.90	0.9	**0.90**	0.93	0.88	0.90
EDUCATION	0.66	0.61	0.63	0.71	0.77	0.74	0.92	0.84	**0.88**
EMPLOYMENT	0.65	0.62	0.63	0.91	0.88	**0.89**	0.92	0.86	0.89
LEGAL	0.16	0.62	0.26	0.67	0.61	0.64	0.72	0.68	**0.70**
MENTAL STATUS	0.56	0.72	0.62	0.85	0.94	**0.89**	0.85	0.94	0.89
MENTAL STATUS EXAM	0.64	0.63	0.64	0.83	0.96	0.89	0.85	0.96	**0.90**
STRENGTHS AND SUPPORTS	0.42	0.82	0.56	0.80	0.92	0.86	0.82	0.92	**0.87**
FORMULATION	0.56	0.71	0.63	0.86	0.78	0.82	0.92	0.82	**0.87**
DIAGNOSES	0.88	0.76	0.81	0.96	0.95	0.96	0.98	0.98	**0.98**
PROGNOSIS	0.66	0.62	0.64	0.84	0.82	0.83	0.90	0.86	**0.88**
TREATMENT PLAN	0.74	0.83	0.78	0.95	0.93	0.94	0.97	0.93	**0.95**
Macro-Average	0.62	0.64	0.62	0.85	0.82	0.83	0.91	0.86	**0.88**
Micro-Average	0.62	0.62	0.62	0.86	0.83	0.84	0.93	0.91	**0.92**

Table 2: Section type identification results (precision, recall and F_1 scores) per section as well as micro and macro averages. Parent sections are in bold.

Seg assumes that a topic change in written text occurs when chains of frequent repetitions of words begin and end. It rewards shorter chains over longer ones and further rewards chains with more repeated terms. Finally, the lexical cohesion between two chains is evaluated using a cosine similarity. The second method is TopicTiling—an augmentation of the well-known TextTiling algorithm (Hearst, 1994). TopicTiling (Riedl and Biemann, 2012) is LDA-based and represents segments as dense vectors of terms contained in dominant topics (as opposed to sparse term vectors).

5.3 Results

For the section labeling problem, our model equaled or outperformed both baselines in all the sections. Table 2 shows the precision, recall, and F_1 scores for the two baselines and our model. The *DIAGNOSIS* section saw the best performance due to a rule-based approach. Similarly, *CURRENT MEDICATIONS* achieved high scores due to the use of dictionaries. All three models performed the worst in identifying the *LEGAL* section. We suspect that this is due to the low prevalence of this section and its content in the dataset. Similarly, sections with lower prevalence saw lower performance than others. Both baselines performed well in identifying the *IDENTIFYING DATA* and *DIAGNOSIS* sections due to their highly distinctive language. Our model performed better for all implicit subsections, and significantly better for two (i.e., *PREGNANCY* and *BIRTH AND DEVELOPMENTAL HISTORY*. Finally, our model performed exactly the same as the Flat HMM baseline for the three parent types, as our model reduces to the Flat HMM in these cases and because the flat HMM model assumes a fixed general ordering of the sections.

Since the report sections vary in size, we computed both macro- and micro-averaged precision, recall, and F-measure (last two rows in Table 2). Our model's micro-averaged F-measure is above 90% which is significantly higher than both the Flat-HMM and the independent bigram baselines performing at 85% and 62% respectively. Similar to Li et al. (2010), both our HHMM and the Flat-HMM baseline seemed to neither overfit nor underfit, which is indicated by higher micro-averaged compared to the macro-averaged scores.

As for the boundary detection problem, and similar to the evaluation in Sherman and Liu (2008), we performed two experiments for the baselines since both baselines require a parameter representing the number of boundaries (number of topics minus one). In the first experiment we allowed the parameter to be chosen by LCSeg and TopicTiling, respectively, while in the second experiment, we provide the algorithms with the correct number of boundaries (i.e., number of sections minus one). Our model however, needs no prior information regarding the number of sections present in a given report. Table 3 shows the W_d and P_k scores for all three approaches. Our system again outperformed both baselines indicated by lower W_d and P_k error rates overall. Both baselines performed better when the number of boundaries is known—an expected result. In fact, TopicTiling outperformed our approach by a small margin when provided with the correct parameter value. We note, however, that when running open loop on new text, the number of sections will be unknown, so this result does not reflect how we envision the approach being used.

# of Boundaries	Algorithm	P_k	W_d
System Choice	LCSeg	0.29	0.37
	TopicTiling	0.27	0.33
Provided	LCSeg	0.25	0.33
	TopicTiling	0.20	0.25
	HHMM	**0.20**	**0.26**

Table 3: Section boundary identification results.

6 Related Work

As discussed above, our work simultaneously solves two problems within a psychiatric evaluation report: identifying section types and identifying section boundaries. The first problem has been referred to as argumentative zoning (Teufel et al., 1999; Li et al., 2010; Denny et al., 2009), while the second is a type of text segmentation problem (Hearst, 1994; Riedl and Biemann, 2012). Argumentative zoning refers to classifying text sections into mutually exclusive categories. Work on this task is mostly centered around identifying scientific article sections (e.g., abstract, introduction, methodology, etc.) (Teufel, 1999).

Our work is a combination and extension of Li et al. (2010)'s work on identifying section types within clinical notes and Sherman and Liu (2008)'s work on text segmentation of meeting minutes. Both approaches integrated n-gram language models into HMMs. The former modeled HMM emissions at the section level using bigrams, while the later modeled the emissions at the sentence level and used unigrams and trigrams. Other approaches followed similar strategies in segmenting story text and in creating generative models for detecting story boundaries (Mulbregt et al., 1998; Yamron et al., 1998). More recently, Yu et al. (2016) used a hybrid deep neural network combined with a Hidden Markov Model (DNN-HMM) to segment speech transcripts from broadcast news to a sequence of stories.

More broadly, there has been some work on applying NLP in the mental health domain. However, due to lack of readily available clinical data (e.g. clinical reports), researchers have focused on non-clinical sources (e.g., social media) (Chapman et al., 2011). Several algorithms were developed to detect specific emotions from suicide notes and online journals (Pestian et al., 2012; Strapparava and Mihalcea, 2008), while twitter data was used to detect distress and suicide ideation (Homan et al., 2014; O'Dea et al., 2015). Additionally, twitter data was used to measure mood valence and detect depression (Sadilek et al., 2013; De Choudhury et al., 2013; Coppersmith et al., 2015). Facebook data was used to measure emotion contagion and to predict post-partum depression (Coviello et al., 2014; De Choudhury et al., 2014). Instead of social media and publicly available, non-clinical data Althoff et al. (2016) used counseling conversations gathered using a messaging service and developed discourse analysis methods to measure the correlation of outcomes with various linguistic aspects.

7 Contributions

To the best of our knowledge, our work represents the only attempt at detecting the position and type of psychiatric report sections. In this paper we present an approach that applies and extends earlier work on document section discovery and segmentation. We collected a corpus of psychiatric documents and created a unified hierarchy of section labels. We built an n-gram-based HHMM model that successful detects the order of sections as well as their boundaries within a given report. We evaluated our model's performance over two separate tasks, namely the section ordering task and the section boundary identification. Our model outperformed baselines for both of those tasks. Finally, our approach further confirms that learning the section ordering of a psychiatric report yields better performance for boundary identification and text segmentation.

Acknowledgments

We thank our colleagues at the Cognition, Narrative, and Culture Laboratory (Cognac Lab), especially Mohammed Aldawsari and Victor Yarlot for their contributions to the overall idea, approach, and evaluation.

References

Tim Althoff, Kevin Clark, and Jure Leskovec. 2016. Large-scale analysis of counseling conversations: An application of natural language processing to mental health. *Transactions of the Association for Computational Linguistics*, 4:463–476.

American Psychiatric Association. 2006. *American Psychiatric Association Practice Guidelines for the Treatment of Psychiatric Disorders: Compendium 2006*. American Psychiatric Association Publishing, Washington, DC.

American Psychiatric Association. 2018. What is psychiatry? Available from: `https://www.psychiatry.org/patients-families/what-is-psychiatry`. (Accessed on Jul 1, 2018).

Regina Barzilay and Lillian Lee. 2004. Catching the drift: Probabilistic content models, with applications to generation and summarization. In *Proceedings of the 2004 North American Chapter of the Association for Computational Linguistics: Human Language Technologies Conference (HLT-NAACL)*, pages 113–120.

Doug Beeferman, Adam Berger, and John Lafferty. 1999. Statistical models for text segmentation. *Machine Learning*, 34(1):177–210.

Hung H. Bui, Dinh Q. Phung, and Svetha Venkatesh. 2004. Hierarchical hidden markov models with general state hierarchy. In *Proceedings of the 19th National Conference on Artifical Intelligence*, AAAI'04, pages 324–329, San Jose, California.

Wendy W Chapman, Prakash M Nadkarni, Lynette Hirschman, Leonard W D'Avolio, Guergana K Savova, and Ozlem Uzuner. 2011. Overcoming barriers to NLP for clinical text: The role of shared tasks and the need for additional creative solutions. *Journal of the American Medical Informatics Association*, 18(5):540–543.

Glen Coppersmith, Mark Dredze, Craig Harman, Kristy Hollingshead, and Margaret Mitchell. 2015. CLPsych 2015 shared task: Depression and PTSD on twitter. In *Proceedings of the 2nd Workshop on Computational Linguistics and Clinical Psychology: From Linguistic Signal to Clinical Reality (CLPsych)*, pages 31–39.

Lorenzo Coviello, Yunkyu Sohn, Adam D. I. Kramer, Cameron Marlow, Massimo Franceschetti, Nicholas A. Christakis, and James H. Fowler. 2014. Detecting emotional contagion in massive social networks. *PLOS ONE*, 9(3):1–6.

Munmun De Choudhury, Scott Counts, Eric J. Horvitz, and Aaron Hoff. 2014. Characterizing and predicting postpartum depression from shared facebook data. In *Proceedings of the 17th ACM Conference on Computer Supported Cooperative Work & Social Computing*, CSCW '14, pages 626–638, Baltimore, MD.

Munmun De Choudhury, Michael Gamon, Scott Counts, and Eric Horvitz. 2013. Predicting depression via social media. In *Proceedings of the 7th International AAAI Conference on Weblogs and Social Media (ICWSM)*, volume 13, pages 1–10, Boston, MA.

Dina Demner-Fushman, Wendy W Chapman, and Clement J McDonald. 2009. What can natural language processing do for clinical decision support? *Journal of Biomedical Informatics*, 42(5):760–772.

Joshua C. Denny, Anderson Spickard, III, Kevin B. Johnson, Neeraja B. Peterson, Josh F. Peterson, and Randolph A. Miller. 2009. Evaluation of a method to identify and categorize section headers in clinical documents. *Journal of the American Medical Informatics Association*, 16(6):806–815.

eMedicineHealth. 2018. Medications and drugs listing. `https://www.emedicinehealth.com/medications-drugs/article_em.htm`. (Accessed on Feb 18, 2018).

Shai Fine, Yoram Singer, and Naftali Tishby. 1998. The hierarchical hidden markov model: Analysis and applications. *Machine Learning*, 32(1):41–62.

Michel Galley, Kathleen McKeown, Eric Fosler-Lussier, and Hongyan Jing. 2003. Discourse segmentation of multi-party conversation. In *Proceedings of the 41st Annual Meeting on Association for Computational Linguistics (ACL)*, volume 1, pages 562–569, Sapporo, Japan.

Karen Goldfinger and Andrew M Pomerantz. 2013. *Psychological Assessment and Report Writing*. Sage, Thousand Oaks, CA.

Gary Groth-Marnat. 2009. *Handbook of Psychological Assessment*. John Wiley & Sons, Hoboken, NJ.

Marti A. Hearst. 1994. Multi-paragraph segmentation of expository text. In *Proceedings of the 32nd Annual Meeting on Association for Computational Linguistics*, ACL '94, pages 9–16, Las Cruces, NM.

Christopher Homan, Ravdeep Johar, Tong Liu, Megan Lytle, Vincent Silenzio, and Cecilia Ovesdotter Alm. 2014. Toward macro-insights for suicide prevention: Analyzing fine-grained distress at scale. In *Proceedings of the Workshop on Computational Linguistics and Clinical Psychology: From Linguistic Signal to Clinical Reality*, pages 107–117.

George Hripcsak, Suzanne Bakken, Peter D Stetson, and Vimla L Patel. 2003. Mining complex clinical data for patient safety research: A framework for event discovery. *Journal of Biomedical Informatics*, 36(1-2):120–130.

Kush Jain, Priya Khatri, and Garima Indolia. 2015. Chunked n-grams for sentence validation. *Procedia Computer Science*, 57:209–213.

Ying Li, Sharon Lipsky Gorman, and Noémie Elhadad. 2010. Section classification in clinical notes using supervised hidden markov model. In *Proceedings of the 1st ACM International Health Informatics Symposium IHI*, pages 744–750, Arlington, VA.

S. Liu, Wei Ma, R. Moore, V. Ganesan, and S. Nelson. 2005. Rxnorm: Prescription for electronic drug information exchange. *IT Professional*, 7(5):17–23.

Paul van Mulbregt, Ira Carp, Lawrence Gillick, Steve Lowe, and Jon Yamron. 1998. Text segmentation and topic tracking on broadcast news via a hidden markov model approach. In *Fifth International Conference on Spoken Language Processing*.

Bridianne O'Dea, Stephen Wan, Philip J. Batterham, Alison L. Calear, Cecile Paris, and Helen Christensen. 2015. Detecting suicidality on twitter. *Internet Interventions*, 2:183–188.

John P. Pestian, Pawel Matykiewicz, Michelle Linn-Gust, Brett South, Ozlem Uzuner, Jan Wiebe, K. Bretonnel Cohen, John Hurdle, and Christopher Brew. 2012. Sentiment analysis of suicide notes: A shared task. *Biomedical Informatics Insights*, 5s1:BII.S9042.

Lev Pevzner and Marti A Hearst. 2002. A critique and improvement of an evaluation metric for text segmentation. *Computational Linguistics*, 28(1):19–36.

L. R. Rabiner. 1989. A tutorial on hidden markov models and selected applications in speech recognition. *Proceedings of the IEEE*, 77(2):257–286.

R Reeves and R Rosner. 2016. *Forensic Psychiatry and Forensic Psychology: Forensic Psychiatric Assessment*. Elsevier.

Jeffrey C Reynar. 1998. *Topic Segmentation: Algorithms and Applications*. Ph.D. thesis, University of Pennsylvania, Philadelphia, PA.

Martin Riedl and Chris Biemann. 2012. Topictiling: A text segmentation algorithm based on lda. In *Proceedings of ACL 2012 Student Research Workshop*, ACL '12, pages 37–42, Jeju Island, Korea.

Adam Sadilek, Christopher Homan, Walter S Lasecki, Vincent Silenzio, and Henry Kautz. 2013. Modeling fine-grained dynamics of mood at scale. *WSDM, Rome, Italy*, pages 3–6.

M. Sherman and Yang Liu. 2008. Using hidden markov models for topic segmentation of meeting transcripts. In *Proceedings of the 2008 IEEE Spoken Language Technology Workshop*, pages 185–188.

Carlo Strapparava and Rada Mihalcea. 2008. Learning to identify emotions in text. In *Proceedings of the ACM Symposium on Applied Computing (SAC)*, SAC '08, pages 1556–1560, Fortaleza, Ceara, Brazil.

Simone Teufel. 1999. *Argumentative zoning: Information Extraction from Scientific Text*. Ph.D. thesis, University of Edinburgh, Edinburgh, Scotland, UK.

Simone Teufel, Jean Carletta, and Marc Moens. 1999. An annotation scheme for discourse-level argumentation in research articles. In *Proceedings of the ninth conference on European chapter of the Association for Computational Linguistics (EACL)*, pages 110–117, Bergen, Norway.

J. P. Yamron, I. Carp, L. Gillick, S. Lowe, and P. van Mulbregt. 1998. A hidden markov model approach to text segmentation and event tracking. In *Proceedings of the IEEE International Conference on Acoustics, Speech and Signal Processing (ICASSP)*, volume 1, pages 333–336 vol.1.

Jia Yu, Xiong Xiao, Lei Xie, Chng Eng Siong, and Haizhou Li. 2016. A dnn-hmm approach to story segmentation. In *INTERSPEECH*, San Francisco, USA.

Iterative development of family history annotation guidelines using a synthetic corpus of clinical text

Taraka Rama* **Pål H. Brekke[†]** **Øystein Nytrø[‡]** **Lilja Øvrelid***

*University of Oslo, Department of Informatics
[†]Oslo University Hospital, Department of Cardiology, Center for Cardiological Innovation
[‡]Norwegian University of Science and Technology, Department of Computer Science
`tarakark@ifi.uio.no,pabrek@ous-hf.no,nytroe@ntnu.no,liljao@ifi.uio.no`

Abstract

In this article, we describe the development of annotation guidelines for family history information in Norwegian clinical text. We make use of incrementally developed synthetic clinical text describing patients' family history relating to cases of cardiac disease and present a general methodology which integrates the synthetically produced clinical statements and guideline development. We analyze inter-annotator agreement based on the developed guidelines and present results from experiments aimed at evaluating the validity and applicability of the annotated corpus using machine learning techniques. The resulting annotated corpus contains 477 sentences and 6030 tokens. Both the annotation guidelines and the annotated corpus are made freely available and as such constitutes the first publicly available resource of Norwegian clinical text.

1 Introduction

The limited availability of clinical text corpora constitutes a major challenge for the development of clinical NLP tools. Such text originates in the (electronic) health record (EHR), and access to and use of the EHR is governed by strict data privacy and health service regulations, which usually restricts secondary use and prohibits re-distribution and sharing with the larger NLP community. Among notable exceptions are anonymized health record texts published as part of the i2b2 challenges (Uzuner and Stubbs, 2015) and the CLEF corpus (Roberts et al., 2008b). For languages other than English the situation is even more difficult, and despite notable annotation efforts (Dalianis et al., 2012), the underlying corpora are largely unavailable.

Clinical texts are radically different in form and function from other biomedical texts: They are communicative, conveying information between health service providers, terse (in that the patient is implicit), and very specialized according to the role of the narrative and profession of the author (Allvin et al., 2010; Røst et al., 2008). In this work, the targeted narrative of family history corresponds to the anamnesis recorded by the cardiologist when interviewing the patient as part of a consultation. However, lacking a corpus of family history statements, we decided to develop a synthetic corpus (Lohr et al., 2018; Boag et al., 2018).

Development of most NLP tools requires manually annotated data and the design of annotation guidelines is crucial for consistent and high quality data suitable for machine learning and classification. Development of annotation guidelines is a time consuming process which in the case of clinical data often also requires access to domain experts (clinicians). The question of how to involve the clinician in the annotation process and make the best use of their domain knowledge is therefore highly relevant.

This article describes the systematic development of annotation guidelines for family history information in Norwegian clinical text. We make use of incrementally developed synthetic clinical text describing patients' family history relating to cases of cardiac diseases. The domain expert is an integral part of this methodology and generates synthetic examples that challenge the guidelines and further participates both in the annotation and development of guidelines. In doing so, the domain knowledge of the clinician informs the annotation process systematically. Measures of inter-annotator agreement is actively used to improve the annotation guideline, as well as to extend the synthetic corpus and range of annotated concepts.

In the rest of the paper, we describe the methodology for corpus generation and annotation guideline design in more detail and provide an overview of our current state of progress in the fam-

111

Proceedings of the 9th International Workshop on Health Text Mining and Information Analysis (LOUHI 2018), pages 111–121
Brussels, Belgium, October 31, 2018. ©2018 Association for Computational Linguistics

ily history domain. We analyse inter-annotator agreement based on the developed guidelines and present results from experiments aimed at evaluating the validity and applicability of the purpose-made annotated corpus using machine learning.

2 Family history in clinical text

A family history is an important part of the medical record. It helps the clinician in identifying risk factors, in diagnosing conditions that have genetic components, and in identifying family members that should be offered genetic counselling or medical follow up. Specific patterns of disease or symptoms in a family suggest modes of inheritance, and could be helpful in the diagnosis of an unrecognised disease or syndrome. For example, if only men in the family are affected, one might expect an X-linked trait, or if approximately half of the offspring in a generation seem to be affected, it would suggest an autosomal dominant disease. In the cases where a pathological mutation has already been identified, the pedigree is used to plan further genetic screening or counselling. Figure 1 shows an example pedigree with a typical autosomal dominant inheritance pattern.

For some diseases, the course of events in the patient's family are important in judging the patient's own risk of serious events. In patients with hereditary hypertrophic cardiomyopathy, the European Society of Cardiology recommends using an online risk calculator to estimate a patient's 5 year risk of sudden cardiac death (SCD). Among the seven factors included in the underlying model – a strong contributor to individual risk – is a history of SCD in first degree relatives (Elliott et al., 2014).

Family histories occur as descriptive text in the EHR, but acknowledging that computational reasoning about family history have substantial benefits in research, diagnosis and decision support where many tools has been developed for interactive pedigree input (Welch et al., 2018). The underlying objective of our NLP challenge is to be able to infer the pedigree of a patient from text. However, even checking consistency of family history information represented in OWL proves to be a challenge (Stevens et al., 2014). A potential outcome of our work would be to transform statements about pedigree into tabular formats directly usable in risk calculators and for bioinformatics application like genome-wide anal-

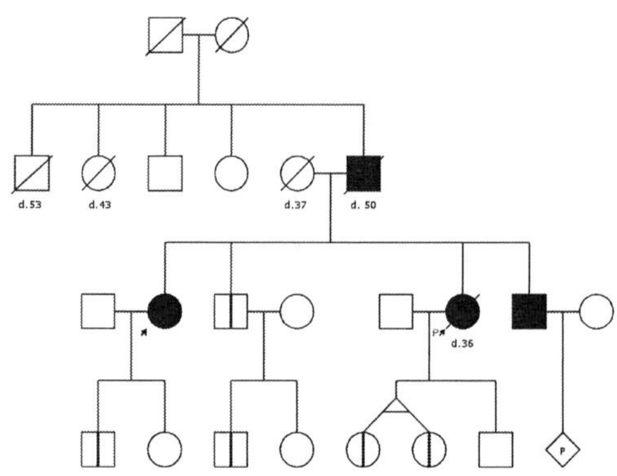

Figure 1: An example pedigree chart with a typical autosomal dominant inheritance pattern. Horizontal rows represent generations, lines represent relationships, lines of descent and sibship. Squares are male, circles female, and diamond shape is unknown gender. A symbol with a 'P' inside denotes a pregnancy. Diagonal lines through symbols denote deceased individuals and the text below their age at the time of death (eg. 'd. 43' means died when 43 years old). Filled symbols represent individuals with manifest disease, symbols with a vertical line are healthy gene carriers who may develop disease later. The small arrow denotes the current patient ("self") and the arrow with the 'P' is the proband or index patient where the genetic analysis of the family started (Bennett et al., 2008).

ysis (Hiekkalinna et al., 2005).

2.1 Previous work

There has been some previous work aimed at extracting family history information from clinical text. Bill et al. (2014) annotate 284 sentences from the publicly available MTSamples corpus of synthetically produced English clinical text for information about family members and clinical observations with some additional attributes (vital status, negation and age of death). However, they do not provide any measures of inter-annotator agreement. Polubriaginof et al. (2015) compared the information contained in structured and free-text descriptions of family history information and found that the free-text descriptions were more comprehensive.

In another work, Goryachev et al. (2008) developed a pipeline of rule based systems to detect family members and diagnosis concepts; and, then assign the family diagnosis to a specific family number. The authors run standard NLP tools such as sentence splitter and part-of-speech taggers on discharge summary notes. The pipeline system is related to Friedlin and McDonald (2006) in only identifying diagnosis concepts that are present in standard medical dictionaries and do not perform relation extraction as performed in this paper.

Both rule based systems (Abacha and Zweigenbaum, 2011) and machine learning methods such as Roberts et al. (2008a) and Minard et al. (2011) use multi-class SVMs to perform relation extraction from clinical reports. Our work in this paper is closest to the work of Roberts et al. (2008a) who manually annotated cancer narratives for entities and relations and, then, trained and tested a one-vs-rest SVM classifier for training and testing. In this paper, we employ widely used features in general purpose named entity recognition (Hong, 2005; Miwa and Sasaki, 2014) to train the SVM models.

3   Incremental annotation guideline and synthetic corpus development

One immediate goal of this work is to develop a tool for the extraction of family history information from Norwegian clinical text. Due to the unavailability of the real health records describing family histories, we developed a methodology for annotation guideline development which makes use of an incrementally developed synthetic corpus. The textual data contained in the corpus was produced by a clinician who has extensive experience with clinical work and genetic cardiology. The data consists of statements that summarize the family history of a patient and will typically correspond to a small part of a patient journal. The descriptions were made by performing web searches for images of "autosomal dominant pedigree", and pseudo-randomly describing parts of the displayed pedigrees while assigning invented but realistic medical events. No real patient histories are reproduced, but coincidental similarities must be expected. The text does not contain any personal identification information.

The first step in a semantic annotation of text is to decide upon the entities and the relations that are interesting to extract or characterize. Biomedicine employs terminologies and classifications that may be used for annotation (Savova et al., 2010). In our domain of family history, we started with family members and relationships, and largely ignored medical conditions apart from death or known (cardiac) disease in general.

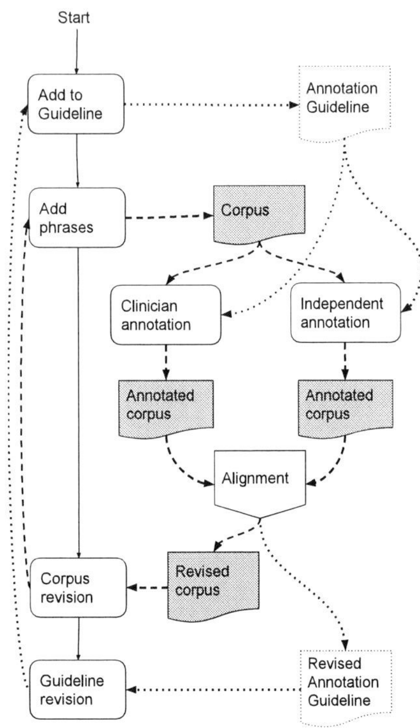

Figure 2: Incremental development of corpus and annotation guidelines

The guideline developers consisted of a clinician and three computational linguists and/or computer scientists. We usually maintained two roles: The clinician would produce a set of representative sentences and along with one of the others propose an annotation scheme for these. Then, the clinician would annotate while another independent person not involved in the design of the annotation scheme would make an *independent annotation*. The results were compared and discrepancies were recorded. We (sometimes artificially) could identify both *semantic* and *pragmatic* discrepancies. Semantic discrepancy would signify a misunderstanding of the underlying domain and required amending the ontology, whereas the pragmatic discrepancy would uncover an underspecified or incomplete annotation rule which could be further specified by adding more examples to the corpus. The drivers and amendments in this quizlike game is shown in the table 1.

Figure 2 shows the double loops of corpus production and guideline development. As shown, the family history statements were produced iteratively. In the initial round, the clinician was

	Driver	Amendment
Guideline iteration	Semantic discrepancy	Add concept or revise guideline
Corpus Iteration	Pragmatic discrepancy	Add sentence to corpus

Table 1: Drivers and amendments guiding the development of annotation guidelines.

asked to produce a set of representative statements about SCD-related family history. Example 1 below shows a sentence from the corpus.

(1) *Indekspasienten er hans onkel på*
Index-patient   is his   uncle on
farssiden,   som hatt hjertestans   og
father's-side, who had cardiac-arrest and
fått implantert ICD.
had implanted ICD.
'The index patient is his uncle on the father's side, who had cardiac arrest and implanted ICD.

Following the initial iterations and discussions with the clinician the need to account for i) relations to groups of family members, ii) temporal statements, and iii) negation emerged. During this iteration the clinician was therefore tasked with the generation of statements that challenged the current guidelines, whilst still producing representative family statements. Example 2 shows an example sentence containing a temporal statement and example 3 shows another type of temporal statement describing the age of the family member at the time of diagnosis.

(2) *Han har kjent hjertebank de siste*
He   has felt   heart-palps the last
fire-fem   månedene.
four-five months
'He has been feeling heart palpitations during the last four-five months'

(3) *Broren   fikk diagnosen i femti-årene.*
Brother-the got diagnosis i fifty-years
'The brother was diagnosed in his fifties'

After arriving at a fairly stable set of guidelines, a large portion of the data set (320 sentences) was doubly annotated. Following this, disagreements were resolved in a round of consolidation between the annotators. The final portion of the data set (91 sentences) was then annotated doubly and the resulting inter-annotator agreement on these data sets is reported here in Section 4.5.

All annotation was performed using the Brat web-based annotation tool (Stenetorp et al., 2012). The data was manually segmented and tokenized prior to annotation.

4 Annotation guidelines

The following section presents an overview of the resulting annotation guidelines. The annotation of the corpus distinguishes semantically relevant clinical *entities* and shows how these relate to each other in the text via a set of *relations*. Figure 4 shows a graphical overview of the annotation schema, where rectangles indicate core clinical entities, ovals indicate modifier entities, and all possible relations are indicated by directed arcs.

4.1 Clinical entities

Clinical entities are marked with one of the following entity types:

- `Family` describes various family members (e.g. *onkelen* 'the uncle', *bestefar* 'grandfather').

- `Self` is used only for the patient under consideration (e.g. *pasienten* 'the patient', *hun* 'she').

- `Index` entities designate the property of being the index patient or *proband*, i.e. the first identified family member with disease *indekspasienten* 'the index patient'.

- `Condition` entities describe a range of clinical conditions such as diseases (*koronarsykdom* 'coronary disease'), diagnoses, various types of mutations, test results (*testet negativt* 'tested negative'), treatments (*hjertetransplantert* 'heart-transplanted'), and vital state (*død* 'dead', *frisk* 'healthy').

- `Event` entities describe clinical events (e.g. *hjertestans* 'cardiac arrest' and *synkope* 'syncope').

The distinction between conditions and events relate to the temporal extension of the entity described: an event is something that happens and then is over, but a condition is a prolonged state of the patient, for instance, the patient has a heart attack (`Event`), but from this point on she is considered to have heart disease (`Condition`).

In addition to the main clinical entities described above, the annotation guidelines also distinguish a set of modifier entities that further de-

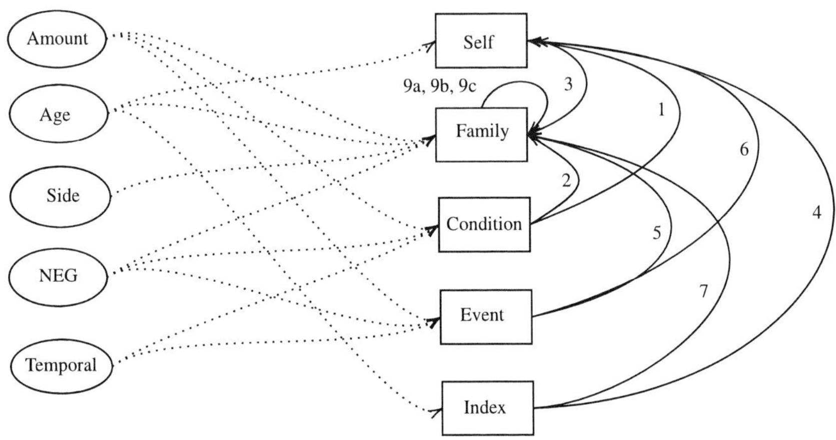

Figure 3: Schematic diagram showing the possible relations between entities. The different relations are marked with a number to avoid cluttering. `Holder`: 1, 2, 4, 5, 6, 7, 8; `Modifier`: Dotted lines; `Related_to`: 3, 9a; `Subset`: 9b; `Partner`: 9c.

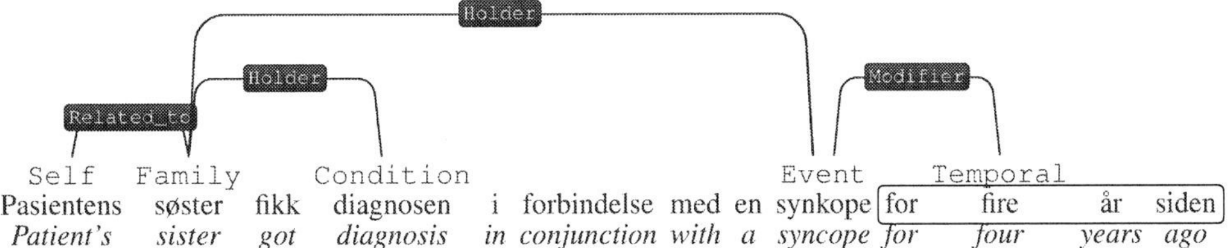

Figure 4: Annotation of clinical entities and relations for an example sentence from the corpus.

scribe the clinical entities for a number of properties that are relevant for semantic interpretation of family history information:

- `Side` entities describe the side of the family and thus modify `Family` entities (e.g. *farssiden* 'paternal side').

- `Age` entities describe the age of a family member *40 år gammel* '40 years old'.

- `Negation` entities mark lexical items that signal negation, so-called *negation cues* in the terminology of Morante and Daelemans (2012). These may be negative adverbs, such as e.g., *ikke* 'not', *aldri* 'never', or negative determiners/pronouns *ingen* 'nobody'. Note that in contrast to Morante and Daelemans (2012), we do not annotate morphological negation cues (e.g. **im**-*possible*). In this version of the guidelines, we treat negation as encompassing uncertainty. The main reason for this is that just like the presence of negation, it marks missing information in the family history.

- `Amount` modifiers describe quantifiers that describe numerical properties of clinical entities, e.g. *to* 'two', *mange* 'many'.

- `Temporal` modifiers typically position `Condition`/`Event` entities in time, e.g. *i sommer* 'this summer', *for tre år siden* 'three years ago'. These are similar to temporal expressions (so-called *timexes*) in previous temporal annotation schemes (Ferro et al., 2002; Saurí et al., 2006).

4.2 Family history relations

In addition to the clinical entities described above, we further annotate a number of relationships between entities in our annotation scheme. Example 4 shows a fully annotated example containing entities and their relations for an sentence from the corpus. The relations are binary undirected relations of the following types:

- `Holder` relations are always between `Condition`/`Event` entity on the one hand and its holder, a `Family`/`Self`/`Index` entity.

- `Modifier` relations hold between modifier entities (e.g. `Side`, `Negation`) and clinical entities (e.g. `Family`, `Condition`).

- `Related_to` relations specify relations between family members and always hold between entities of the `Family` type.

- `Subset` relations specify relations between family members, where one is a subset of the other, e.g. in statements such as *Hun har to brødre, den ene har mutasjonen* 'She has two brothers, one of them has the mutation', where *den ene* 'one of them' would be connected to the `Family` entity *brødre* 'brothers' with a `Subset`-relation.

- `Partner` relations specify relations between entities of the `Family` type, used to identify couples (husbands and wives, civil partnerships) that are able to provide offspring. The assumption is no kinship.

4.3 Span of annotations

In general, annotation should pick out the minimal span in the text which denotes the entity or property in question. This will most often be a single word (*onkel* 'uncle', *mutasjon* 'mutation') but will in some cases also include more than one word (*plutselig hjertedød* 'sudden cardiac death', *voksende hjerte* 'growing heart'). Genitive modifiers of an entity, e.g. *farens* 'father's' in *farens søster* 'the father's sister' or *Søsteren til faren* 'the sister of the father' should not be included in the annotation span. Rather, these are annotated as two separate entities related by a `Related_to` relation. The span of `Family` entities usually encompass only the family term itself (*onkel* 'uncle', *søster* 'sister'), however, when the family term is described using a pronominal element (*hun* 'she', *den ene* 'one (of them)') this should be annotated as a family entity. When both are present (*den ene broren* 'the one brother') only the family term is annotated.

Temporal expressions will often be more complex and should include both numerical expressions denoting amount (*tre* 'three', *flere* 'several'), temporal units such as month/year, as well as expressions denoting temporal ordering or duration (*i* 'in', *siden* 'since' as in *tre år siden* 'three years since', *i tre år* 'for three years'). Initial iterations of annotation showed that agreement for this category was low due to differences in annotation span. We therefore introduced the generalization

Entities	Number	Spans
Family	1704	96
Condition	681	135
Event	542	115
Self	509	–
Amount	273	9
Temporal	214	178
Negation	131	33
Age	57	34
Side	36	3
Index	7	–

Relations	Number	Spans
Holder	880	–
Modifier	687	–
Related_to	389	–
Subset	108	–
Partner	14	–

Table 2: Distribution of entities and relations in the data annotated by the clinician. The Spans column shows the number of entities that span across words. Both the entities and relations are sorted in decreasing order of number of occurrences.

that temporal annotation should make use of a replacement rule where the full constituent replaced by a temporal pronoun corresponding to English *then* is annotated. This means that unlike e.g. Ferro et al. (2002), our temporal annotations will include prepositions (e.g. *i tre år* 'for three years').

4.4 Statistics

The resulting annotated corpus contains 477 sentences and 6030 tokens. In table 2 we present the distribution of the entities and relations in the corpus. We see that `Condition` and `Event` entities are fairly equally distributed in the corpus. Temporal modifiers span more than one word in a majority of cases. Whereas `Holder`-relations are the most common type of relation in the corpus, there are only 14 cases of the `Partner` relation.

4.5 Inter-Annotator Agreement

As described in Section 3, two final rounds of annotation with different second annotators (in addition to the clinician, here dubbed A1 and A2) were used to complete the annotation guidelines. We measured the inter-annotator agreement at two levels. At the first level, IAA is based on match of the entities spans and their labels. At the second level, IAA is based on the relationship matches between the matched spans. Therefore, the relationship agreement measurement is stricter than the entity level agreement measurement. We examine token level agreement where we treat the clinician's notes as gold standard and compute the per token F-measures i.e., Precision, Recall, and F_1-score. We measure the inter-annotator agreement using micro F_1-score. The Precision, Recall, and F_1-scores of the agreement is provided in table 3.

Annotator	Precision	Recall	F_1-score
A1, 320	0.743	0.648	0.692
	0.645	0.559	0.599
A2, 91	0.821	0.797	0.809
	0.752	0.678	0.713

Table 3: Each row shows the number of sentences annotated by each annotator. The first and second rows shows the Precision, Recall, and F_1-score for entities and relations. All the results are in comparison to the texts annotated by the clinician.

We find that the round of consolidation and improvement of the guidelines was useful and improves the IAA scores for both entities and relations. When we compare the annotations of the clinician (A0) and the second additional annotator (A2), we find that there are still a number of remaining discrepancies. Some of these are what we termed semantic discrepancies above in Section 3 above, annotation decisions that require domain knowledge. For instance, in several places A2 annotates clinical conditions that are not marked by the clinician, e.g. marking *symptomer* 'symptoms' as a `Condition`. There are also examples where additional distinctions should probably be added to the guidelines, in particular with respect to annotation of temporal and negation-related information, both examples of complex annotation tasks by themselves. For instance, A2 annotates the phrase *under en flytur til Spania* 'during a flight to Spain' as Temporal, where A0 does not. With respect to negation, the distinction between negation and uncertainty causes differences in annotation spans, where A0 annotates *husker ikke* 'does not remember' as `NEG`, whereas A2 annotates only *ikke* 'not'.

5 Preliminary experiments

In this section, we perform entity classification and relation extraction experiments to verify the viability of our annotation. We train and test SVM model on the data annotated by the domain expert in five-fold cross-validation fashion. The domain expert annotated dataset has 477 sentences and we performed five-fold cross-validation to train and test our model. In all our experiments, we split the sentences into five folds and extracted entities and relations. Then, we treated each of five folds as test dataset and trained on the other four folds in an iterative fashion.

5.1 Entity detection

In this experiment, we trained and tested a linear classifier (SVM model) for entity classification. We treat entity classification as a multi-class classification problem where there are 11 classes including the O entity that denotes unmarked lexical units. Our model is a linear SVM model that is trained on the following features:

- Lexical: Current word, words in a context window size of 2.

- Universal POS tags: Current word, words in a context window size of 2.

- Entity tags: The two previous entity tags where the model uses the gold entity tags to train but uses the previous predicted entity tags to predict the current tag.

We also experimented with lowercasing a word and orthographic features such as prefixes and suffixes of length 3 which did not improve the performance of the SVM model. We evaluate the performance of the SVM model using weighted F_1 score to account for class imbalance. On an average, these feature templates yielded 5000 features across the five cross-validation experiments. All the Universal POS tags are obtained through the CoNLL17 Baseline model (Zeman et al., 2017) trained on the publicly available Universal Dependencies Norwegian Bokmål treebank (Øvrelid and Hohle, 2016). We used the majority class "O" as the baseline in our experiments. The results of our experiments are given in table 4. It has to be noted that these results are not comparable to the IAA scores presented in table 3, which are calculated only over entities and completely disregard the remaining tokens. Moreover, the IAA

System	Precision	Recall	F_1-score
Baseline	0.34	0.582	0.429
SVM	0.843	0.843	0.841

Table 4: The average of the weighted F_1-scores across the five folds. On an average, there are 6030 training instances and 1507 test instances.

score is computed only on parts of the annotated data whereas the SVM models are trained and tested on the whole of the data annotated by A0. The SVM model performs better than the majority class baseline model across all the measures. The SVM model made errors at distinguishing Condition entities from Event entities and Age from Temporal entities. Most of the errors occurred when the SVM model misclassified the rest of the classes as "O".

5.2 Relation extraction

In this subsection, we performed a relation detection and classification experiment. In this experiment, we treat a relation defined between exactly two entities to belong to one of the six relations where five of them are given in table 2 and the sixth relation is "No_Relation". We train and test an SVM model in a five-fold cross-validation fashion. Apart from entity labels, we experimented with increasingly complex set of features:

- Lexical: Words belonging to the entities are treated as two separate features.

- POS tags: Universal POS tags of the entities' lexical tokens as separate features.

- Dependency features: The dependency label of a entity word's incoming arc as a feature.

If a entity is spanning across multiple words, we concatenate the per-word feature and treated them as a single feature when training and testing the SVM model. The results of the experiments are given in table 5. Our results suggest that word based features themselves yield a performance which is close to the model with more complex features. Incremental inclusion of POS tags and dependency labels increases the performance of the SVM model, whereas the inclusion of predicted entity labels does not improve the performance of the SVM model. We experimented if including the gold standard labels would improve the performance of the SVM model. We find that

the quality of entity labels does improve the performance of the model.

Finally, we present the confusion matrix for the best fold is presented in table 6. The SVM model makes most of the errors when it misclassifies one of the five annotated relations as "No_Relation" and vice-versa. The classifier errs when distinguishing between "Related_to", "Partner" and "Subset" relations. Finally, the classifier makes errors when distinguishing between the Norwegian indefinite determiner *en* which is unmarked and the quantifier *en*.

Features	Precision	Recall	F_1-score
Words	0.716	0.732	0.719
+POS tags	0.73	0.738	0.731
+Dependency labels	0.743	0.746	0.743
+Entity labels (Predicted)	0.743	0.745	0.743
+Entity labels (Gold)	0.771	0.767	0.768

Table 5: Average of the weighted F_1-scores on five fold cross-validation. On an average, there are 5530 training instances and 1461 test instances.

	Holder	Modifier	No_Relation	Partner	Related_to	Subset
Holder	127	1	54	0	1	0
Modifier	2	82	66	0	0	0
No_Relation	65	100	1045	1	30	15
Partner	0	0	0	2	3	0
Related_to	7	0	22	0	58	2
Subset	0	0	11	0	4	13

Table 6: Confusion matrix for the SVM model at the task of relation extraction on the best performing fold.

6 Discussion

The validity of our study is limited by using synthetic data. While the clinician producing the clinical text works in genetic cardiology, and writes similar patient histories in his clinical practice, the synthetic data can not be expected to be fully representative of real clinical notes from a large patient cohort. The analysis pertaining to the synthetic data should be thought of as an illustration of one iteration of the cycle described in 2, and the objective of the iterative process is a stepwise, guided, design of an annotation guideline in a setting where the target text data is unavailable. The same process could be used with a real corpus, where specific new examples would present challenges driving guideline development. The only difference is that in our case, a specialist produced text, instead of finding representative text.

The guideline development workflow itself may also be improved or expanded by storing a representation of the input data (the pedigree) and linking it to the resulting synthetic text description, which would allow further downstream comparison of extraction results to the actual source material.

7 Conclusion

This article has investigated the development of annotation guidelines for family history information in Norwegian by leveraging synthetically produced clinical text. Inter-annotator agreement scores show that the annotation schema can be applied fairly consistently and that it may also be generalized to unseen text using machine learning. Both the annotation guidelines and the annotated corpus will be made freely available and as such constitutes the first freely available resource of Norwegian clinical text.

In the near future, we will apply the annotation schema to real clinical texts. The family history is but a minor part of a patient record, and segmentation as shown in Bill et al. (2014) is needed. Analysis of the annotation disagreements along with the experimental results also highlighted part of the schema that will need to be further refined, e.g. the analysis of temporality and our treatment of uncertainty. We will develop the method for incremental and systematic annotation guideline development further. The method will be put to test when we iteratively improve the current guideline in order to capture real patient pedigree information from the EHR.

Acknowledgments

We are grateful to three anonymous reviewers for constructive comments on the first version of the paper. This work is funded by the Norwegian Research Council and more specifically by the BigMed project, an IKTPLUSS Lighthouse project.

References

Asma Ben Abacha and Pierre Zweigenbaum. 2011. Automatic extraction of semantic relations between medical entities: a rule based approach. *Journal of biomedical semantics*, 2(5):S4.

Helen Allvin, Elin Carlsson, Hercules Dalianis, Riitta Danielsson-Ojala, Vidas Daudaravičius, Martin Hassel, Dimitrios Kokkinakis, Heljä Lundgren-Laine, Gunnar Nilsson, Øystein Nytrø, et al. 2010. Characteristics and analysis of finnish and swedish clinical intensive care nursing narratives. In *Proceedings of the NAACL HLT 2010 Second Louhi Workshop on Text and Data Mining of Health Documents*, pages 53–60. Association for Computational Linguistics.

Robin L Bennett, Kathryn Steinhaus French, Robert G Resta, and Debra Lochner Doyle. 2008. Standardized human pedigree nomenclature: update and assessment of the recommendations of the national society of genetic counselors. *Journal of genetic counseling*, 17(5):424–433.

Robert Bill, Serguei Pakhomov, Elizabeth S Chen, Tamara J Winden, Elizabeth W Carter, and Genevieve B Melton. 2014. Automated extraction of family history information from clinical notes. In *AMIA Annual Symposium Proceedings*, volume 2014, page 1709. American Medical Informatics Association.

Willie Boag, Tristan Naumann, and Peter Szolovits. 2018. Towards the creation of a large corpus of synthetically-identified clinical notes. *CoRR*, abs/1803.02728.

Hercules Dalianis, Martin Hassel, Aron Henriksson, and Maria Skeppstedt. 2012. Stockholm EPR Corpus: A Clinical Database Used to Improve Health Care. In *Proceedings of the Fourth Swedish Language Technology Conference*, pages 17–18.

Perry M Elliott, Aris Anastasakis, Michael A Borger, Martin Borggrefe, Franco Cecchi, Philippe Charron, Albert Alain Hagege, Antoine Lafont, Giuseppe Limongelli, Heiko Mahrholdt, William J McKenna, Jens Mogensen, Petros Nihoyannopoulos, Stefano Nistri, Petronella G Pieper, Burkert Pieske, Claudio Rapezzi, Frans H Rutten, Christoph Tillmanns, Hugh Watkins, Additional Contributor, Constantinos O'Mahony, ESC Committee for Practice Guidelines (CPG), Jose Luis Zamorano, Stephan Achenbach, Helmut Baumgartner, Jeroen J Bax, Héctor Bueno, Veronica Dean, Christi Deaton, Çetin Erol, Robert Fagard, Roberto Ferrari, David Hasdai, Arno W Hoes, Paulus Kirchhof, Juhani Knuuti, Philippe Kolh, Patrizio Lancellotti, Ales Linhart, Petros Nihoyannopoulos, Massimo F Piepoli, Piotr Ponikowski, Per Anton Sirnes, Juan Luis Tamargo, Michal Tendera, Adam Torbicki, William Wijns, Stephan Windecker, Document Reviewers, David Hasdai, Piotr Ponikowski, Stephan Achenbach, Fernando Alfonso, Cristina Basso, Nuno Miguel Cardim, Juan Ramón Gimeno, Stephane Heymans, Per Johan Holm, Andre Keren, Paulus Kirchhof, Philippe Kolh, Christos Lionis, Claudio Muneretto, Silvia Priori, Maria Jesus Salvador, Christian Wolpert, Jose Luis Zamorano, Matthias Frick, Farid Aliyev, Svetlana Komissarova, Georges Mairesse, Elnur Smajić, Vasil Velchev, Loizos Antoniades,

Ales Linhart, Henning Bundgaard, Tiina Heliö, Antoine Leenhardt, Hugo A Katus, George Efthymiadis, Róbert Sepp, Gunnar Thor Gunnarsson, Shemy Carasso, Alina Kerimkulova, Ginta Kamzola, Hady Skouri, Ghada Eldirsi, Ausra Kavoliuniene, Tiziana Felice, Michelle Michels, Kristina Hermann Haugaa, Radosław Lenarczyk, Dulce Brito, Eduard Apetrei, Leo Bokheria, Dragan Lovic, Robert Hatala, Pablo Garcia Pavía, Maria Eriksson, Stéphane Noble, Elizabeta Srbinovska, Murat Özdemir, Elena Nesukay, and Neha Sekhri. 2014. 2014 esc guidelines on diagnosis and management of hypertrophic cardiomyopathy: the task force for the diagnosis and management of hypertrophic cardiomyopathy of the european society of cardiology (esc). *European heart journal*, 35(39).

L. Ferro, L. Gerber, I. Mani, B. Sundheim, and G. Wilson. 2002. Instruction manual for the annotation of temporal expressions. Technical report, MITRE, Washington C3 Cen- ter, McLean, Virginia.

Jeff Friedlin and Clement J McDonald. 2006. Using a natural language processing system to extract and code family history data from admission reports. In *AMIA Annual Symposium Proceedings*, volume 2006, page 925. American Medical Informatics Association.

Sergey Goryachev, Hyeoneui Kim, and Qing Zeng-Treitler. 2008. Identification and extraction of family history information from clinical reports. In *AMIA Annual Symposium Proceedings*, volume 2008, page 247. American Medical Informatics Association.

Tero Hiekkalinna, Joseph D. Terwilliger, Sampo Sammalisto, Leena Peltonen, and Markus Perola. 2005. Autogscan: Powerful tools for automated genome-wide linkage and linkage disequilibrium analysis. *Twin Research and Human Genetics*, 8(1):16–21.

Gumwon Hong. 2005. Relation extraction using support vector machine. In *International Conference on Natural Language Processing*, pages 366–377. Springer.

Christina Lohr, Sven Buechel, and Udo Hahn. 2018. Sharing copies of synthetic clinical corpora without physical distribution — a case study to get around IPRs and privacy constraints featuring the German JSYNCC corpus. In *Proceedings of the Eleventh International Conference on Language Resources and Evaluation*, pages 1259 – 1266, Miyazaki, Japan.

Anne-Lyse Minard, Anne-Laure Ligozat, and Brigitte Grau. 2011. Multi-class svm for relation extraction from clinical reports. In *Proceedings of the International Conference Recent Advances in Natural Language Processing 2011*, pages 604–609.

Makoto Miwa and Yutaka Sasaki. 2014. Modeling joint entity and relation extraction with table representation. In *Proceedings of the 2014 Conference on Empirical Methods in Natural Language Processing (EMNLP)*, pages 1858–1869.

Roser Morante and Walter Daelemans. 2012. Conandoyle-neg: Annotation of negation cues and their scope in conan doyle stories. In *Proceedings of the Eighth International Conference on Language Resources and Evaluation (LREC-2012)*. European Language Resources Association (ELRA).

Lilja Øvrelid and Petter Hohle. 2016. Universal Dependencies for Norwegian. In *Proceedings of the International Conference on Language Resources and Evaluation (LREC)*.

Fernanda Polubriaginof, Nicholas P Tatonetti, and David K Vawdrey. 2015. An assessment of family history information captured in an electronic health record. In *AMIA Annual Symposium Proceedings*, volume 2015, page 2035. American Medical Informatics Association.

Angus Roberts, Robert Gaizauskas, and Mark Hepple. 2008a. Extracting clinical relationships from patient narratives. In *Proceedings of the Workshop on Current Trends in Biomedical Natural Language Processing*, pages 10–18. Association for Computational Linguistics.

Angus Roberts, Robert Gaizauskas, Mark Hepple, George Demetriou, Yikun Guo, Andrea Setzer, and Ian Roberts. 2008b. Semantic annotation of clinical text: The clef corpus. In *Proceedings of the LREC 2008 workshop on building and evaluating resources for biomedical text mining*, pages 19–26.

Thomas Brox Røst, Ola Huseth, Øystein Nytrø, and Anders Grimsmo. 2008. Lessons from developing an annotated corpus of patient histories. *JCSE*, 2(2):162–179.

R. Saurí, J. Littman, B. Knippen, R. Gaizauskas, A. Setzer, and J. Pustejovsky. 2006. TimeML annotation guidelines version 1.2. 1. Technical report, LDC.

Guergana K Savova, James J Masanz, Philip V Ogren, Jiaping Zheng, Sunghwan Sohn, Karin C Kipper-Schuler, and Christopher G Chute. 2010. Mayo clinical text analysis and knowledge extraction system (ctakes): architecture, component evaluation and applications. *Journal of the American Medical Informatics Association*, 17(5):507–513.

Pontus Stenetorp, Sampo Pyysalo, Goran Topić, Tomoko Ohta, Sophia Ananiadou, and Jun'ichi Tsujii. 2012. brat: a web-based tool for nlp-assisted text annotation. In *Proceedings of the Demonstrations Session at EACL 2012*, pages 102 – 107, Avignon, France.

Robert Stevens, Nicolas Matentzoglu, Uli Sattler, and Margaret Stevens. 2014. A family history knowledge base in OWL 2. In *Informal Proceedings of the 3rd International Workshop on OWL Reasoner Evaluation (ORE 2014) co-located with the Vienna Summer of Logic (VSL 2014), Vienna, Austria, July 13,*

2014., volume 1207 of *CEUR Workshop Proceedings*, pages 71–76. CEUR-WS.org.

Özlem Uzuner and Amber Stubbs. 2015. Practical applications for natural language processing in clinical research: The 2014 i2b2/uthealth shared tasks. *Journal of biomedical informatics*, 58(Suppl):S1.

Brandon M. Welch, Kevin Wiley, Lance Pflieger, Rosaline Achiangia, Karen Baker, Chanita Hughes-Halbert, Heath Morrison, Joshua Schiffman, and Megan Doerr. 2018. Review and comparison of electronic patient-facing family health history tools. *Journal of Genetic Counseling*, 27(2):381–391.

Daniel Zeman, Martin Popel, Milan Straka, Jan Hajic, Joakim Nivre, Filip Ginter, Juhani Luotolahti, Sampo Pyysalo, Slav Petrov, Martin Potthast, Francis Tyers, Elena Badmaeva, Memduh Gokirmak, Anna Nedoluzhko, Silvie Cinkova, Jan Hajic jr., Jaroslava Hlavacova, Václava Kettnerová, Zdenka Uresova, Jenna Kanerva, Stina Ojala, Anna Missilä, Christopher D. Manning, Sebastian Schuster, Siva Reddy, Dima Taji, Nizar Habash, Herman Leung, Marie-Catherine de Marneffe, Manuela Sanguinetti, Maria Simi, Hiroshi Kanayama, Valeria de-Paiva, Kira Droganova, Héctor Martínez Alonso, Çağrı Çöltekin, Umut Sulubacak, Hans Uszkoreit, Vivien Macketanz, Aljoscha Burchardt, Kim Harris, Katrin Marheinecke, Georg Rehm, Tolga Kayadelen, Mohammed Attia, Ali Elkahky, Zhuoran Yu, Emily Pitler, Saran Lertpradit, Michael Mandl, Jesse Kirchner, Hector Fernandez Alcalde, Jana Strnadová, Esha Banerjee, Ruli Manurung, Antonio Stella, Atsuko Shimada, Sookyoung Kwak, Gustavo Mendonca, Tatiana Lando, Rattima Nitisaroj, and Josie Li. 2017. CoNLL 2017 Shared Task: Multilingual Parsing from Raw Text to Universal Dependencies. In *Proceedings of the CoNLL 2017 Shared Task: Multilingual Parsing from Raw Text to Universal Dependencies*, Vancouver, Canada.

A   Supplemental material

The annotated data and the code used in this paper is available at: `https://github.com/ltgoslo/NorSynthClinical`.

CAS: French Corpus with Clinical Cases

Natalia Grabar
UMR CNRS 8163 – STL
F-59000 Lille, France
natalia.grabar@univ-lille.fr

Vincent Claveau, Clément Dalloux
CNRS, IRISA, Rennes, France
vincent.claveau@irisa.fr
clement.dalloux@irisa.fr

Abstract

Textual corpora are extremely important for various NLP applications as they provide information necessary for creating, setting and testing these applications and the corresponding tools. They are also crucial for designing reliable methods and reproducible results. Yet, in some areas, such as the medical area, due to confidentiality or to ethical reasons, it is complicated and even impossible to access textual data representative of those produced in these areas. We propose the CAS corpus built with clinical cases, such as they are reported in the published scientific literature in French. We describe this corpus, currently containing over 397,000 word occurrences, and the existing linguistic and semantic annotations.

1   Introduction

Textual corpora are extremely important for various NLP applications as they provide information necessary for creating, setting and testing these applications and the corresponding tools. Yet, in some areas, due to confidentiality or to ethical reasons, it is complicated and even impossible to access representative textual data. Medical and legal areas correspond to such examples: in the legal area, information on lawsuits and trials remain confidential, while in the medical area, the medical secret must be respected. In both situations, personal data cannot be used. For several years now, anonymization and de-identification methods and tools have been made available and provide competitive and reliable results (Ruch et al., 2000; Sibanda and Uzuner, 2006; Uzuner et al., 2007; Grouin and Zweigenbaum, 2013) reaching up to 90% precision and recall. But even de-identified data may be difficult to be freely accessed and used for the research purpose because there is a risk of re-identification of people, and more particularly of patients (Meystre et al., 2014; Grouin

et al., 2015) because several medical histories are unique, or because of other reasons. Hence, the application of the de-identification tools on personal data often does not permit to make these data freely available and usable within the research context.

Yet, there is a real need for the development of methods and tools for several applications suited for such restricted areas. For instance, in the medical area, it is important to have suitable tools for information retrieval and extraction, for the recruiting of patients for clinical trials, and for performing several other important tasks such as indexing, study of temporality, negation, etc. (Embi et al., 2005; Hamon and Grabar, 2010; Uzuner et al., 2011; Fletcher et al., 2012; Sun et al., 2013; Campillo-Gimenez et al., 2015; Kang et al., 2017). Another important issue is related to the reliability of tools and to the reproducibility of study results across similar data from different sources. The scientific research and clinical community are indeed increasingly coming under criticism for the lack of reproducibility in the biomedical area (Chapman et al., 2011; Collins and Tabak, 2014; Cohen et al., 2016), as well as in other areas. First step towards the reproducibility of results is the availability of freely usable tools and corpora. In our work, we are mainly concerned by building freely available corpora from the medical area.

The purpose of our work is to introduce the CAS corpus with French medical data, containing clinical cases such as those published in scientific literature or used for the education and training of medical students. In what follows, we first present some works on creation of medical corpora stressing more particularly on corpora freely available for the research (Section 2). We then introduce and describe the CAS corpus in French (Section 3) and its current annotations. We conclude with some directions for the future work (Section 4).

Proceedings of the 9th International Workshop on Health Text Mining and Information Analysis (LOUHI 2018), pages 122–128
Brussels, Belgium, October 31, 2018. ©2018 Association for Computational Linguistics

2 Freely available clinical corpora

Within the medical area, we can distinguish two main types of medical corpora: scientific and clinical. *Scientific corpora* are issued from scientific publications and reporting. Such corpora are becoming increasingly available for the research thanks to the recent and less recent initiatives dedicated to the open publication, such as those promoted by the NLM (National Library of Medicine) through the PUBMED portal[1] and specifically dedicated to the biomedical area, and by the HAL[2] and ISTEX[3] initiatives, which provide generic portals for accessing scientific publications from various areas, including medicine. Such corpora describe the research works, their motivation, methods, results and issues on precise research questions. Other portals may also provide access to scientific literature following specific purposes, like indexing of reliable literature, such as proposed by HON (Boyer et al., 1997), CIS-MEF (Darmoni et al., 1999), and other similar initiatives (Risk and Dzenowagis, 2001). Thanks to some research works, there are also scientific corpora which provide precise annotations and categorizations. These are mainly built for the purposes of challenges (Kelly et al., 2013; Goeuriot et al., 2014) but may also be provided from works of researchers, such as POS-tag (Tsuruoka et al., 2005) and negation (Szarvas et al., 2008) annotated corpora. As for *clinical corpora*, they are related to hospital and clinical events of patients. Such corpora typically describe medical history of patients and the medical care they are undergoing. It is complicated to obtain free access to this kind of medical data and, for this reason, there are very few clinical corpora freely available for the research. In our work, we are mainly interested in clinical corpora: the proposed literature review of the existing work is aimed at clinical corpora which are freely available for the research. We present here the main existing clinical corpora.

MIMIC (Medical Information Mart for Intensive Care) corpora, now in their version III, provide the largest available set of structured and unstructured clinical data in English. MIMIC III is a single-center database comprising information relating to patients admitted to critical care units at a large tertiary care hospital. These data include vital signs, medications, laboratory measurements, observations and notes charted by care providers, fluid balance, procedure codes, diagnostic codes, imaging reports, hospital length of stay, survival data, and more. The database supports applications including academic and industrial research, quality improvement initiatives, and higher education coursework (Johnson et al., 2016). These data are widely used by researchers, for instance for the prediction of mortality (Anand et al., 2018; Feng et al., 2018), for the diagnosis identification and coding (Perotte et al., 2014; Li et al., 2018), for the study of temporality (Che et al., 2018) or for the identification of similar clinical notes (Gabriel et al., 2018) to cite just a few of such works. Data from these corpora are also used in challenges, such as I2B2, N2C2 and CLEF-eHEALTH.

I2B2 (Informatics for Integrating Biology and the Bedside)[4] is an NIH-funded initiative promoting the development and test of NLP tools for healthcare improvement. In order to enhance the ability of NLP tools to process fine grained information from clinical records, I2B2 challenges provide sets of fully deidentified clinical notes enriched with specific annotations (Uzuner, 2008; Uzuner et al., 2011; Sun et al., 2013), such as: deidentification, smoking status, medication-related information, semantic relations between entities, or temporality. The clinical corpora and their annotations built for the I2B2 NLP challenges are available now for the general research purposes.

N2C2 (National NLP Clinical Challenges)[5], held for the first time in 2018, is dedicated to the inclusion of patients in clinical trials and the detection of adverse-drug events.

CLEF-eHEALTH challenges[6] held in 2013 and 2014 provide annotations for the detection of disorders and normalization of abbreviations, in 2016 the focus was done on structuring of Australian free-text nurse notes, and in 2016 and 2017 death reports in French, provided by the CépiDc[7], have been processed for the extraction of death causes.

Finally, medical data, close to those handled in the clinical context, can be found in the clinical trials protocols. One example is the corpus of clinical trials annotated with information on numerical

---

[1] https://www.ncbi.nlm.nih.gov/pubmed
[2] https://hal.archives-ouvertes.fr/
[3] https://www.istex.fr/

[4] https://www.i2b2.org/NLP/DataSets/Main.php
[5] https://n2c2.dbmi.hms.harvard.edu/
[6] https://sites.google.com/site/shareclefehealth/
[7] http://www.cepidc.inserm.fr/

*A term female infant was born by vaginal delivery with normal birth weight, body length and APGAR score, from a 42-year-old mother with 13 previous pregnancies resulting in 3 miscarriages and 10 live births. The mother had no history of antenatal medical illness nor of exposure to smoking, drinking and other drugs. At birth, general and systemic examination revealed a round face, single palmar crease, left precordial systolic murmur. Two hours after birth a deterioration of the general condition occurred, with generalized hypotonia, cyanosis, poor feeding. The blood count revealed white blood cell count of 35.6*10 /µL with 20.6*10 /µL, 57.9% monocytes, normal neutrophils, lymphocytes and eosinophils count, hemoglobin levels of 19.1 g/dl and 27*10 /µL platelets count. The acute phase reactants were negative. Because she maintained the altered general condition and the platelets ranged between 17-18 *10 /µL, on the 8th day after birth she was referred to our unit for proper diagnosis and treatment. Physical examination showed a phenotype suggestive for Down syndrome, later confirmed by karyotyping (47, XX + 21). She was lethargic, tachypneic and a systolic heart murmur was observed. The liver was 2 cm below the right costal margin, along with a slight enlargement of the spleen. The laboratory tests on the first day of admission in our unit revealed a white blood count of 15.8*10 /µL, with an abnormal monocyte count (increased absolute and percentile count: 5.66*10 /µL, respectively 35.5%), normal absolute neutrophil count (5.53*10 /µL), a hemoglobin level of 15.9 g/dl and severe thrombocytopenia (15*10 /µL). The biochemical parameters including electrolytes, uric acid, creatinine, bilirubin, liver enzymes were normal. The serum lactate dehydrogenase was raised. The bacterial culture work-up and titers of antibodies against toxoplasmosis, cytomegalovirus, Epstein Barr virus, hepatitis C, HIV were negative. The peripheral blood smear presented atypical cells. The bone marrow aspiration showed hemodiluted aspirate with blast cells. Immunophenotyping revealed 23% blast cells, positive for megakaryocytic markers (CD42b, CD41, CD61), myeloid markers (CD33), progenitor cell markers (CD117, CD34) and T cell marker - CD7 positive. MPO and HLA/DR were negative. The mutational status of AM-LETO, PML-RARα, FLT3 and NPM1 fusion genes came out absent. The positive diagnosis was acute megakaryoblastic leukemia (AMKL).*

The echocardiography found a patent foramen ovale. The infant underwent chemotherapy according to the Down syndrome-specific AML chemotherapy protocol, consisting in four cycles of treatment: the first two cycles (induction phase) included combinations of cytarabine and liposomal daunorubicin and the last two cycles (consolidation phase): etoposide, cytarabine and mitoxantrone. Our patient aquired clinical and hematological remission without serious adverse events.

Figure 1: Example of clinical case

values in English (Claveau et al., 2017), and on negation in French and Brazilian Portuguese (Dalloux et al., 2018).

3 The CAS corpus

3.1 Content of the corpus

We present the CAS corpus in French. It contains clinical cases such as published in scientific literature and training material. Cases from these different sources are included in the corpus. Usually, the source data are available as pdf files. Their conversion in the text format is automatic but then needs to be fully checked out in order to correct potential segmentation errors (remove the paratext specific to a given journal, verify the conversion of columns, of end of lines and pages, etc.).

Similarly to clinical documents, the content of clinical cases depends on the clinical situations which are illustrated and on the disorders, but also on the purpose of the presented cases (description of diagnoses, treatments or procedures, expected audience, etc.).

Figure 1 presents an example of clinical case in English. Such data are de-identified by the auhors and their publication is done with the written permission of patients. The case reports can be related to any medical situation (diagnosis, treatment, procedure, follow-up...) and to any disorder. Publication of clinical cases usually has didactic purposes: train medical students, report on unusual or new clinical situations, present novel treatment or imaging issue... A typical structure of publications with clinical cases starts with the introduction to the clinical situation, then one or more clinical cases are presented to support the situation. Schemas, imaging, examination results,

word	PoS	lemma	uncert. cue	uncert. scope	CUI	neg cue	neg scope
L'	B-determiner	le	O	O	O	O	O
adolescent	B-common_noun	adolescent	O	O	B-C0205653	O	O
parait	B-present_verb_form	paraître	B-u-1	O	O	O	O
triste	B-adjective	triste	O	B-u-1	O	O	O
et	B-coordination_conjunction	et	O	O	O	O	O
ne	B-adverb	ne	O	O	O	B-n-1	O
parle	B-present_verb_form	parler	O	O	O	O	B_n-1
pas	B-adverb	pas	O	O	O	I-n-1	O
.	B-ending_punctuation_mark	.	O	O	O	O	O

Table 1: Example of the annotated sentence from the corpus (B-u-x stands for the beginning of the uncertainty cue or scope number x, B-n-y for the negation cue or scope number y)

patient history, lab results, clinical evolution, treatment, etc. can also be provided for the illustration of clinical cases. Finally, these clinical cases are discussed. Hence, such cases may present an extensive description of medical problems. Such publications gather medical information related to clinical discourse (clinical cases) and to scientific discourse (introduction and discussion). Related scientific literature is also provided.

As we can see from Figure 1, the clinical part of publications on clinical cases may be very similar to clinical documents: it describes patients, and proposes their diagnosis based on examination, imaging, and biological and genetic information. Besides, numerical values and abbreviations are also present. Misspellings, which are quite frequent in clinical documents, may be missing in publications on clinical cases.

3.2 Annotation of the corpus

Currently, the corpus contains linguistic and semantic annotations.

At the linguistic level, the corpus is PoS-tagged and lemmatized with a tool developed in-house and available as a web-service at `https://anonymized_url`. Then, several layers of semantic annotation are performed automatically:

- *Concept Unique Identifiers (CUI)* corresponding to French terms from the UMLS (Lindberg et al., 1993) for single or multi-word terms. For multi-word terms, the annotations exploits the IOB (Inside-Outside-Begin) format. For instance, the two-word term *vitamine B12* is encoded as follows:
  ```
  ...        O
  vitamine   B-C0042845
  B12        I-C0042845
  ...        O
  ```

In the current version of the corpus, in case of several concurrent CUIs, only the longest, and supposedly more precise, CUIs are kept. For instance, *carence en vitamine B12 (deficiency in B12 vitamin)* (C0042847) will be preferred to *vitamine B12* (C0042845);

- *Negation.* Negation indicates whether a given disorder, procedure or treatment are present or not in the medical history and care of a given patient. For this reason, its annotation and detection are important. We adopt the approach proposed by Fancellu et al. (2016) and adapted for French by Dalloux et al. (2018) based on Machine Learning techniques trained on annotated data. This follows a two-step process: (1) the negation markers are detected with a specifically trained CRF; (2) the scope of each detected marker is found with a neural network (Bi-LSTM with a CRF layer). On the French and English data tested, the detection of negation gives up to 0.98 for the cues and 0.86 for their scope;

- *Uncertainty.* Uncertainty is also an integral part of medical discourse and should be taken into account for a more precise computing of the status of disorders, procedures and treatments. A set of markers has been built manually. It contains simple and complex lexical markers like *probablement, certainement (probably, certainly)* and morphological cues like conditional verbs (*indiquerait, proviendrait (should indicate, may be caused by)*). These markers and cues are projected on the corpus and their scope are found by heuristic rules. Detection of uncertainty gives about

type	# annotations
CUI	47,708
uncertainty	4,723
negations	4,620

Table 2: Statistics on annotations

0.90 F-measure for the cues and 0.80 for the scope.

Since there may be several markers of negation and uncertainty in a sentence, they are numbered with their scopes accordingly.

In Table 1, we present an excerpt from the corpus with all the aforementioned linguistic and semantic annotations for the sentence *L'adolescent paraît triste et ne parle pas.* (*The teenager seems to be sad and doesn't speak.*)

3.3 Annotation statistics

Overall, the corpus currently contains 20,363 sentences and over 397,000 word occurrences excluding punctuation marks. Table 2 indicates the number of units automatically recognized for each category.

4 Conclusion

We presented a new corpus in French which provides medical data close to those produced in the clinical context: description of clinical cases and their discussion. Overall, the corpus currently contains over 397,000 word occurrences excluding punctuation marks. The corpus is currently annotated with several layers of information: linguistic (PoS-tagging, lemmas) and semantic (the UMLS concepts, uncertainty, negation and their scopes). The corpus will be enriched with more clinical cases published. Other annotation layers will be added and their correctness cross-validated by human annotators. The enriched version of the corpus will undergo a more detailed description, such as statistics on age and gender of patients, their diseases, or the sources of publications.

Besides, similar corpora will be built for other languages. For instance, the repository of clinical cases in English is available on a dedicated website *Archive of Clinical Cases*[8] respecting the Creative Commons License.

The very purpose of our work is to make these annotated corpora freely available for research. We expect that this may encourage development of robust NLP tools for medical free-text documents in French and other languages.

Acknowledgements

This work was partly funded by the French government support granted to the CominLabs LabEx managed by the ANR in Investing for the Future program under reference ANR-10-LABX-07-01.

The authors would like to thank Cyril Grouin for the discussions on existing medical corpora and the reviewers for their helpful comments.

References

RS Anand, P Stey, S Jain, DR Biron, H Bhatt, K Monteiro, E Feller, Ranney ML, Sarkar IN, and Chen ES. 2018. Predicting mortality in diabetic icu patients using machine learning and severity indices. In *AMIA Jt Summits Transl Sci Proc*, pages 310–319.

Celia Boyer, O Baujard, Vincent Baujard, S Aurel, M Selby, and RD Appel. 1997. Health on the net automated database of health and medical information. *Int J Med Inform*, 47(1-2):27–9.

B Campillo-Gimenez, C Buscail, O Zekri, B Laguerre, E Le Prisé, R De Crevoisier, and M Cuggia. 2015. Improving the pre-screening of eligible patients in order to increase enrollment in cancer clinical trials. *Trials*, 16(1):1–15.

Wendy W Chapman, Prakash M Nadkarni, Lynette Hirschman, Leonard W D'Avolio, Guergana K Savova, and Ozlem Uzuner. 2011. Overcoming barriers to nlp for clinical text: the role of shared tasks and the need for additional creative solutions. *J Am Med Inform Assoc*, 18(5):540–543.

Z Che, S Purushotham, K Cho, D Sontag, and Y Liu. 2018. Recurrent neural networks for multivariate time series with missing values. *Sci Rep*, 8(1):6085.

Vincent Claveau, Lucas Emanuel Silva Oliveira, Guillaume Bouzillé, Marc Cuggia, Claudia Maria Cabral Moro, and Natalia Grabar. 2017. Numerical eligibility criteria in clinical protocols: annotation, automatic detection and interpretation. In *AIME (Artifical Intelligence in Medicine in Europe)*.

K. Bretonnel Cohen, Jingbo Xia, Christophe Roeder, and Lawrence E. Hunter. 2016. Reproducibility in natural language processing: A case study of two r libraries for mining pubmed/medline. In *LREC Int Conf Lang Resour Eval*, pages 6–12.

FS Collins and LA Tabak. 2014. Nih plans to enhance reproducibility. *Nature*, 505:612–613.

Clément Dalloux, Vincent Claveau, Natalia Grabar, and Claudia Moro. 2018. Portée de la négation : détection par apprentissage supervisé en français et portugais brésilien. In *TALN 2018*, pages 1–6.

---

[8]http://www.clinicalcases.eu

SJ Darmoni, JP Leroy, F Baudic, M Douyère, J Piot, and B Thirion. 1999. CISMeF: cataloque and index of french speaking health resources. In *Stud Health Technol Inform*, pages 493–6.

PJ Embi, A Jain, J Clark, and CL Harris. 2005. Development of an electronic health record-based clinical trial alert system to enhance recruitment at the point of care. In *Ann Symp Am Med Inform Assoc (AMIA)*, pages 231–35.

Federico Fancellu, Adam Lopez, and Bonnie Webber. 2016. Neural networks for negation scope detection. In *An Meeting of the Ass for Comp Linguistics*, volume 1.

M Feng, JI McSparron, DT Kien, DJ Stone, DH Roberts, RM Schwartzstein, A Vieillard-Baron, and LA Celi. 2018. Transthoracic echocardiography and mortality in sepsis: analysis of the mimic-iii database. *Intensive Care Med*, 44(6):884–892.

B Fletcher, A Gheorghe, D Moore, S Wilson, and S Damery. 2012. Improving the recruitment activity of clinicians in randomised controlled trials: A systematic review. *BMJ Open*, 2(1):1–14.

RA Gabriel, TT Kuo, J McAuley, and CN Hsu. 2018. Identifying and characterizing highly similar notes in big clinical note datasets. *J Biomed Inform*, 82:63–69.

Lorraine Goeuriot, Liadh Kelly, Wei Li, Joao Palotti, Pavel Pecina, Guido Zuccon, Allan Hanbury, Gareth Jones, and Henning Müller. 2014. Share/clef ehealth evaluation lab 2014, task 3: User-centred health information retrieval. In *CLEF*, Lecture Notes in Computer Science (LNCS), pages 43–61. Springer.

Cyril Grouin, Nicolas Griffon, and Aurélie Névéol. 2015. Is it possible to recover personal health information from an automatically de-identified corpus of French EHRs? In *Proc of LOUHI*, Lisbon, Portugal.

Cyril Grouin and Pierre Zweigenbaum. 2013. Automatic de-identification of french clinical records: Comparison of rule-based and machine-learning approches. In *Stud Health Technol Inform, Proc of MedInfo*, volume 192, pages 476–80, Copenhagen, Denmark.

T Hamon and N Grabar. 2010. Linguistic approach for identification of medication names and related information in clinical narratives. *J Am Med Inform Assoc*, 17(5):549–54.

Alistair E.W. Johnson, Tom J. Pollard, Lu Shen, Li wei H. Lehman, Mengling Feng, Mohammad Ghassemi, Benjamin Moody, Peter Szolovits, Leo Anthony Celi, and Roger G. Mark. 2016. MIMIC-iii, a freely accessible critical care database. *Scientific Data*, 3(160035):1–9.

Tian Kang, Shaodian Zhang, Youlan Tang, Gregory W Hruby, Alexander Rusanov, Noemie Elhadad, and Chunhua Weng. 2017. EliIE: An open-source information extraction system for clinical trial eligibility criteria. *J Am Med Inform Assoc*, 24(6):1062–1071.

Liadh Kelly, Lorraine Goeuriot, Hanna Suominen, Danielle L. Mowery, Sumithra Velupillai, Wendy W. Chapman, Guido Zuccon, and Joao Palotti. 2013. Overview of the share/clef ehealth evaluation lab 2013. In *CLEF*, Lecture Notes in Computer Science (LNCS). Springer.

M Li, Z Fei, M Zeng, F Wu, Y Li, Y Pan, and J Wang. 2018. Automated ICD-9 coding via a deep learning approach. In *IEEE/ACM Trans Comput Biol Bioinform*.

DA Lindberg, BL Humphreys, and AT McCray. 1993. The unified medical language system. *Methods Inf Med*, 32(4):281–291.

Stephane Meystre, Shuying Shen, Deborah Hofmann, and Adi Gundlapalli. 2014. Can physicians recognize their own patients in de-identified notes? In *Stud Health Technol Inform 205*, pages 778–82.

Adler Perotte, Rimma Pivovarov, Karthik Natarajan, Nicole Weiskopf, Frank Wood, and Noémie Elhadad. 2014. Diagnosis code assignment: models and evaluation metrics. *J Am Med Inform Assoc*, 21:231–237.

Ahmad Risk and J Dzenowagis. 2001. Review of internet information quality initiatives. *Journal of Medical Internet Research*, 3(4).

Patrick Ruch, Robert H. Baud, Anne-Marie Rassinoux, Pierrette Bouillon, and Gilbert Robert. 2000. Medical document anonymization with a semantic lexicon. In *Ann Symp Am Med Inform Assoc (AMIA)*, pages 729–733, Los Angeles, CA.

T Sibanda and O Uzuner. 2006. Role of local context in de-identification of ungrammatical, fragmented test. In *NAACL-HLT 2006*, New York, USA.

Weiyi Sun, Anna Rumshisky, and Özlem Uzuner. 2013. Evaluating temporal relations in clinical text: 2012 i2b2 challenge. *JAMIA*, 20(5):806–813.

G Szarvas, V Vincze, R Farkas, and J Csirik. 2008. The BioScope corpus: annotation for negation, uncertainty and their scope in biomedical texts. In *BIONLP*, pages 38–45.

Y Tsuruoka, Y Tateishi, JD Kim, T Ohta, J McNaught, S Ananiadou, and J Tsujii. 2005. Developing a robust part-of-speech tagger for biomedical text. *LNCS*, 3746:382–392.

O Uzuner. 2008. Second i2b2 workshop on natural language processing challenges for clinical records. In *Ann Symp Am Med Inform Assoc (AMIA)*, pages 1252–3.

O Uzuner, Y Luo, and P Szolovits. 2007. Evaluating the state-of-the-art in automatic de-identification. *J Am Med Inform Assoc*, 14:550–563.

Özlem Uzuner, Brett R South, Shuying Shen, and
Scott L DuVall. 2011. 2010 i2b2/va challenge on
concepts, assertions, and relations in clinical text. *J
Am Med Inform Assoc*, 18(5):552–556.

Analysis of Risk Factor Domains in Psychosis Patient Health Records

Eben Holderness[1,2], Nicholas Miller[1,2], Philip Cawkwell[1], Kirsten Bolton[1],
James Pustejovsky[2], Marie Meteer[2] and Mei-Hua Hall[1]

[1]Psychosis Neurobiology Laboratory, McLean Hospital, Harvard Medical School
[2]Department of Computer Science, Brandeis University
```
{eholderness, mhall}@mclean.harvard.edu
    nicholas.anthony.miller@gmail.com
    {pcawkwell, kbolton}@partners.org
    {jamesp, mmeteer}@cs.brandeis.edu
```


Abstract

Readmission after discharge from a hospital is disruptive and costly, regardless of the reason. However, it can be particularly problematic for psychiatric patients, so predicting which patients may be readmitted is critically important but also very difficult. Clinical narratives in psychiatric electronic health records (EHRs) span a wide range of topics and vocabulary; therefore, a psychiatric readmission prediction model must begin with a robust and interpretable topic extraction component. We created a data pipeline for using document vector similarity metrics to perform topic extraction on psychiatric EHR data in service of our long-term goal of creating a readmission risk classifier. We show initial results for our topic extraction model and identify additional features we will be incorporating in the future.

1 Introduction

Psychotic disorders typically emerge in late adolescence or early adulthood (Kessler et al., 2007; Thomsen, 1996) and affect approximately 2.5-4% of the population (Perälä et al., 2007; Bogren et al., 2009), making them one of the leading causes of disability worldwide (Vos et al., 2015). A substantial proportion of psychiatric inpatients are readmitted after discharge (Wiersma et al., 1998). Readmissions are disruptive for patients and families, and are a key driver of rising healthcare costs (Mangalore and Knapp, 2007; Wu et al., 2005). Reducing readmission risk is therefore a major unmet need of psychiatric care. Developing clinically implementable machine learning tools to enable accurate assessment of risk factors associated with readmission offers opportunities to inform the selection of treatment interventions and implement appropriate preventive measures.

In psychiatry, traditional strategies to study readmission risk factors rely on clinical observa-

tion and manual retrospective chart review (Olfson et al., 1999; Lorine et al., 2015). This approach, although benefitting from clinical expertise, does not scale well for large data sets, is effort-intensive, and lacks automation. An efficient, more robust, and cheaper NLP-based alternative approach has been developed and met with some success in other medical fields (Murff et al., 2011). However, this approach has seldom been applied in psychiatry because of the unique characteristics of psychiatric medical record content.

There are several challenges for topic extraction when dealing with clinical narratives in psychiatric EHRs. First, the vocabulary used is highly varied and context-sensitive. A patient may report "feeling 'really great and excited'" – symptoms of mania – without any explicit mention of keywords that differ from everyday vocabulary. Also, many technical terms in clinical narratives are multiword expressions (MWEs) such as 'obsessive body image', 'linear thinking', 'short attention span', or 'panic attack'. These phrasemes are comprised of words that in isolation do not impart much information in determining relatedness to a given topic but do in the context of the expression.

Second, the narrative structure in psychiatric clinical narratives varies considerably in how the same phenomenon can be described. Hallucinations, for example, could be described as "the patient reports auditory hallucinations," or "the patient has been hearing voices for several months," amongst many other possibilities.

Third, phenomena can be directly mentioned without necessarily being relevant to the patient specifically. Psychosis patient discharge summaries, for instance, can include future treatment plans (e.g. "Prevent relapse of a manic or major depressive episode.", "Prevent recurrence of psychosis.") containing vocabulary that at the word-level seem strongly correlated with readmission

129

Proceedings of the 9th International Workshop on Health Text Mining and Information Analysis (LOUHI 2018), pages 129–138
Brussels, Belgium, October 31, 2018. ©2018 Association for Computational Linguistics

risk. Yet at the paragraph-level these do not indicate the presence of a readmission risk factor in the patient and in fact indicate the absence of a risk factor that was formerly present.

Lastly, given the complexity of phenotypic assessment in psychiatric illnesses, patients with psychosis exhibit considerable differences in terms of illness and symptom presentation. The constellation of symptoms leads to various diagnoses and comorbidities that can change over time, including schizophrenia, schizoaffective disorder, bipolar disorder with psychosis, and substance use induced psychosis. Thus, the lexicon of words and phrases used in EHRs differs not only across diagnoses but also across patients and time.

Taken together, these factors make topic extraction a difficult task that cannot be accomplished by keyword search or other simple text-mining techniques.

To identify specific risk factors to focus on, we not only reviewed clinical literature of risk factors associated with readmission (Alvarez-Jimenez et al., 2012; Addington et al., 2010), but also considered research related to functional remission (Harvey and Bellack, 2009), forensic risk factors (Singh and Fazel, 2010), and consulted clinicians involved with this project. Seven risk factor domains – Appearance, Mood, Interpersonal, Occupation, Thought Content, Thought Process, and Substance – were chosen because they are clinically relevant, consistent with literature, replicable across data sets, explainable, and implementable in NLP algorithms.

In our present study, we evaluate multiple approaches to automatically identify which risk factor domains are associated with which paragraphs in psychotic patient EHRs.[1] We perform this study in support of our long-term goal of creating a readmission risk classifier that can aid clinicians in targeting individual treatment interventions and assessing patient risk of harm (e.g. suicide risk, homicidal risk). Unlike other contemporary approaches in machine learning, we intend to create a model that is clinically explainable and flexible across training data while maintaining consistent performance.

To incorporate clinical expertise in the identification of risk factor domains, we undertake an annotation project, detailed in section 3.1. We identify a test set of over 1,600 EHR paragraphs

---

[1]This study has received IRB approval.

which a team of three domain-expert clinicians annotate paragraph-by-paragraph for relevant risk factor domains. Section 3.2 describes the results of this annotation task. We then use the gold standard from the annotation project to assess the performance of multiple neural classification models trained exclusively on Term Frequency – Inverse Document Frequency (TF-IDF) vectorized EHR data, described in section 4. To further improve the performance of our model, we incorporate domain-relevant MWEs identified using all in-house data.

2 Related Work

McCoy et al. (2015) constructed a corpus of web data based on the Research Domain Criteria (RDoC)(Insel et al., 2010), and used this corpus to create a vector space document similarity model for topic extraction. They found that the 'negative valence' and 'social' RDoC domains were associated with readmission. Using web data (in this case data retrieved from the Bing API) to train a similarity model for EHR texts is problematic since it differs from the target data in both structure and content. Based on reconstruction of the procedure, we conclude that many of the informative MWEs critical to understanding the topics of paragraphs in EHRs are not captured in the web data. Additionally, RDoC is by design a generalized research construct to describe the entire spectrum of mental disorders and does not include domains that are based on observation or causes of symptoms. Important indicators within EHRs of patient health, like appearance or occupation, are not included in the RDoC constructs.

Rumshisky et al. (2016) used a corpus of EHRs from patients with a primary diagnosis of major depressive disorder to create a 75-topic LDA topic model that they then used in a readmission prediction classifier pipeline. Like with McCoy et al. (2015), the data used to train the LDA model was not ideal as the generalizability of the data was narrow, focusing on only one disorder. Their model achieved readmission prediction performance with an area under the curve of .784 compared to a baseline of .618. To perform clinical validation of the topics derived from the LDA model, they manually evaluated and annotated the topics, identifying the most informative vocabulary for the top ten topics. With their training data, they found the strongest coherence occurred in topics involving substance use, suicidality, and anxiety disorders. But given the unsupervised nature of the LDA clustering algorithm, the topic coherence they observed is not guaranteed across data sets.

3 Data

Our target data set consists of a corpus of discharge summaries, admission notes, individual encounter notes, and other clinical notes from 220 patients in the OnTrack[TM] program at McLean Hospital. OnTrack[TM] is an outpatient program, focusing on treating adults ages 18 to 30 who are experiencing their first episodes of psychosis. The length of time in the program varies depending on patient improvement and insurance coverage, with an average of two to three years. The program focuses primarily on early intervention via individual therapy, group therapy, medication evaluation, and medication management. See Table 1 for a demographic breakdown of the 220 patients, for which we have so far extracted approximately 240,000 total EHR paragraphs spanning from 2011 to 2014 using Meditech, the software employed by McLean for storing and organizing EHR data.

These patients are part of a larger research cohort of approximately 1,800 psychosis patients, which will allow us to connect the results of this EHR study with other ongoing research studies incorporating genetic, cognitive, neurobiological, and functional outcome data from this cohort.

We also use an additional data set for training our vector space model, comprised of EHR texts queried from the Research Patient Data Registry (RPDR), a centralized regional data repository of clinical data from all institutions in the Partners HealthCare network. These records are highly comparable in style and vocabulary to our target data set. The corpus consists of discharge summaries, encounter notes, and visit notes from approximately 30,000 patients admitted to the system's hospitals with psychiatric diagnoses and symptoms. This breadth of data captures a wide range of clinical narratives, creating a comprehensive foundation for topic extraction.

After using the RPDR query tool to extract EHR paragraphs from the RPDR database, we created a training corpus by categorizing the extracted para-

Mean Age (2014)	20.7
Gender (Male)	79%
Race	
Asian	6%
Black	7%
Caucasian	77%
Latino	5%
Multiracial	5%
Insurance (Public)[2]	5.5%
30-day Inpatient Readmission Rate	14%

Table 1: Demographic breakdown of the target cohort.

graphs according to their risk factor domain using a lexicon of 120 keywords that were identified by the clinicians involved in this project. Certain domains – particularly those involving thoughts and other abstract concepts – are often identifiable by MWEs rather than single words. The same clinicians who identified the keywords manually examined the bigrams and trigrams with the highest TF-IDF scores for each domain in the categorized paragraphs, identifying those which are conceptually related to the given domain. We then used this lexicon of 775 keyphrases to identify more relevant training paragraphs in RPDR and treat them as (non-stemmed) unigrams when generating the matrix. By converting MWEs such as 'shortened attention span', 'unusual motor activity', 'wide-ranging affect', or 'linear thinking' to non-stemmed unigrams, the TF-IDF score (and therefore the predictive value) of these terms is magnified. In total, we constructed a corpus of roughly 100,000 paragraphs consisting of 7,000,000 tokens for training our model.

3.1 Annotation Task

In order to evaluate our models, we annotated 1,654 paragraphs selected from the 240,000 paragraphs extracted from Meditech with the clinically relevant domains described in Table 2. The annotation task was completed by three licensed clinicians. All paragraphs were removed from the surrounding EHR context to ensure annotators were not influenced by the additional contextual information. Our domain classification models consider each paragraph independently and thus we designed the annotation task to mirror the information available to the models.

The annotators were instructed to label each

Domain	Description	Example Paragraph	Example Keywords
Appearance	Physical appearance, gestures, and mannerisms	"A well-appearing, clean young woman appearing her stated age, pleasant and cooperative. Eye contact was good."	disheveled, clothing, groomed, wearing, clean
Thought Content	Suicidal/homicidal ideation, obsessions, phobias, delusions, hallucinations	"No SI[3], No HI[4], No hallucinations, Ideas of reference, Paranoid delusions"	obsession, delusion, grandiose, ideation, suicidal, paranoid
Interpersonal	Family situation, friendships, and other social relationships	"Pt. overall appears to be functioning very well despite this conflict with a romantic interest of hers."	boyfriend, relationship, peers, family, parents, social
Mood	Feelings and overall disposition	"Pt. indicates that his mood is becoming more 'depressed.'"	anxious, calm, depressed, labile, confused, cooperative
Occupation	School and/or employment	"Pt. followed through with decision to leave college at this point in time."	boss, employed, job, school, class, homework, work
Thought Process	Pace and coherence of thoughts. Includes linear, goal-directed, perseverative, tangential, and flight of ideas	"Disorganized (Difficult to communicate with patient.), Paucity of thought, Thought-blocking."	linear, tangential, prosody, blocking, goal-directed, perseverant
Substance	Drug and/or alcohol use	"Patient used marijuana once which he believes triggered the current episode."	cocaine, marijuana, ETOH[5], addiction, narcotic
Other	Any paragraph that does not fall into any of the other seven domains	"Maintain mood stabilization, prevent future episodes of mania, improve self-monitoring skills."	–

Table 2: Annotation scheme for the domain classification task.

paragraph with one or more of the seven risk factor domains. In instances where more than one domain was applicable, annotators assigned the domains in order of prevalence within the paragraph. An eighth label, 'Other', was included if a paragraph was ambiguous, uninterpretable, or about a domain not included in the seven risk factor domains (e.g. non-psychiatric medical concerns and lab results). The annotations were then reviewed by a team of two clinicians who adjudicated collaboratively to create a gold standard. The gold standard and the clinician-identified keywords and MWEs have received IRB approval for release to the community. They are available as supplementary data to this paper.

3.2 Inter-Annotator Agreement

Inter-annotator agreement (IAA) was assessed using a combination of Fleiss's Kappa (a variant of Scott's Pi that measures pairwise agreement for annotation tasks involving more than two annotators) (Fleiss, 1971) and Cohen's Multi-Kappa as proposed by Davies and Fleiss (1982). Table 3 shows IAA calculations for both overall agreement and agreement on the first (most important) domain only. Following adjudication, accuracy scores were calculated for each annotator by evaluating their annotations against the gold standard.

Overall agreement was generally good and aligned almost exactly with the IAA on the first domain only. Out of the 1,654 annotated paragraphs, 671 (41%) had total agreement across all three annotators. We defined total agreement for the task as a set-theoretic complete intersection of domains for a paragraph identified by all annotators. 98% of paragraphs in total agreement involved one domain. Only 35 paragraphs had total disagreement, which we defined as a set-theoretic null intersection between the three annotators. An analysis of the 35 paragraphs with total disagreement showed that nearly 30% included the term "blunted/restricted". In clinical terminology, these terms can be used to refer to appearance, affect, mood, or emotion. Because the paragraphs being annotated were extracted from larger clinical narratives and examined independently of any surrounding context, it was difficult for the annotators to determine the most appropriate domain. This lack of contextual information resulted in each annotator using a different 'default' label: Appearance, Mood, and Other. During adjudication, Other was decided as the most appropriate label unless the paragraph contained additional content that encompassed other domains, as it avoids making unnecessary assumptions.

---

[3] Suicidal ideation
[4] Homicidal ideation
[5] Ethyl alcohol and ethanol

Labels	Fleiss's Kappa	Cohen's Multi-Kappa	Mean Accuracy
Overall	0.575	0.571	0.746
First Domain Only	0.536	0.528	0.805

Table 3: Inter-annotator agreement

Network	MLP	RBF
Input Layer		
Nodes	100	100
Dropout	0.2	0.2
Activation	ReLU[6]	ReLU
Hidden Layer		
Nodes	100	350
Dropout	0.5	0.0
Activation	ReLU	RBF
Output Layer		
Nodes	7	7
Activation	Sigmoid	Linear
Optimizer	Adam[7]	Adam
Loss Function	Categorical Cross Entropy	Mean Squared Error
Training Epochs	30	50
Batch Size	128	128

Table 4: Architectures of our highest-performing MLP and RBF networks.

A Fleiss's Kappa of 0.575 lies on the boundary between 'Moderate' and 'Substantial' agreement as proposed by Landis and Koch (1977). This is a promising indication that our risk factor domains are adequately defined by our present guidelines and can be employed by clinicians involved in similar work at other institutions.

The fourth column in Table 3, Mean Accuracy, was calculated by averaging the three annotator accuracies as evaluated against the gold standard. This provides us with an informative baseline of human parity on the domain classification task.

4   Topic Extraction

Figure 1 illustrates the data pipeline for generating our training and testing corpora, and applying them to our classification models.

We use the TfidfVectorizer tool included in the scikit-learn machine learning toolkit (Pedregosa et al., 2011) to generate our TF-IDF vector space models, stemming tokens with the Porter Stemmer tool provided by the NLTK library (Bird et al., 2009), and calculating TF-IDF scores for unigrams, bigrams, and trigrams. Applying Singular Value Decomposition (SVD) to the TF-IDF matrix, we reduce the vector space to 100 dimensions, which Zhang et al. (2011) found to improve classifier performance.

Starting with the approach taken by McCoy et al. (2015), who used aggregate cosine similarity scores to compute domain similarity directly from their TF-IDF vector space model, we extend this method by training a suite of three-layer multilayer perceptron (MLP) and radial basis function (RBF) neural networks using a variety of parameters to compare performance. We employ the Keras deep learning library (Chollet et al., 2015) using a TensorFlow backend (Abadi et al.) for this task. The architectures of our highest performing MLP and RBF models are summarized in Table 4. Prototype vectors for the nodes in the hidden layer of our RBF model are selected via k-means clustering (MacQueen et al., 1967) on each domain paragraph megadocument individually. The RBF transfer function for each hidden layer node is assigned the same width, which is based off the maximum Euclidean distance between the centroids that were computed using k-means.

To prevent overfitting to the training data, we utilize a dropout rate (Srivastava et al., 2014) of 0.2 on the input layer of all models and 0.5 on the MLP hidden layer.

Since our classification problem is multiclass, multilabel, and open-world, we employ seven nodes with sigmoid activations in the output layer, one for each risk factor domain. This allows us to identify paragraphs that fall into more than one of the seven domains, as well as determine paragraphs that should be classified as Other. Unlike the traditionally used softmax activation function, which is ideal for single-label, closed-world classification tasks, sigmoid nodes output class like-

---

[6] Rectified Linear Units, $f(x) = max(0, x)$ (Nair and Hinton, 2010)

[7] Adaptive Moment Estimation (Kingma and Ba, 2014)

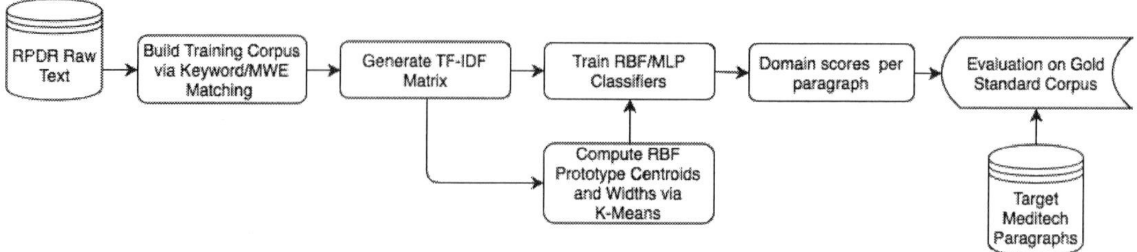

Figure 1: Data pipeline for training and evaluating our risk factor domain classifiers.

lihoods for each node independently without the normalization across all classes that occurs in softmax.

We find that the risk factor domains vary in the degree of homogeneity of language used, and as such certain domains produce higher similarity scores, on average, than others. To account for this, we calculate threshold similarity scores for each domain using the formula min=avg(sim)+α*σ(sim), where σ is standard deviation and α is a constant, which we set to 0.78 for our MLP model and 1.2 for our RBF model through trial-and-error. Employing a generalized formula as opposed to manually identifying threshold similarity scores for each domain has the advantage of flexibility in regards to the target data, which may vary in average similarity scores depending on its similarity to the training data. If a paragraph does not meet threshold on any domain, it is classified as Other.

5 Results and Discussion

Table 5 shows the performance of our models on classifying the paragraphs in our gold standard. To assess relative performance of feature representations, we also include performance metrics of our models without MWEs. Because this is a multilabel classification task we use macro-averaging to compute precision, recall, and F1 scores for each paragraph in the testing set. In identifying domains individually, our models achieved the highest per-domain scores on Substance (F1 $\approx$ 0.8) and the lowest scores on Interpersonal and Mood (F1 $\approx$ 0.5). We observe a consistency in per-domain performance rankings between our MLP and RBF models.

The wide variance in per-domain performance is due to a number of factors. Most notably, the training examples we extracted from RPDR – while very comparable to our target OnTrack[TM]

	Precision	Recall	F1
Aggregate Cosine Similarity Scores	0.602	0.563	0.574
MLP Baseline (No MWEs)	0.611	0.567	0.579
RBF Baseline (No MWEs)	0.603	0.618	0.606
MLP (w/ MWEs)	0.717	0.666	0.681
Appearance	0.886	0.414	0.564
Interpersonal	0.548	0.453	0.496
Mood	0.691	0.430	0.530
Occupation	0.826	0.461	0.592
Substance	0.920	0.703	0.797
Thought Content	0.926	0.590	0.721
Thought Process	0.654	0.617	0.635
Other	0.632	0.798	0.710
RBF (w/ MWEs)	0.684	0.630	0.645
Appearance	0.670	0.490	0.566
Interpersonal	0.410	0.493	0.448
Mood	0.655	0.399	0.496
Occupation	0.720	0.501	0.598
Substance	0.866	0.730	0.792
Thought Content	0.892	0.547	0.678
Thought Process	0.569	0.691	0.624
Other	0.651	0.650	0.651

Table 5: Overall and domain-specific Precision, Recall, and F1 scores for our models. The first row computes similarity directly from the TF-IDF matrix, as in (McCoy et al., 2015). All other rows are classifier outputs.

data – may not have an adequate variety of content and range of vocabulary. Although using keyword and MWE matching to create our training corpus has the advantage of being significantly less labor intensive than manually labeling every paragraph in the corpus, it is likely that the homogeneity of language used in the training para-

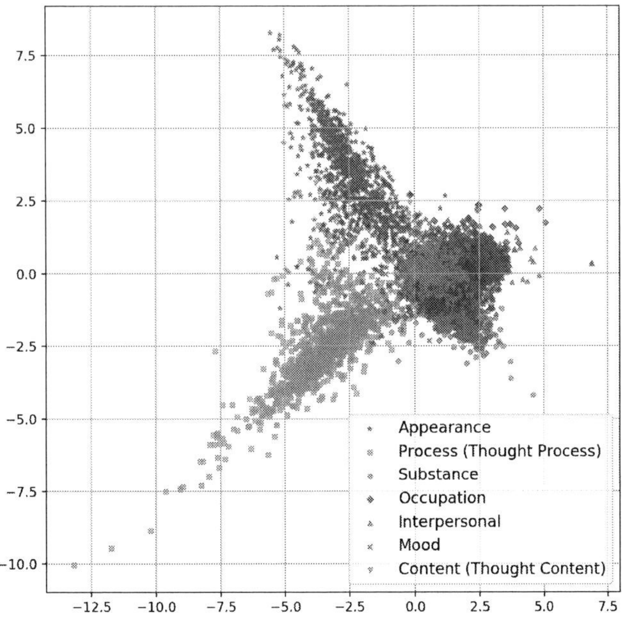

Figure 2: 2-component linear discriminant analysis of the RPDR training data.

graphs is higher than it would be otherwise. Additionally, all of the paragraphs in the training data are assigned exactly one risk factor domain even if they actually involve multiple risk factor domains, making the clustering behavior of the paragraphs more difficult to define. Figure 2 illustrates the distribution of paragraphs in vector space using 2-component Linear Discriminant Analysis (LDA) (Johnson and Wichern, 2004).

Despite prior research indicating that similar classification tasks to ours are more effectively performed by RBF networks (Scheirer et al., 2014; Jain et al., 2014; Bendale and Boult, 2015), we find that a MLP network performs marginally better with significantly less preprocessing (i.e. k-means and width calculations) involved. We can see in Figure 2 that Thought Process, Appearance, Substance, and – to a certain extent – Occupation clearly occupy specific regions, whereas Interpersonal, Mood, and Thought Content occupy the same noisy region where multiple domains overlap. Given that similarity is computed using Euclidean distance in an RBF network, it is difficult to accurately classify paragraphs that fall in

regions occupied by multiple risk factor domain clusters since prototype centroids from the risk factor domains will overlap and be less differentiable. This is confirmed by the results in Table 5, where the differences in performance between the RBF and MLP models are more pronounced in the three overlapping domains (0.496 vs 0.448 for Interpersonal, 0.530 vs 0.496 for Mood, and 0.721 vs 0.678 for Thought Content) compared to the non-overlapping domains (0.564 vs 0.566 for Appearance, 0.592 vs 0.598 for Occupation, 0.797 vs 0.792 for Substance, and 0.635 vs 0.624 for Thought Process). We also observe a similarity in the words and phrases with the highest TF-IDF scores across the overlapping domains: many of the Thought Content words and phrases with the highest TF-IDF scores involve interpersonal relations (e.g. 'fear surrounding daughter', 'father', 'family history', 'familial conflict') and there is a high degree of similarity between high-scoring words for Mood (e.g. 'meets anxiety criteria', 'cope with mania', 'ocd'[8]) and Thought Content (e.g. 'mania', 'feels anxious', 'feels exhausted').

---

[8]Obsessive-compulsive disorder

MWEs play a large role in correctly identifying risk factor domains. Factoring them into our models increased classification performance by 15%, a marked improvement over our baseline model. This aligns with our expectations that MWEs comprised of a quotidian vocabulary hold much more clinical significance than when the words in the expressions are treated independently.

Threshold similarity scores also play a large role in determining the precision and recall of our models: higher thresholds lead to a smaller number of false positives and a greater number of false negatives for each risk factor domain. Conversely, more paragraphs are incorrectly classified as Other when thresholds are set higher. Since our classifier will be used in future work as an early step in a data analysis pipeline for determining readmission risk, misclassifying a paragraph with an incorrect risk factor domain at this stage can lead to greater inaccuracies at later stages. Paragraphs misclassified as Other, however, will be discarded from the data pipeline. Therefore, we intentionally set a conservative threshold where only the most confidently labeled paragraphs are assigned membership in a particular domain.

6 Future Work and Conclusion

To achieve our goal of creating a framework for a readmission risk classifier, the present study performed necessary evaluation steps by updating and adding to our model iteratively. In the first stage of the project, we focused on collecting the data necessary for training and testing, and on the domain classification annotation task. At the same time, we began creating the tools necessary for automatically extracting domain relevance scores at the paragraph and document level from patient EHRs using several forms of vectorization and topic modeling. In future versions of our risk factor domain classification model we will explore increasing robustness through sequence modeling that considers more contextual information.

Our current feature set for training a machine learning classifier is relatively small, consisting of paragraph domain scores, bag-of-words, length of stay, and number of previous admissions, but we intend to factor in many additional features that extend beyond the scope of the present study. These include a deeper analysis of clinical narratives in EHRs: our next task will be to extend our EHR data pipeline by distinguishing between clinically positive and negative phenomena within each risk factor domain. This will involve a series of annotation tasks that will allow us to generate lexicon-based and corpus-based sentiment analysis tools. We can then use these clinical sentiment scores to generate a gradient of patient improvement or deterioration over time.

We will also take into account structured data that have been collected on the target cohort throughout the course of this study such as brain based electrophysiological (EEG) biomarkers, structural brain anatomy from MRI scans (gray matter volume, cortical thickness, cortical surface-area), social and role functioning assessments, personality assessment (NEO-FFI[9]), and various symptom scales (PANSS[10], MADRS[11], YMRS[12]). For each feature we consider adding, we will evaluate the performance of the classifier with and without the feature to determine its contribution as a predictor of readmission.

7 Acknowledgments

This work was supported by a grant from the National Institute of Mental Health (grant no. 5R01MH109687 to Mei-Hua Hall). We would also like to thank the LOUHI 2018 Workshop reviewers for their constructive and helpful comments.

References

Martín Abadi, Paul Barham, Jianmin Chen, Zhifeng Chen, Andy Davis, Jeffrey Dean, Matthieu Devin, Sanjay Ghemawat, Geoffrey Irving, Michael Isard, et al. Tensorflow: a system for large-scale machine learning.

Donald Emile Addington, Cindy Beck, JianLi Wang, Beverly Adams, Cathy Pryce, Haifeng Zhu, Jian Kang, and Emily McKenzie. 2010. Predictors of admission in first-episode psychosis: developing a risk adjustment model for service comparisons. *Psychiatric Services*, 61(5):483–488.

Mario Alvarez-Jimenez, A Priede, SE Hetrick, Sarah Bendall, Eoin Killackey, AG Parker, PD McGorry, and JF Gleeson. 2012. Risk factors for relapse following treatment for first episode psychosis: a systematic review and meta-analysis of longitudinal studies. *Schizophrenia Research*, 139(1-3):116–128.

---

[9]NEO Five-Factor Inventory (Costa and McCrae, 2010)

[10]Positive and Negative Syndrome Scale (Kay et al., 1987)

[11]Montgomery-Asperg Depression Rating Scale (Montgomery and Åsberg, 1979)

[12]Young Mania Rating Scale (Young et al., 1978)

Abhijit Bendale and Terrance Boult. 2015. Towards open world recognition. In *Proceedings of the IEEE Conference on Computer Vision and Pattern Recognition*, pages 1893–1902.

Steven Bird, Ewan Klein, and Edward Loper. 2009. *Natural language processing with Python: analyzing text with the natural language toolkit.* " O'Reilly Media, Inc.".

Mats Bogren, Cecilia Mattisson, Per-Erik Isberg, and Per Nettelbladt. 2009. How common are psychotic and bipolar disorders? a 50-year follow-up of the lundby population. *Nordic journal of psychiatry*, 63(4):336–346.

François Chollet et al. 2015. Keras. https:// keras.io.

PT Costa and Robert R McCrae. 2010. The neo personality inventory: 3. *Odessa, FL: Psychological assessment resources.*

Mark Davies and Joseph L Fleiss. 1982. Measuring agreement for multinomial data. *Biometrics*, pages 1047–1051.

Joseph L Fleiss. 1971. Measuring nominal scale agreement among many raters. *Psychological bulletin*, 76(5):378.

Philip D Harvey and Alan S Bellack. 2009. Toward a terminology for functional recovery in schizophrenia: is functional remission a viable concept? *Schizophrenia Bulletin*, 35(2):300–306.

Thomas Insel, Bruce Cuthbert, Marjorie Garvey, Robert Heinssen, Daniel S Pine, Kevin Quinn, Charles Sanislow, and Philip Wang. 2010. Research domain criteria (rdoc): toward a new classification framework for research on mental disorders.

Lalit P Jain, Walter J Scheirer, and Terrance E Boult. 2014. Multi-class open set recognition using probability of inclusion. In *European Conference on Computer Vision*, pages 393–409. Springer.

Richard A Johnson and Dean W Wichern. 2004. Multivariate analysis. *Encyclopedia of Statistical Sciences*, 8.

Stanley R Kay, Abraham Fiszbein, and Lewis A Opler. 1987. The positive and negative syndrome scale (panss) for schizophrenia. *Schizophrenia bulletin*, 13(2):261–276.

Ronald C Kessler, G Paul Amminger, Sergio Aguilar-Gaxiola, Jordi Alonso, Sing Lee, and T Bedirhan Ustun. 2007. Age of onset of mental disorders: a review of recent literature. *Current opinion in psychiatry*, 20(4):359.

Diederik P Kingma and Jimmy Ba. 2014. Adam: A method for stochastic optimization. *arXiv preprint arXiv:1412.6980.*

J Richard Landis and Gary G Koch. 1977. The measurement of observer agreement for categorical data. *biometrics*, pages 159–174.

Kim Lorine, Haig Goenjian, Soeun Kim, Alan M Steinberg, Kendall Schmidt, and Armen K Goenjian. 2015. Risk factors associated with psychiatric readmission. *The Journal of nervous and mental disease*, 203(6):425–430.

James MacQueen et al. 1967. Some methods for classification and analysis of multivariate observations.

Roshni Mangalore and Martin Knapp. 2007. Cost of schizophrenia in england. *The journal of mental health policy and economics*, 10(1):23–41.

Thomas H McCoy, Victor M Castro, Hannah R Rosenfield, Andrew Cagan, Isaac S Kohane, and Roy H Perlis. 2015. A clinical perspective on the relevance of research domain criteria in electronic health records. *American Journal of Psychiatry*, 172(4):316–320.

Stuart A Montgomery and MARIE Åsberg. 1979. A new depression scale designed to be sensitive to change. *The British journal of psychiatry*, 134(4):382–389.

Harvey J Murff, Fern FitzHenry, Michael E Matheny, Nancy Gentry, Kristen L Kotter, Kimberly Crimin, Robert S Dittus, Amy K Rosen, Peter L Elkin, Steven H Brown, et al. 2011. Automated identification of postoperative complications within an electronic medical record using natural language processing. *Jama*, 306(8):848–855.

Vinod Nair and Geoffrey E Hinton. 2010. Rectified linear units improve restricted boltzmann machines. In *Proceedings of the 27th international conference on machine learning (ICML-10)*, pages 807–814.

Mark Olfson, David Mechanic, Carol A Boyer, Stephen Hansell, James Walkup, and Peter J Weiden. 1999. Assessing clinical predictions of early rehospitalization in schizophrenia. *The Journal of nervous and mental disease*, 187(12):721–729.

Fabian Pedregosa, Gaël Varoquaux, Alexandre Gramfort, Vincent Michel, Bertrand Thirion, Olivier Grisel, Mathieu Blondel, Peter Prettenhofer, Ron Weiss, Vincent Dubourg, et al. 2011. Scikit-learn: Machine learning in python. *Journal of machine learning research*, 12(Oct):2825–2830.

Jonna Perälä, Jaana Suvisaari, Samuli I Saarni, Kimmo Kuoppasalmi, Erkki Isometsä, Sami Pirkola, Timo Partonen, Annamari Tuulio-Henriksson, Jukka Hintikka, Tuula Kieseppä, et al. 2007. Lifetime prevalence of psychotic and bipolar i disorders in a general population. *Archives of general psychiatry*, 64(1):19–28.

A Rumshisky, M Ghassemi, T Naumann, P Szolovits, VM Castro, TH McCoy, and RH Perlis. 2016. Predicting early psychiatric readmission with natural language processing of narrative discharge summaries. *Translational psychiatry*, 6(10):e921.

Walter J Scheirer, Lalit P Jain, and Terrance E Boult. 2014. Probability models for open set recognition. *IEEE transactions on pattern analysis and machine intelligence*, 36(11):2317–2324.

Jay P Singh and Seena Fazel. 2010. Forensic risk assessment: A metareview. *Criminal Justice and Behavior*, 37(9):965–988.

Nitish Srivastava, Geoffrey Hinton, Alex Krizhevsky, Ilya Sutskever, and Ruslan Salakhutdinov. 2014. Dropout: a simple way to prevent neural networks from overfitting. *The Journal of Machine Learning Research*, 15(1):1929–1958.

PH Thomsen. 1996. Schizophrenia with childhood and adolescent onseta nationwide register-based study. *Acta Psychiatrica Scandinavica*, 94(3):187–193.

Theo Vos, Ryan M Barber, Brad Bell, Amelia Bertozzi-Villa, Stan Biryukov, Ian Bolliger, Fiona Charlson, Adrian Davis, Louisa Degenhardt, Daniel Dicker, et al. 2015. Global, regional, and national incidence, prevalence, and years lived with disability for 301 acute and chronic diseases and injuries in 188 countries, 1990–2013: a systematic analysis for the global burden of disease study 2013. *The Lancet*, 386(9995):743–800.

Durk Wiersma, Fokko J Nienhuis, Cees J Slooff, and Robert Giel. 1998. Natural course of schizophrenic disorders: a 15-year followup of a dutch incidence cohort. *Schizophrenia bulletin*, 24(1):75–85.

Eric Q Wu, Howard G Birnbaum, Lizheng Shi, Daniel E Ball, Ronald C Kessler, Matthew Moulis, and Jyoti Aggarwal. 2005. The economic burden of schizophrenia in the united states in 2002. *Journal of Clinical Psychiatry*, 66(9):1122–1129.

RC Young, JT Biggs, VE Ziegler, and DA Meyer. 1978. A rating scale for mania: reliability, validity and sensitivity. *The British Journal of Psychiatry*, 133(5):429–435.

Wen Zhang, Taketoshi Yoshida, and Xijin Tang. 2011. A comparative study of tf* idf, lsi and multi-words for text classification. *Expert Systems with Applications*, 38(3):2758–2765.

Patient Risk Assessment and Warning Symptom Detection Using Deep Attention-Based Neural Networks

Ivan Girardi[1], Pengfei Ji[1,2], An-phi Nguyen[1], Nora Hollenstein[1,2], Adam Ivankay[1],
Lorenz Kuhn[1], Chiara Marchiori[1] and Ce Zhang[2]

[1] IBM Research Zurich, Switzerland
[2] ETH Zurich, Switzerland

`ivg@zurich.ibm.com, pji@student.ethz.ch, uye@zurich.ibm.com,`
`noraho@ethz.ch, aiv@zurich.ibm.com, kuhnl@student.ethz.ch,`
`chi@zurich.ibm.com, ce.zhang@ethz.ch`

Abstract

We present an operational component of a real-world patient triage system. Given a specific patient presentation, the system is able to assess the level of medical urgency and issue the most appropriate recommendation in terms of best *point of care* and *time to treat*. We use an attention-based convolutional neural network architecture trained on 600,000 doctor notes in German. We compare two approaches, one that uses the full text of the medical notes and one that uses only a selected list of medical entities extracted from the text. These approaches achieve 79% and 66% precision, respectively, but on a confidence threshold of 0.6, precision increases to 85% and 75%, respectively. In addition, a method to detect *warning symptoms* is implemented to render the classification task transparent from a medical perspective. The method is based on the learning of attention scores and a method of automatic validation using the same data.

1 Introduction

Several intelligent triage systems have recently been developed that attempt to evaluate automatically the risk related to specific patient conditions and direct patients to the appropriate care provider (Semigran et al., 2015). The work presented here is part of an interactive triage system being developed for industrial applications. The system takes patient demographics and symptoms as input, assesses their current medical conditions and suggests where and by when the patients should seek medical care. A key feature of the system is the detection of warning symptoms, namely, red flags. This is crucial to distinguish potential emergencies from common or less urgent cases and therefore provides the medical rationale behind a given recommendation. In addition, for triage systems that involve a dialogue with patients through multiple question-and-answer interactions (such as Ada

(2018)), warning symptom detection is fundamental to determine the most informative questions to ask patients.

We propose a model that assesses patient risk and detects warning symptoms based on a large volume of doctor notes in German, sometimes even mixed with Swiss German expressions. In this context, assessing patient risk can be regarded as a supervised text classification task, where the content of the medical records represents the feature space, and the recommendations assigned by medical professionals are the ground truth labels. The use of recurrent neural networks (RNN) has been proposed to solve text classification tasks (Tang et al., 2015). However, the proposed RNN models must be modified to be consistent with the requirement that warning symptoms must be detected, because in RNNs it is generally not possible to know which hidden states are most relevant.

To address these challenges, we propose an integrated approach to assess patient risk and detect warning symptoms simultaneously using an attention-based convolutional neural network (ACNN), which is a combination of a convolutional neural network (CNN) and an attention mechanism (Kim, 2014; Yang et al., 2016; Du et al., 2017). To the best of our knowledge, such an integrated approach is applied for the first time to the medical domain.

The main contributions of this paper are twofold. First, we propose a neural network architecture that can be used simultaneously for text classification and the detection of important words. Comparing our model to other neural architectures of similar complexity, we achieve competitive classification results. The model is especially useful to explain the recommendation rationale in classification scenarios, where the given input consists of a set of extracted entities, rather than full text. Second, a formal pipeline to detect

139

Proceedings of the 9th International Workshop on Health Text Mining and Information Analysis (LOUHI 2018), pages 139–148
Brussels, Belgium, October 31, 2018. ©2018 Association for Computational Linguistics

warning symptoms based on learned importance factors is applied in an industrial application. Our model identifies symptoms that indicate a medical emergency. These warning symptoms can then be used by intelligent medical care services or in an ontology.

2 Related Work

2.1 Text Classification with Deep Learning

Traditional text classification approaches represent documents with sparse lexical features, such as n-grams, and use a linear model or kernel methods on this representation (Wang and Manning, 2012; Joachims, 1998). More recently, deep learning technologies have been applied to text categorization problems. RNNs are designed to handle sequences of any length and capture long-term dependencies. Like sequence-based (Tang et al., 2015) and tree-structured (Tai et al., 2015) models, they have achieved remarkable results in document modeling.

Moreover, CNN models have achieved high accuracy on text categorization. For example, Kim (2014) used one convolutional layer (with multiple widths and filters) followed by a max pooling layer over time. Johnson and Zhang (2015) built a model that uses up to six convolutional layers, followed by three fully connected classification layers. Conneau et al. (2016) published a model with a 32-layer character-level CNN, that achieved a significant improvement on a large dataset. Models that combine CNN and RNN components for document classification also yield competitive results on several public datasets (Zhou et al., 2015; Lai et al., 2015).

To the best of our knowledge, not many research efforts have focused on augmenting CNNs for text classification with attention mechanisms. In fact, attention layers are more typically coupled with RNNs in order to better handle long-term dependencies (Yang et al., 2016). Interestingly, Du et al. (2017) used a CNN not as a classifier, but to compute the attention weights to apply to the hidden layers of a RNN. An example of combining attention layers with a CNN is the work by Shen and Huang (2016). However, the authors do not augment the CNN features using attention weights. They use an attention mechanism to compute sentence-level features, which they then concatenate to the convolutional features to ultimately perform the classification.

2.2 Intelligent Triage Systems

Intelligent triage systems inform patients where and when they should seek medical care, based on methods such as expert rules, Bayesian inference and deep learning (Semigran et al., 2015). For example, Symptomate (2018) uses a Bayesian network and a medical database for triage advice. Clinical records written by medical experts have also been used to make triage suggestions with deep learning technologies. Li et al. (2017) uses a shallow CNN model to predict a patient's diseases using the corresponding admission notes. Nigam (2016) applied a LSTM model to the multi-label classification task of assigning ICD-9 labels to medical notes.

3 Methodology

3.1 Data Processing

To build the triage application described here, we used 600,000 case records written in German and collected over the past five years. This is only 50% of the total available data, as we selected only those cases treated by top-ranked doctors. Case records contain demographic data such as age and gender, previous illnesses, and a full-text description of the patient's current medical condition. Potential diagnoses consistent with the symptom description are listed.

The descriptions in the records are expressed in formal medical language as well as in layman's terminology. The notes are not always written in complete sentences and include misspellings, dialect vocabulary, non-standard medical abbreviations and inconsistent punctuation. This is a challenge for the linguistic processing of case files.

The original case records are very unevenly distributed over ten recommendation classes (a combination of a point-of-care and a time-to-treat class). To mitigate this problem and for the purpose of this work, the original classes, (*emergency, urgent*), (*grundversorger, urgent*), (*specialist, urgent*), (*grundversorger, within a day*), (*specialist, within a day*), (*grundversorger, not urgent*), (*specialist, not urgent*), (*telecare, –*), were merged, with the help of healthcare professionals, into three categories: *Urgent Care, General Practice, Telecare*. The categorization of cases is shown in Table 1.

Recommendations	Number of Cases
Urgent Care	270,000
General Practice	104,000
Telecare	244,000

Table 1: Ground truth distribution for the reduced classes. *Urgent Care* = Patient needs to seek medical care within a short time period; *General Practice* = Patient requires medical attention in a physical consultation, but not urgently; *Telecare* = In-person medical appointment not required, instructions over the phone are sufficient.

3.1.1 NLP Pipeline

A natural language processing (NLP) pipeline extracted medically relevant concepts associated with each written case. The pipeline consisted of the following stages: (1) data preprocessing for misspelling correction and abbreviation expansion, (2) named entity recognition (NER) and (3) concept clustering. Acronyms and abbreviations used unambiguously were linked to the corresponding entities directly in the dictionaries. Ambiguous acronyms and abbreviations were resolved, when possible, using algorithms that include context for disambiguation. For NER, we used a rule-based medical entity extraction system built with IBM Watson Explorer, using algorithms based on dictionary look-up and advanced rules. This allowed us to detect 51 entity types in the following categories: *anatomy, physiology, symptoms, diseases, medical procedures, medicines, negated symptoms, negated diseases, ability/inability of, foreign-body objects, negations, patient information, symptom characterization, disease characterization, time expressions*. The distinction between symptoms and diagnosis was made using existing ontologies, where these semantic types were assigned with the help of a team of clinical experts. The dictionaries used in the NER were built partially based on existing German-language medical dictionaries and ontologies (UMLS mapped German terms, ICD10, Meddra, etc.) and partially using the list of words contained in the case records. The dictionaries therefore contain a mapping of technical and layman's terms. The NLP pipeline was designed to detect and resolve the negated mentions of the entities listed above (using German language-specific negation particles or expressions), which are very frequent in this type of records. Only 31 entity types in the categories

symptoms, diseases, ability/inability of, negated symptoms, negated diseases were included in the current final list. The average number of extracted annotations per case was 70 for all entities, but only 17 for the selected entities. Performance was evaluated using the manual annotations of a set of ground truth cases performed by a team of clinical experts. Concept clustering is a hierarchical procedure that allowed us to group annotations describing the same medical concept. The same entity may be expressed in a variety of forms (compound vs. simple nouns, dialect or common language vs. medical terminology). Concept clustering is performed either at the dictionary level or by algorithms based on similarity between lemmas associated with the annotations.

Table 2 lists the concepts extracted from an original case record after preprocessing by the described NLP pipeline.

key	value
Gender	*Male*
Class	*Urgent Care*
Content	*"Seit heute beschwert sich der Patient über heftige Brustschmerzen; Fieber 37,4°C; Schwierigkeiten beim Atmen, leichte Kopfschmerzen."*
Entities	*starke Brustschmerzen Fieber 36-38°C Atembeschwerde leichte Kopfschmerzen*

Table 2: (key,value)-pairs of an original patient case file and extracted entities.

In this paper, we will benchmark the classification approach of using the extracted concepts with respect to the one of using the full text.

3.2 Model Architecture

The overall architecture of the attention-based CNN is shown in Fig. 1. It consists of several components: a word embedding look-up layer obtained using *word2vec* (Mikolov et al., 2013), a CNN-based n-gram encoder, an n-gram level attention layer and several fully-connected layers. By means of word embeddings, each word is represented as a real-valued vector. The word embedding look-up layer is a word embedding table $T \in \mathcal{R}^{n \times k}$, where n is the total vocabulary size and k is the embedding dimension. The parame-

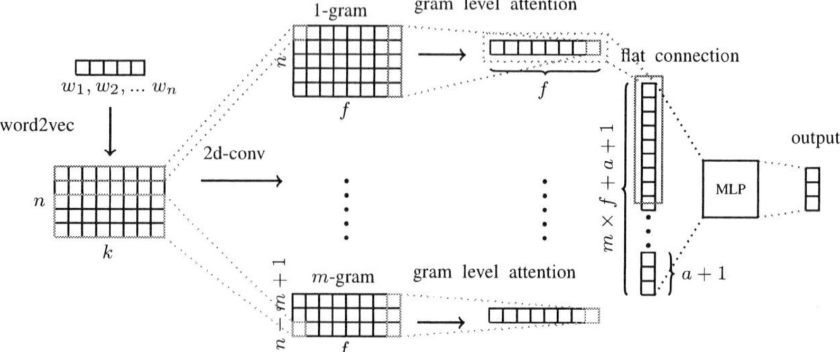

Figure 1: Model Architecture.

ters of the embedding table were fine-tuned during the training phase.

3.2.1 N-Gram Encoder

We used a 2D convolution layer (Kim, 2014) to encode the word sequence into n-gram representations, thus capturing contextual information. For a given document, a 2D convolution filter $w \in \mathcal{R}^{m \times k}$ was applied to a window of m words to produce a new feature. A feature c_i was generated from a window of words $x_{i:i+m-1}$ by

$$c_i = Relu(w \cdot x_{i:i+m-1} + b). \tag{1}$$

This filter was applied to each possible window of words in the sentence $x_{1:m}, x_{2:m+1}, \ldots x_{n-m+1:n}$ to produce a feature map:

$$c = [c_1, c_2, \ldots, c_{n-m+1}], \tag{2}$$

with $c \in \mathcal{R}^{n-m+1}$. By applying multiple filters (denoted f) on $x_{i:i+m-1}$, we obtained a new representation of the document. By setting different values for m, we obtained different n-gram representations of the documents. This operation was useful in our application setting because these layers create local region embeddings by n-grams. Moreover, this allowed us to compute the attention factors for a combination of several symptoms. This in turn enabled us to detect pairs and even triplets of symptoms that are harmless if they appear individually, yet become red flags when they appear together. For example, the individual symptoms *pain in arm* and *sudden nausea* are no cause for concern. However, if a patient experiences both, this might indicate an impending heart attack.

3.2.2 N-Gram Level Attention Layer

For each n-gram representation, we wanted to derive a corresponding fully-connected represen-

tation for the document. As different n-grams are of different importance to the document, we introduced an attention mechanism to extract n-grams that are relevant to the meaning of the document and aggregated the representation of those informative n-grams to form a document vector. The relevant n-grams then became candidates for warning symptoms. More specifically, the attention mechanism was defined such that:

$$u_{it} = \tanh(W_w v_{it} + b_w), \tag{3}$$

where v_{it} refers to the tth row of ith-gram representation. That is, we first fed the n-gram annotations v_{it} through a one-layer neural network to obtain u_{it} as a hidden representation of v_{it}. Then we measured the importance of the word as the similarity of u_{it} with a word-level context vector u_w and obtained a normalized importance weight α_{it} through a softmax function:

$$\alpha_{it} = \frac{exp(u_{it}^T u_w)}{\sum_t exp(u_{it}^T u_w)}. \tag{4}$$

The context vector u_w can be regarded as a high-level representation of a fixed query "what is the most informative word?" used in memory networks (Sukhbaatar et al., 2015; Kumar et al., 2016). Context vector u_w was randomly initialized and jointly learned during the training process. Thereafter, we computed the document vector s_i as a weighted sum of the n-gram annotations based on the weights:

$$s_i = \sum_t \alpha_{it} v_{it}. \tag{5}$$

Finally, all n-gram document level representations were flattened into a one-dimensional vector (flat connection layer in Fig. 1) plus patient gender and age ($a + 1$ in Fig. 1). This vector was then fed into a multilayer perceptron (MLP) for classification.

3.3 Warning Symptom Detection

Warning symptoms, or red flags, indicate the need for urgent medical care. The ACNN model is able to distinguish the importance of each symptom in the final classification. Thereafter, we calculated the attention score for each symptom as follows:

$$score(s_j) = \frac{\sum_{c_i \in C} \Phi(c_i, s_j) f(c_i, s_j)}{occur(s_j)}, \quad (6)$$

$$f(c_i, s_j) = \frac{att(s_j)}{\max_{s_k \in c_i} att(s_k)}, \quad (7)$$

where $\Phi(c_i, s_j)$ is equal to 1 if symptom s_j is contained in case record c_i and zero elsewhere; C is the set of urgent care cases in the data; $occur(s_i)$ is the total occurrences of symptom s_i; $att(s_k)$ are the attention weights returned by the ACNN. The attention weights gave us a measurement of the warning level of the symptoms.

This procedure was applied for all classes to detect the most important symptoms that drive the model's prediction. As expected for the other classes, the model assigns high attention weights to non-warning symptoms.

4 Results

4.1 Patient Risk Assessment Experiment

4.1.1 Training Details

We conducted a detailed evaluation of this model on both the original *full-text* dataset and a dataset of a few selected medical entities (see Section 3.1.1 for details) denoted for simplicity as a *symptoms* dataset. The machine learning framework where all the neural network models have been implemented was based on TensorFlow and Keras. The vocabulary size, average document size and maximum document length are 134,000, 62.9 and 959 words for the full-text dataset; and 20,000, 14.15 and 94 for the *symptoms* dataset. We used 90% of the data for training, 5% for validation, and 5% for test randomly sampled. Both datasets were preprocessed by removing stop words and low-occurrence words and zero-padding the documents. We learned 200-dimensional word embeddings on our datasets with *word2vec* over 25 iterations. The embeddings were different for each dataset.

We tuned our parameters on a 30,000 validation set and report the result on another 30,000 test set. For model-specific parameters, we used grid search to find the optimal values. We used a cross-entropy loss function with 256-mini-batch updating and Adam optimizer for five epochs. The learning rate was between 0.001 and 0.003; regularization was performed by weight decay of 0.0001 and a dropout of 0.8 was applied to every MLP layer. The attention vector size was set up to 100, and the window size was set from 1 to 5. For each n-gram extraction, we used up to 128 filters for 2D convolution.

4.1.2 Model Comparison

In this section, we compare our system to the following approaches:

CLSTM (Zhou et al., 2015) applies a CNN model on text and feeds consecutive window features directly to a LSTM model.

Kim CNN (Kim, 2014) uses 2D convolution windows to extract an n-gram representation followed by max-pooling.

BiGRU Attention Network (Yang et al., 2016) consists of RNNs applied on both word and sentence level to extract a hidden state. An attention mechanism is applied after the bidirectional gated recurrent units.

The results on the datasets with the *full text* and the *symptoms* only are shown in Tables 3 and 4, respectively. All the analyzed models show similar performance in the classification task. For all models, the performance decreases as we move from the full text dataset to the symptoms dataset because the medical and contextual information also diminishes by taking into account only the extracted symptom concepts.

4.1.3 Result Analysis

In this section, we compare our ACNN model with the state-of-the-art deep learning models to obtain a benchmark on our triage use case. We also describe how our approach, a combination of convolutional neural networks and attention mechanisms, equals the performance of existing models with the advantage of being explainable.

Kim CNN uses 2D convolution windows to extract n-gram representations. Max pooling was then applied to each of the filter outputs. A single value was retained for each feature map. This might work well for short sentences containing only a few "leading" words indicating the cate-

Model	$P(f_1)$	$R(f_1)$	$F(f_1)$	$P(f_2)$	$R(f_2)$	$F(f_2)$	$P(f_3)$	$R(f_3)$	$F(f_3)$
KIM CNN	82.3	80.5	81.9	69.2	65.4	68.4	82.2	86.1	83.0
CLSTM	78.2	82.6	79.0	66.9	62.4	66.0	83.4	80.7	82.8
BiGRU Attention Net	74.9	80.2	75.9	62.6	59.0	61.9	80.8	76.6	79.9
ACNN	80.5	81.1	80.7	67.6	60.9	66.1	82.0	84.8	82.6

Table 3: Prediction results in % for the different architectures on *full text*, where $P(f_k)$, $R(f_k)$, $F(f_k)$ are precision, recall and f-score divided by class, and where f_1, f_2, f_3 are urgent care, general practice and telecare, respectively. Similar values were obtained by conducting several experiments and averaging the results.

Model	$P(s_1)$	$R(s_1)$	$F(s_1)$	$P(s_2)$	$R(s_2)$	$F(s_2)$	$P(s_3)$	$R(s_3)$	$F(s_3)$
KIM CNN	70.5	73.6	71.1	55.2	40.6	51.5	66.5	70.6	67.3
CLSTM	70.0	71.6	70.3	53.8	40.1	50.4	65.4	70.8	66.4
BiGRU Attention Net	69.2	72.5	69.9	53.0	43.1	50.7	66.7	68.4	67.0
ACNN	72.5	68.2	71.6	51.9	47.8	51.0	65.5	72.0	66.5

Table 4: Same as Table 3 but on *symptoms* dataset, where s_1, s_2, s_3 are urgent care, general practice and telecare cases, respectively.

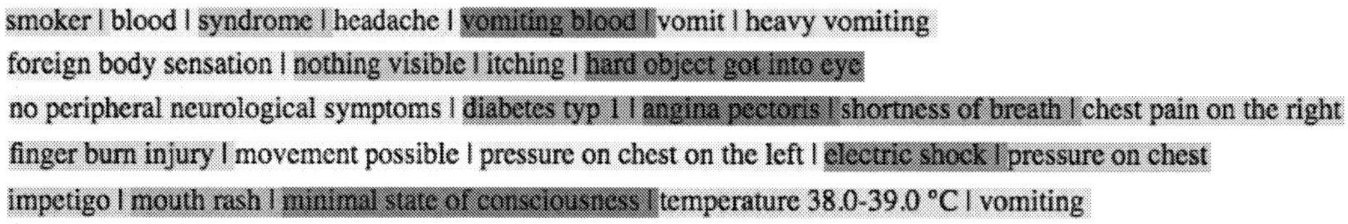

Figure 2: Visualization of attention factors from neural network used to explain recommendation rationale. Each line represents the (translated) symptoms extracted for a patient case file. The darker the color, the higher the attention factor for a symptom.

gory. For longer documents, however, all information about n-grams is lost apart from the strongest signal. The presence of highly important symptoms in clinical data is the reason why this model performs well especially for urgent care and telecare classes. This hypothesis is supported by the number of symptoms with large attention scores found in the ACNN model for these classes.

The BiGRU Attention Network applies an attention layer after bidirectional GRU components. For a given word in a sentence, it encodes information about the word context in that sentence. However, compared to a 2D convolution window, only a single context window is used. It is not trivial to choose the optimal window size. Thus, it is difficult to detect warning symptom pairs or triplets. For 2D convolution in our model, identifying such pairs or triplets would be more straightforward because attention factors are also learned for 2 and 3-grams. Another limitation of GRU models is that they rely on fully sequential data. In our use case, however, the data is composed of several separate phrases, words or incomplete sentences.

Our ACNN combines the merits of 2D convolution and attention mechanisms by stacking 2D convolution layers to extract contextual information and an attention mechanism to assign importance factors to different symptoms and combinations thereof.

4.2 Warning Symptom Detection

Owing to the lack of ground truth, we used the following evaluation method to detect warning symptoms with the ACNN. First, we measured the recall of the ACNN on urgent care cases containing only symptom concepts. Then, a new dataset was created by removing from each case record the 1-gram with the highest attention score, calculated as described in Section 3.3. For urgent cases, we expected the removed 1-grams to be highly important signals of medical urgency, hence warning symptoms. For instance, *starke Brustschmerzen* would be removed from the case described in Section 3.1.1. We then compared the ACNN recall for the urgent cases on the new dataset (Attention Drop) with respect to the recall on the original symptoms dataset (Baseline). This procedure is performed on all the classes for validation. The decrease in recall demonstrates the importance of the detected warning symptoms in order to classify urgent cases correctly. To verify that the detected warning symptoms are indeed highly informative, we furthermore generated datasets in which either random symptoms (Random Drop) or symptoms that appear most frequently in urgent cases (Frequency Drop) are dropped.

As shown in Table 5, dropping the attention-detected warning symptoms led to the largest decrease in performance. The difference became even more distinct if two symptoms instead of one were removed from the cases.

Performance also decreased for the urgent care and general practice classes, whereas almost a flat behavior was found for telecare class, as expected. In the latter case, random, frequency, attention drops showed the same results because several features had the same attention scores. Manual inspection of the symptoms with the highest attention scores further supports these results. The darker the color of the symptom in Figure 2, the higher its attention factor in the model. In the examined samples, darker colors did indeed correlate with symptoms that made patients require urgent care, such as *vomiting blood* and *electric shock*.

With single or double removals for the *full-text* dataset, a much lower decrease in performance was observed because of the higher number of features per case.

4.3   Explainable Deep Learning

In current research, but especially in medical industry applications, *transparent* or *explainable* machine learning models are becoming increasingly important. Some machine learning models have become so complex, they are black boxes. End users need to understand why a certain recommendation was made.

In our application, the attention mechanism on which we based our warning (and non-warning) symptom detection represents a transparent method of reasoning why a given case belongs to a certain class.

For instance, by analyzing the patient symptoms with the highest attention scores, it becomes apparent *why* a case would be predicted to be urgent,

general practice or telecare. Table 6 shows some examples with high/low attention scores computed using 1-gram attention values for urgent care, general practice and telecare classes. As can be seen, the symptoms with the highest score in the urgent cases are the most severe, whereas the symptoms in the telecare cases are less severe. In other words, symptoms with a high/low score for a given class are the most/least relevant ones for that class. As expected, if the model predicts an urgent (non-urgent) class, the model assigns a higher weight to warning (non-warning) symptoms. The computation of 1-gram feature scores results in 2,000 (3,600), 734 (3,700), 1,500 (3,800) features with scores of > 0.8 (< 0.2) for s_1, s_2 and s_3, respectively. The use of an attention layer on n-gram representations allowed us to compute feature relevance including correlations between pairs, triplets, etc. An example of scores of feature pairs obtained by extracting the attention weights for the 2-grams is shown in Tables 7 and 8. Strong correlation between feature pairs is found for the cases where the score of the pair is much higher than those of the single features. The computation of 2-gram feature scores results in 12,000 (28,000), 4,800 (13,000), 10,000 (24,000) features with scores of > 0.8 (< 0.2) for s_1, s_2 and s_3, respectively.

4.4   Confidence

To reach higher performance in an operative triage application, we define a confidence score in the classification based on which the system decides whether to trust the recommendation. In Table 9 and Table 10 we show the same results obtained in Tables 3 and 4, respectively, discarding all test cases in which the predicted probability of the classifier was lower than 0.6. With the chosen threshold, we discarded roughly 30% cases. Overall a performance improvement of between 5% and 10% is observed. In future work, we plan to apply additional techniques, e.g., based on hierarchical decision trees, to minimize medical risk even further.

5   Conclusion

We have described an attention-based CNN model to assess patient risk and to detect warning symptoms, which will be used in an industrial application for medical triage. We achieved a precision of 79% on the *full-text* dataset and

Dataset	P(s_1)	R(s_1)	F(s_1)	P(s_2)	R(s_2)	F(s_2)	P(s_3)	R(s_3)	F(s_3)
Baseline	72.5	68.2	71.6	51.9	47.8	51.0	65.5	72.0	66.7
Random Drop	71.7	65.7	70.4	49.9	44.2	48.6	65.8	72.8	67.1
Frequency Drop	71.7	65.7	70.4	50.9	46.0	49.8	66.0	73.5	67.4
Attention Drop	70.3	61.3	68.2	48.0	40.9	46.4	66.3	74.6	67.8
2 Random Drops	70.8	62.8	69.0	47.6	40.2	45.9	66.0	73.3	67.3
2 Frequency Drops	70.3	61.2	68.3	49.0	42.6	47.6	66.6	75.6	68.2
2 Attention Drops	67.5	53.6	64.2	44.0	34.8	41.8	67.0	77.0	68.8

Table 5: Different datasets and the model's precision, recall and f-score in %, where s_1, s_2, s_3 are urgent care, general practice and telecare *symptoms* dataset, respectively.

s_1	score	s_2	score	s_3	score
shortness of breath *(Atemnot)*	1.0	intermittent shoulder pain *(intermittierende Schulterschmerzen)*	1.0	back distortion *(Rückenzerrung)*	1.0
pain after accident *(Schmerzen nach Unfall)*	1.0	severely itchy wound *(stark juckende Wunde)*	1.0	abrasion on the back *(Schürfung am Rücken)*	1.0
foreign body in esophagus *(Fremdkörper im Ösophagus)*	1.0	purulent nasal discharge *(eitriger Nasenausfluss)*	1.0	toenail injury *(Zehennagelverletzung)*	1.0
severe rectal bleed *(blutet stark rektal)*	1.0	neck abscess *(Abszess am Nacken)*	1.0	itching forehead *(Juckreiz an der Stirn)*	1.0
itching back *(Juckreiz am Rücken)*	0.05	no pain when walking *(beim Laufen keine Schmerzen)*	0.03	stabbed with knife *(Messerstich)*	0.03
pain in thumb *(Daumenschmerz)*	0.04	throat is normal *(Hals normal)*	0.01	ear bleeding *(Ohrenblutung)*	0.02
nail injury *(Nagelverletzung)*	0.03	blister on tongue *(Blase auf der Zunge)*	0.01	hardened lower abdomen *(verhärteter Unterbauch)*	0.01
wart on foot *(Warze am Fuss)*	0.01	can drink normally *(kann normal trinken)*	0.003	difficulty breathing *(Schwierigkeiten beim Atmen)*	0.005

Table 6: Symptoms (translated from German into English) scores divided by class using 1-gram attention values (only a few examples with high/low scores shown here). The corresponding German terms are given in parentheses.

(f_i, f_j)	score of f_i	score of f_j	score of (f_i, f_j)
(acute abdominal pain, severe abdominal pain) *(akute Bauchschmerzen, starke Bauchschmerzen)*	0.86	0.39	1.0
(loss of consciousness, head injury) *(Bewusstseinsverlust, Schädelverletzungen)*	0.32	0.55	1.0
(epigastric pain, colic) *(Oberbauchschmerzen, Kolik)*	0.35	0.26	1.0
(pneumonia, respiratory tract inflammation) *(Pneumonie, Atemwegentzündung)*	0.45	0.24	0.92
(severe vomiting, dehydration) *(starkes Erbrechen, Dehydration)*	0.44	0.56	0.87
(very high blood pressure, hypertensive crisis) *(Blutdruck stark erhöht, hypertensive Krise)*	0.49	0.78	0.86

Table 7: Symptoms (translated from German into English) scores using 2-gram attention values (only a few high scores shown here) for s_1. The corresponding German terms are given in parentheses.

66% on the *symptoms* set. On a confidence threshold of 0.6, precision increases to 85% and 75%, respectively. The learned attention weights allowed us to compute the symptom relevance, i.e., the attention score, which is then used to extract warning symptoms more precisely and to make the recommendation rationale transparent.

(f_i, f_j)	score of f_i	score of f_j	score of (f_i, f_j)
(chronic back pain, back pain)	0.29	0.08	1.0
(*chronische Rückenschmerzen, Rückenschmerzen*)			
(patella pain, knee pain)	0.26	0.21	0.79
(*Schmerzen an der Kniescheibe, Knieschmerzen*)			
(chronic anemia, food allergy)	0.18	0.17	0.66
(*chronische Anämie, Nahrungsmittelallergie*)			
(rheumatoid arthritis, joint pain)	0.13	0.13	0.59
(*rheumatoide Arthritis, Gelenkschmerzen*)			
(colonoscopy, blood in stool)	0.22	0.22	0.59
(*Darmspiegelung, Blut im Stuhlgang*)			
(fatigue, chronic anemia)	0.07	0.18	0.58
(*Müdigkeit, chronische Anämie*)			
(non-swollen lymph nodes, viral infection)	0.20	0.31	1.0
(*keine Lymphknotenschwellung, virale Entzündung*)			
(conjunctivitis, slight redness)	0.32	0.17	1.0
(*Konjunktivitis, leichte Rötung*)			
(abnormally frequent urination, no complication)	0.26	0.63	1.0
(*hufiges Urinieren, keine Komplikation*)			
(bladder infection, no pregnancy)	0.29	0.21	0.96
(*Harnblasenentzündung, keine Schwangerschaft*)			
(local reaction, itchiness)	0.31	0.17	0.92
(*Lokalreaktion, Juckreiz*)			
(gastroenteritis, no travel abroad)	0.42	0.23	0.90
(*Gastroenteritis, kein Auslandaufenthalt*)			

Table 8: Symptoms (translated from German into English) scores divided by class using 2-gram attention values (only a few high scores shown here) for s_2 (upper), s_3 (lower) panel. The corresponding German terms are given in parentheses.

Model	P(f_1)	R(f_1)	F(f_1)	P(f_2)	R(f_2)	F(f_2)	P(f_3)	R(f_3)	F(f_3)
KIM CNN	87.8	86.3	87.5	73.0	78.6	74.0	90.6	90.0	90.4
CLSTM	84.4	88.4	85.2	76.2	66.5	74.0	88.4	87.6	88.3
BiGRU Attention Net	76.5	82.2	77.6	65.1	60.7	64.2	82.5	78.2	81.6
ACNN	85.3	87.3	85.6	77.3	65.1	74.5	87.1	89.5	87.6

Table 9: Same as Table 3 applying a threshold to the probabilities of 0.6.

Model	P(s_1)	R(s_1)	F(s_1)	P(s_2)	R(s_2)	F(s_2)	P(s_3)	R(s_3)	F(s_3)
KIM CNN	75.6	86.0	77.5	65.0	45.5	59.8	77.5	72.0	76.3
CLSTM	76.5	83.2	77.8	66.8	44.5	60.7	75.2	75.2	75.2
BiGRU Attention Net	73.2	76.8	73.4	58.4	45.1	55.1	70.5	72.5	70.9
ACNN	77.0	81.5	77.9	62.8	60.0	60.3	75.7	74.9	75.5

Table 10: Same as Table 4 applying a threshold to the probabilities of 0.6.

Acknowledgements. We deeply acknowledge D. Dykeman, D. Ortiz-Yepes and K. Thandiackal.

References

Ada. 2018. https://ada.com.

Alexis Conneau, Holger Schwenk, Loïc Barrault, and Yann Lecun. 2016. Very deep convolutional networks for natural language processing. *arXiv preprint arXiv:1606.01781*.

Jiachen Du, Lin Gui, Ruifeng Xu, and Yulan He. 2017. A convolutional attention model for text classification. In *National CCF Conference on Natural Language Processing and Chinese Computing*, pages 183–195. Springer.

Thorsten Joachims. 1998. Text categorization with support vector machines: Learning with many relevant features. *Machine learning: ECML-98*, pages 137–142.

Rie Johnson and Tong Zhang. 2015. Semi-supervised convolutional neural networks for text categorization via region embedding. In *Advances in neural information processing systems*, pages 919–927.

Yoon Kim. 2014. Convolutional neural networks for sentence classification. *arXiv preprint arXiv:1408.5882*.

Ankit Kumar, Ozan Irsoy, Peter Ondruska, Mohit Iyyer, James Bradbury, Ishaan Gulrajani, Victor Zhong, Romain Paulus, and Richard Socher. 2016. Ask me anything: Dynamic memory networks for natural language processing. In *International Conference on Machine Learning*, pages 1378–1387.

Siwei Lai, Liheng Xu, Kang Liu, and Jun Zhao. 2015. Recurrent convolutional neural networks for text classification. In *AAAI*, volume 333, pages 2267–2273.

Christy Li, Dimitris Konomis, Graham Neubig, Pengtao Xie, Carol Cheng, and Eric Xing. 2017. Convolutional neural networks for medical diagnosis from admission notes. *arXiv preprint arXiv:1712.02768*.

Tomas Mikolov, Ilya Sutskever, Kai Chen, Greg Corrado, and Jeffrey Dean. 2013. Distributed representations of words and phrases and their compositionality. In *Advances in neural information processing systems*, pages 3111–3119.

Priyanka Nigam. 2016. Applying deep learning to ICD-9 Multi-label Classification from Medical Records. Technical report, Stanford University.

Hannah L Semigran, Jeffrey A Linder, Courtney Gidengil, and Ateev Mehrotra. 2015. Evaluation of symptom checkers for self diagnosis and triage: audit study. *bmj*, 351:h3480.

Yatian Shen and Xuanjing Huang. 2016. Attention-based convolutional neural network for semantic relation extraction. In *COLING 2016, 26th International Conference on Computational Linguistics, Proceedings of the Conference: Technical Papers, December 11-16, 2016, Osaka, Japan*, pages 2526–2536.

Sainbayar Sukhbaatar, Jason Weston, Rob Fergus, et al. 2015. End-to-end memory networks. In *Advances in neural information processing systems*, pages 2440–2448.

Symptomate. 2018. https://symptomate.com.

Kai Sheng Tai, Richard Socher, and Christopher D Manning. 2015. Improved semantic representations from tree-structured long short-term memory networks. *arXiv preprint arXiv:1503.00075*.

Duyu Tang, Bing Qin, and Ting Liu. 2015. Document modeling with gated recurrent neural network for sentiment classification. In *EMNLP*, pages 1422–1432.

Sida Wang and Christopher D Manning. 2012. Baselines and bigrams: Simple, good sentiment and topic classification. In *Proceedings of the 50th Annual Meeting of the Association for Computational Linguistics: Short Papers-Volume 2*, pages 90–94. Association for Computational Linguistics.

Zichao Yang, Diyi Yang, Chris Dyer, Xiaodong He, Alexander J Smola, and Eduard H Hovy. 2016. Hierarchical attention networks for document classification. In *HLT-NAACL*, pages 1480–1489.

Chunting Zhou, Chonglin Sun, Zhiyuan Liu, and Francis Lau. 2015. A C-LSTM neural network for text classification. *arXiv preprint arXiv:1511.08630*.

Syntax-based Transfer Learning for the Task
of Biomedical Relation Extraction

Joël Legrand[1], Yannick Toussaint[1], Chedy Raïssi[1] and Adrien Coulet[1,2]

[1] Université de Lorraine, CNRS, Inria, LORIA, 54000 Nancy, France
[2] Stanford University, Stanford Center for Biomedical Informatics Research, California
joel.legrand@loria.fr

Abstract

Transfer learning (TL) proposes to enhance machine learning performance on a problem, by reusing labeled data originally designed for a related problem. In particular, domain adaptation consists, for a specific task, in reusing training data developed for the same task but a distinct domain. This is particularly relevant to the applications of deep learning in Natural Language Processing, because those usually require large annotated corpora that may not exist for the targeted domain, but exist for side domains. In this paper, we experiment with TL for the task of Relation Extraction (RE) from biomedical texts, using the TreeLSTM model. We empirically show the impact of TreeLSTM alone and with domain adaptation by obtaining better performances than the state of the art on two biomedical RE tasks and equal performances for two others, for which few annotated data are available. Furthermore, we propose an analysis of the role that syntactic features may play in TL for RE.

1 Introduction

A bottleneck problem for training deep learning-based architecture on text is the availability of large enough annotated training corpora. This is especially an issue in highly specialized domains such as those of biomedicine. TL approaches address this problem by leveraging existing labeled data originally designed for related tasks or domains (Weiss et al., 2016). However, adaptation between dissimilar domains may lead to negative transfer, *i.e.* transfer that decreases the performance for the target domain. In this article, we apply a TL strategy using the TreeLSTM model for the task of biomedical Relation Extraction (RE). We propose an analysis of the syntactic features of source and target domain corpora to provide elements of interpretation for the improvements we obtained.

Figure 1: Example of relationship typed as *Weak Confidence Association* between two named entities: a *SNP (single nucleotide polymorphism)* and a *Phenotype*, from the SNPPhenA corpus.

Relation Extraction (RE) aims at identifying in raw and unstructured text all the instances of a pre-defined set of relations between identified entities. A relationship takes the form of an edge between two or more named entities as illustrated in Figure 1. We are considering here binary RE that can be seen as a classification task by computing a score for each possible relation type, given a sentence and two identified entities.

Deep learning methods have demonstrated good ability for RE (Zeng et al., 2014), but one of their drawbacks is that, in order to obtain reasonable performances, they generally require a large amount of training data, *i.e.*, text corpora where entities and relationships between them are annotated. The assembly of this kind of domain- and task-specific corpora, such as those of interest in biomedicine, is time consuming and expensive because it involves complex entities (*e.g.*, genomic variations, complex phenotypes), complex relationships (which may be hypothetical, contextualized, negated, *n*-ary) and requires trained annotators. This explains why only few and relatively small (*i.e.*, few hundreds of sentences) corpora are available for some biomedical RE tasks, making these resources particularly valuable. Distinct approaches, such as TL or *distant supervision* (Mintz et al., 2009) have been particularly explored to overcome this limit. With the latter approach, existing relationships available in knowledge- or data-bases are used to enrich the training set, with-

Proceedings of the 9th International Workshop on Health Text Mining and Information Analysis (LOUHI 2018), pages 149–159
Brussels, Belgium, October 31, 2018. ©2018 Association for Computational Linguistics

out considering more labeled corpora .

Domain adaptation is a type of TL that allows taking advantage of data annotated for a *source* domain to improve the performances in a related *target* domain (Weiss et al., 2016). However, even if the source and target domain share the same language (*i.e.*, English), thus a common syntax, TL between domains may lead to negative transfer since specific source domains may use specific vocabularies as well as specific formulations that are inadequate to the target domain. Hence, we need to better understand and characterize what makes a source corpus potentially helpful, or harmful, with regard to a RE task.

The contribution of this paper is twofold. First, we show that, compared to a baseline Convolutional Neural Network (CNN)-based model, a syntax-based model (*i.e.*, the TreeLSTM model) can better benefit from a TL strategy, even with very dissimilar additional source data. We conduct our experiments with two biomedical RE tasks and relatively small associated corpora, SNPPhenA (Bokharaeian et al., 2017) and EU-ADR (van Mulligen et al., 2012) as target corpora and three larger RE corpora, Semeval 2013 DDI (Herrero-Zazo et al., 2013), ADE-EXT (Gurulingappa et al., 2012), reACE (Hachey et al., 2012) as source corpora. Second, we propose a syntax-based analysis, using both quantitative criteria and qualitative observations, to better understand the role of syntactic features in the TL behavior.

2   Related work

2.1   Deep Learning Models for Relation Extraction

Deep learning models, based on continuous word representations have been proposed to overcome the problem of sparsity inherent to NLP (Huang and Yates, 2009). In Collobert et al. (2011), the authors proposed a unified CNN architecture to tackle various NLP problems traditionally handled with statistical approaches. They obtained state-of-the-art performances for several tasks, while avoiding the hand design of task specific features.

Zeng et al. (2014) showed that CNN models can also be applied to RE. In this study, they learn a vectorial sentence representation, by applying a CNN model over word and word position embeddings, which is used to feed a softmax classifier (Bishop, 2007). To improve the performance of RE, authors, such as Xu et al. (2015) and Yang

et al. (2016), consider elements of syntax within the embedding provided to the model.

Beside CNN models that incorporate syntactic knowledge in their embeddings, other approaches proposed neural networks (NN) in which the topology is adapted to the syntactic structure of the sentence. In particular, Recursive Neural Network (RNN) have been proposed to adapt to tree structures resulting from constituency parsing (Socher et al., 2013; Legrand and Collobert, 2014). In this vein, Tai et al. (2015) introduced a TreeLSTM, a generalization of LSTM for tree-structured network topologies, which allows processing trees with arbitrary branching factors.

The first model to use RNN for RE was proposed by Liu et al. (2015). The authors introduced a CNN-based model applied on the shortest dependency path between two entities, augmented with a RNN-based feature designed to model subtrees attached to the shortest path. Miwa and Bansal (2016) introduced a variant of the TreeLSTM that allows, like the model used in this paper, to take the whole dependency tree into account and not only the shortest path between two entities.

In this paper, we compare two deep learning strategies for RE: (1) the MultiChannel CNN (MCCNN) model (Quan et al., 2016), which has been successfully applied to the task of protein-protein interaction extraction without using any syntactic feature as input and (2) the TreeLSTM model (Tai et al., 2015), which is designed for considering dependency trees. These two models are detailed in section 3.

2.2   Transfer learning

TL allows to overcome the lack of training data for a given *target* task by transferring knowledge from *source* data not originally designed for that purpose (Weiss et al., 2016). One can distinguish *multitask learning* in which performances on a given task are improved using information contained in the training signals of auxiliary related tasks (Caruana, 1997) from *domain adaptation* in which only one task is considered but its application domains differ (Ben-David et al., 2010). While the former is a form of inductive transfer in which the auxiliary task introduces an inductive bias during training, the latter is a form of transductive transfer.

Domain adaptation approaches have been proposed for RE, including kernel based methods

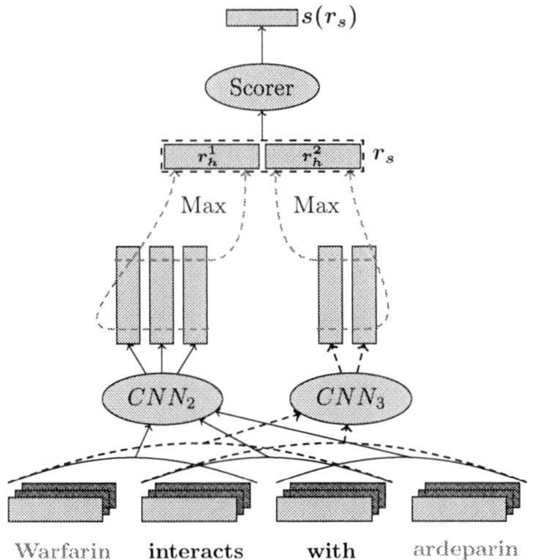

Figure 2: The MCCNN model with three channels, two CNN kernels of size 2 (CNN_2) and 3 (CNN_3). Red words correspond to the entities.

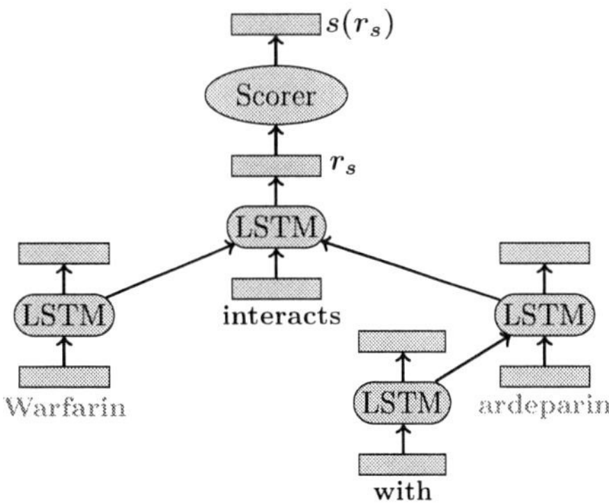

Figure 3: The TreeLSTM model. Each node takes as input the representation of its children. Red words correspond to the entities.

such as Plank and Moschitti (2013) focusing on unsupervised domain adaptation (*i.e.*, without any labeled target data) and deep learning based ones such as (Fu et al., 2017; Zhao et al., 2017) focusing on domain adversarial learning (an approach which ensures that the feature distributions over the source and target domains are made similar using an extra domain classifier at train time). Differently, our approach is a case of multi-source domain adaptation (*i.e.*, implying that we have labeled data, both in target and source corpora) and does not involve adversarial training.

Negative transfer occurs when the information learned from a source domain and task has a negative impact on the performances of the target task. Despite the fact that negative transfer is a major issue in TL, to our knowledge only few works have been conducted to overcome this problem (Weiss et al., 2016). Most of them use a relatedness metrics to select the elements of the source that are the most related to the target. For instance, Seah et al. (2013) defined a positive transferability measure that allows removing irrelevant source data. Ge et al. (2014) also focused on domain adaptation from multiple sources. They proposed a method to avoid negative learning caused by unrelated or irrelevant source domains, using a weighting mechanism based on a relatedness metrics between the source and target data.

In this work, we experiment with a domain adaptation method on the RE task using the TreeL-STM model, with relatively small biomedical corpora as target corpora and, larger biomedical or general domain corpora as source corpora. We also provide elements of interpretation of the impact of syntactic dependency structures on TL. In this matter, and unlike Seah et al. (2013) or Ge et al. (2014), the relatedness measures used in this work emphasizes the key role of syntax in TL with TreeLSTM.

3 Models

We compare in this article the performances of the MCCNN and TreeLSTM models. Both models compute a fixed-size vector representation for a whole sentence by composing input embeddings. A score is computed for each possible type of relationship (*e.g.*, negative, positive or speculative) between two identified entities.

In this section, we first introduce the embedding input layer, which is common to both approaches (*i.e.*, MCCNN and TreeLSTM); Then, we detail how each approach composes sequences of embedding in order to compute a unique vectorial sentence representation; Finally, we present the scoring layer, which is common to both approaches.

3.1 Input layer

Both models are fed with *word embeddings* (*i.e.*, continuous vectors) of dimension d_w, along with extra *entity embeddings* of size d_e. These embeddings are concatenated to form the input of the model. Formally, given a sentence of N words, $w_1, w_2, \ldots, w_N$, each word $w_i \in \mathcal{W}$ is first embedded in a d_w-dimensional vector space by ap-

	Corpus name	Subcorpus	Train Size		Test Size		#Entity Types	#Relation Types
			sent.	rel.	sent.	rel.		
Target corpora	SNPPhenA	–	362	935	121	365	2	3
	EU-ADR	drug-disease	244	176	–	–	4	3
		drug-target	247	310			4	3
		target-disease	355	262			4	3
Source corpora	SemEval 2013 DDI	DrugBank	5,675	3,805	973	889	4	4
		MEDLINE	1,301	232	326	95	4	4
	ADE-EXT	–	5,939	6,701	–	–	2	1
	reACE	–	5,984	2,486	–	–	4	5

Table 1: Main characteristics of our target and source corpora. Two corpora are divided into subcorpora. The sizes of the training and test corpora are reported in term of number of sentences (sent.) and annotated relationships (rel.). EU-ADR, ADR-EXT and reACE have no proper test corpus.

plying a lookup-table operation: $LT_W(w_i) = W_{w_i}$, where the matrix $W \in R^{d_w \times |\mathcal{W}|}$ represents the parameters to be trained in this lookup-table layer. The dictionary $\mathcal{W}$ is composed of all the words of the given corpus. Each column $W_{w_i} \in R^{d_w}$ corresponds to the vector embedding of the $w_i{}^{th}$ word in our dictionary $\mathcal{W}$.

Besides, entity embeddings (coming from a simple 3-elements dictionary) enable to distinguish between words which compose either the first entity, the second entity or are not part of any entity. They are respectively called *first entity*, *second entity* and *other* embeddings. Finally, word and entity embeddings are concatenated to form the input corresponding to a given word. Let's denote x_i the concatenated input corresponding to the ith word.

3.2 Composition layers

Both models take the embeddings as input and output a fixed-size representation r_s of size d_s, which corresponds to the whole sentence with two identified entities. Accordingly, one sentence with more than two entities will lead to one embedding for each pair of entities. This section details the two models used in this study.

3.2.1 MCCNN

The MCCNN models applies a variable kernel size CNN to multiple input channels of word embeddings. Inspired by the three-channel RGB image processing models, it considers different embedding channels (i.e., different word embeddings versions for each word) allowing to capture different aspects of input words.

More formally, given an input sequence $x_1, \ldots, x_N$, applying a kernel to the ith window of size k is done using the following formula:

$$C = h\left(\sum_{j=1}^{N-k+1} W[x_i, \ldots, x_{i+k-1}]^j + b \right)$$

where $[\ ]^j$ denotes the concatenation of inputs from channel j, $W \in \mathcal{R}^{(d_w+d_e) \times d_h}$ and $b \in \mathcal{R}^{d_h}$ are the parameters, d_h is the size of the hidden layer, h is a pointwise non-linear function such as the hyperbolic tangent and c is the number of input channels. For each kernel, a fixed size representation $r_h \in \mathcal{R}^{d_h}$ is then obtained by applying a max-pooling over time (here, the time means the position in the sentence).: $r_h = \max C$

We denote K the number of kernels with different sizes. A sentence representation $r_s \in \mathcal{R}^{d_s}$ (with $d_s = K * d_h$) is finally obtained by concatenating the output corresponding to the K kernels $r_s = [r_h^1, \ldots, r_h^k]$, where r_h^k corresponds to the output of the kth kernel. Figure 2 illustrates the structure of a two-channel CNN, with two kernels of size 2 and 3, on a four-words sentence.

3.2.2 TreeLSTM

The TreeLSTM model, and more specifically its *Child-Sum* version, (Tai et al., 2015) processes the dependency tree associated with an input sentence in a bottom-up manner. This model is suitable for processing dependency trees since it handles trees with arbitrary branching factors and no order between children of a node. This is done by recursively processing the nodes of the tree, using at each iteration, the representations of the children of the current node as input. The transition function for a node j and a set of children $C(j)$ can be found in the original paper (Tai et al., 2015) using $x_j \in \mathcal{R}^{d_w+d_e}$ as input for node j. The TreeLSTM outputs a sentence representation $r_s \in \mathcal{R}^{d_s}$ corresponding to the output state o_j of the top tree node (*i.e.*, the *root* node of the dependency tree that spans all the others). Figure 3 illustrates the structure of the TreeLSTM computed for a four-words sentence.

3.3 Scoring layer

Both the MCCNN and TreeLSTM models output a unique vector representation $r_s \in \mathcal{R}^{d_s}$ that takes the entire sentence into account, as well as two identified entities. This representation is used to feed a single layer NN classifier, which outputs a score vector with one score for each possible type of relationship. This vector is obtained using the formula: $s(r_s) = W^{(s)}r_s + b^{(s)}$, where $W^{(s)} \in \mathcal{R}^{d_s \times |S|}$ and $b^{(s)} \in \mathcal{R}^{|S|}$ are the trained parameters of the scorer, $|S|$ is the number of possible relation types. The scores are interpreted as probabilities using a softmax layer (Bishop, 2007).

4 Datasets

We explore how RE tasks that focus on a type of relationship associated with scarce resources may take advantage from larger corpora developed for distinct domains. To this purpose, we selected *(i)* two small *target* biomedical corpora and *(ii)* three larger *source* corpora. All are publicly available and detailed in the following section. Table 3 summarizes their main characteristics.

4.1 Target corpora

SNPPhenA (Bokharaeian et al., 2017) is a corpus of abstracts of biomedical publications, obtained from PubMed (Fiorini et al., 2017), annotated with two types of entities: *single nucleotide polymorphisms* (SNPs) and *phenotypes*. Relationships between them are annotated and classified in 3 types: *positive*, *negative* and *neutral*.

EU-ADR (van Mulligen et al., 2012) is a corpus of PubMed abstracts annotated with *drugs*, *diseases* and drug targets (*proteins/genes* or *gene variants*) entities. It is composed of 3 subcorpora of 100 abstracts each, encompassing annotations of either target-disease, target-drug or drug-disease relationships. Annotated relationships are classified in 3 types: *positive*, *speculative* and *negative associations* (PA, SA and NA respectively). In (Bravo et al., 2015), performances are assessed over the TRUE class, which is composed of the PA, SA and NA types, in contrast with the FALSE class.

4.2 Source corpora

SemEval 2013 DDI (Drug-Drug Interaction) (Herrero-Zazo et al., 2013) consists of texts from DrugBank and MEDLINE annotated with drugs. Drug are categorized in 4 categories: *drug*, *brand*, *group* and *drug_n* (*i.e.*, active substances not approved for human use). Relationships are classified in 4 types: *mechanism*, *effect*, *advice* and *int* (default category, when no detail is provided).

ADE-EXT (Adverse Drug Effect corpus, extended) (Gurulingappa et al., 2012) consists of MEDLINE case reports, annotated with *drugs* and *conditions* (*e.g.*, diseases, signs and symptoms), along with untyped relationships between them.

reACE (Edinburgh Regularized Automatic Content Extraction) (Hachey et al., 2012) consists of English broadcast news and newswire annotated with *organization*, *person*, *fvw* (facility, vehicle or weapon) and *gpl* (geographical, political or location) entities along with relationships between them. Relationships are classified in five types: *general-affiliation*, *organisation-affiliation*, *part-whole*, *personal-social* and *agent-artifact*.

5 Experiments

5.1 Training and Experimental Settings

Our models were trained by minimizing the log-likelihood over the training data. All parameters (weights, biases and embeddings) were iteratively updated via backpropagation for the MCCNN and backpropagation Through Structure (Goller and Kuchler, 1996) for the TreeLSTM. Hyper-parameters were tuned using a 10-fold cross-validation by selecting the values leading to the best averaged performance, and fixed for the remaining experiments. Word embeddings were pre-trained on ~3.4 million PubMed abstracts (corresponding to all those published between Jan. 1, 2014 and Dec. 31, 2016) using the method described in Lebret and Collobert (2014).

MCCNN model. Following Kim (2014) both channels were initialized with pre-trained word embeddings, but gradients were backpropagated only through one of the channels. Hyper-parameters were fixed to $d_w = 100$, $d_e = 10$, $d_h = 100$ for each of the 2 channels, $d_s = 2 \times d_h = 200$. We used two kernels of size 3 and 5 respectively. We applied a dropout regularization after the embedding layers (Srivastava et al., 2014) with a dropout probability fixed to 0.25.

Test Corpus	Model	Train corpus	P	R	F	σ_F
SNPPhenA	TreeLSTM	SNPPhenA alone	58.9	73.8	65.5	4.1
		+ SemEval 2013 DDI	65.2	71.1	**68.0**	4.7
		+ ADE-EXT	62.8	72.1	67.2	3.4
		+ reACE	61.8	74.3	67.1	3.6
	MCCNN	SNPPhenA alone	55.1	75.0	**63.3**	4.8
		+ SemEval 2013 DDI	55.3	74.4	**63.3**	4.9
		+ ADE-EXT	56.1	73.2	63.2	4.8
		+ reACE	53.2	70.9	60.6	4.1
EU-ADR drug-disease	TreeLSTM	EU-ADR drug-disease alone	74.8	84.1	79.1	12.3
		+ SemEval 2013 DDI	74.8	90.6	**82.0**	13.1
		+ ADE-EXT	73.9	88.2	80.4	13.7
		+ reACE	74.3	91.1	79.3	14.3
	MCCNN	EU-ADR drug-disease alone	73.3	94.7	**80.2**	14.2
		+ SemEval 2013 DDI	72.6	87.9	76.6	14.3
		+ ADE-EXT	73.0	85.5	76.0	14.5
		+ reACE	74.1	91.5	79.2	13.8
EU-ADR drug-target	TreeLSTM	EU-ADR drug-target alone	72.4	90.6	80.2	10.9
		+ SemEval 2013 DDI	71.9	95.5	**82.5**	8.5
		+ ADE-EXT	70.2	96.7	80.9	9.2
		+ reACE	70.4	96.5	80.8	9.3
	MCCNN	EU-ADR drug-target alone	74.5	92.3	**81.0**	9.3
		+ SemEval 2013 DDI	74.9	88.8	80.0	10.6
		+ ADE-EXT	76.3	87.4	80.3	10.1
		+ reACE	73.4	92.1	80.5	7.8
EU-ADR target-disease	TreeLSTM	EU-ADR target-disease alone	77.0	89.7	82.7	6.4
		+ SemEval 2013 DDI	77.4	91.6	**83.9**	8.2
		+ ADE-EXT	77.7	89.5	83.3	6.9
		+ reACE	75.9	91.7	83.0	7.7
	MCCNN	EU-ADR target-disease alone	76.9	91.8	**82.6**	7.7
		+ SemEval 2013 DDI	77.6	90.6	82.5	7.1
		+ ADE-EXT	75.5	87.4	81.8	10.1
		+ reACE	77.1	91.2	82.0	6.8

Table 2: Results of our TL strategy in terms of precision (P), recall (R) and f-measure (F). σ_F is the standard deviation of the f-measure. The $+$ in the column *Train corpus* indicates that we trained our model using the target corpus plus one additional source corpus.

TreeLSTM model. Dependency trees were derived from parsing trees obtained using the Charniak-Johnson parser trained on GENIA and PubMed data (McClosky and Charniak, 2008). Hyper-parameters were fixed to $d_w = 100$, $d_e = 10$, $d_h = 200$ and $d_s = 200$. We applied a dropout regularization after every TreeLSTM unit and after the embedding layers. The dropout probability was fixed to 0.25. All the parameters are initialized randomly except the word embeddings.

We evaluated performances in terms of precision (P), recall (R) and f-measure (F). For multi-label classifications, we report the macro-average performance[1]. For SNPPhenA, we performed a cross-validation using 10% of the corpus for the validation and the provided test corpus for testing (which is about 30% the size of the training cor-pus). Because no test corpus is provided with EU-ADR, we performed a 10-fold cross-validation using 10% of the corpus for the validation and 10% for the test of our models.

5.2 Transfer learning experiment

In this subsection, we present our TL strategy and its results. Following a standard practice in deep learning, the transfer learning is done by training models in parallel while using shared representations, as illustrated by (Collobert et al., 2011). In other terms, for each experiment, the same network, initialized with random weights, is used for each corpus (i.e., same embedding layer and TreeLSTM weights), except for the scorer, which is adapted to each corpus as the number and types of relationships may change. During the training phase, using a standard stochastic gradient descent procedure (Robbins and Monro, 1985), we randomly pick training sentences from the mixed corpus (i.e., target + one source training corpora).

---

[1]The macro-average metric is less impacted by classes with few test instances (and thus a high variance). For this reason, it is more representative of the performance of our model.

Test corpus	Work (train corpus)	P	R	F
SNPPhena	**Bokharaeian et al. (2017)** (SNPPhenA)	56.6	59.8	58.2
	This work (SNPPhenA + SemEval 2013 DDI)	64.5	75.2	**69.4**
EU-ADR drug-disease	**Bravo et al. (2015)** (EU-ADR drug-disease)	70.2	93.2	79.3
	This work (EU-ADR drug-disease + SemEval 2013 DDI)	74.8	90.6	**82.0**
EU-ADR drug-target	**Bravo et al. (2015)** (EU-ADR drug-target)	74.2	97.4	**83.3**
	This work (EU-ADR drug-target + SemEval 2013 DDI)	73.5	95.6	83.1
EU-ADR target-disease	**Bravo et al. (2015)** (EU-ADR target-disease)	75.1	97.7	**84.6**
	This work (EU-ADR target-disease + SemEval 2013 DDI)	78.7	91.4	**84.6**

Table 3: Performance comparison between the state of the art (Bokharaeian et al., 2017; Bravo et al., 2015) and this work in terms of precision (P), recall (R) and F-measure (F). Results reported for this work are ensembles of the 5 best models obtained.

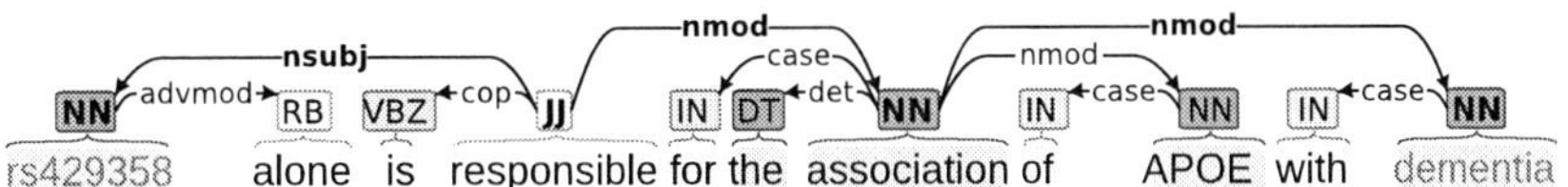

Figure 4: Dependency parse tree of a sentence from SNPPhena expressing a relation between the entities *rs429358* and *dementia*. The shortest dependency path between the two entities is shown in bold.

This training procedure is done, starting from different random initialization for each fold of our cross-validation. Table 2 presents the results of the TL study. Each results is an average of 100 experiment (10 experiments for each fold starting from different random initialization). We observed that for the TreeLSTM model, additional source corpora consistently improved the performances. More interestingly, this phenomenon occurs even for corpora of distinct types of entities such as the combination of SNPPhenA and SemEval 2013 DDI and, to a lesser extend, with the corpus that is outside of the biomedical domain, reACE. We note that the pre-trained embeddings were obtained using biomedical sources. This may affect the TL performance with reACE that is not of the biomedical domain. Also, we did not observed any benefit of the TL strategy for the MCCNN model, which performances decrease slightly in comparison with the baseline experiments.

5.3 Comparison with the state of the art

Table 3 presents a comparison of performances obtained with our approach *versus* two state-of-the-art systems applied to the RE tasks associated respectively with SNPPhenA (Bokharaeian et al., 2017) and EU-ADR (Bravo et al., 2015). Our results are obtained using, for each fold, an ensemble of the 5 best models for this fold, according to the validation. The ensembling was done by averaging the scores $s(r_s)$ of each individual model, following Legrand and Collobert (2014). We report the 10-folds average performance. Both state-of-the-art systems use a combination of a shallow linguistic kernel with a kernel that exploits deep syntactic features. Our approach outperforms the performances reported for SNPPhenA and one EU-ADR subtasks and lead to similar performances for the two remaining EU-ADR subtasks.

6 On the role of syntactic features in transfer learning

Empirical results suggest that the TreeLSTM model is more positively-influenced by syntactic similarity between source and target corpora than by domain closeness. Indeed, the TreeLSTM model explicitly includes the syntactic structure of the sentences in the network topology. Thus, a source corpus, such as reACE, that share neither entity nor vocabulary with the target corpus proved to be helpful. We propose in the following an analysis of the role of the syntactic features. We also provide real examples illustrating similarities between corpora and comment them.

Syntactic features. We propose three comparisons based on patterns extracted from shortest paths between two entities in dependency graphs. Shortest path proved to be effective for RE (Bunescu and Mooney, 2005; Cellier et al., 2010). From a shortest path (as between *rs429358* and *dementia* in Figure 4), we extract 3 different patterns. The first one is made with the part-of-speech (POS) and dependency tags (DT): for example, in Figure 4, *"NN nsubj *JJ* nmod NN nmod NN"*[2]. The second and the third patterns are built by keeping only either the POS or the DT. The patterns associated with our running example are then: *"NN *JJ* NN NN"* and *"nsubj ** nmod nmod"*. For a given pattern, the *syntactic similarity* score is obtained using the following procedure: Given 2 corpora, (1) we first extract all the shortest path pattern that appear between two related entities. (2) For each corpus, we compute the pattern distribution (*i.e.*, the list of patterns, along with their frequency) by normalizing over all the patterns in the corpus. (3) The score is then computed with the cosine similarity between the pattern distributions of two corpora. Table 4 shows the cosine similarity measures between target and source corpora for the three different pattern distributions. We observe that, for the two target corpora, the performance gain obtained using the TL strategy using a given source corpus can be related to the cosine similarity with this corpus: the higher cosine similarity lead to the best transfer TL.

| | | Source corpora | | |
		DDI	ADE	reACE
		POS + DT		
	SNPPhena	0.53	0.22	0.13
	EU-ADR	0.24	0.20	0.09
		POS only		
	SNPPhena	0.80	0.70	0.35
Source corpora	**EU-ADR**	0.77	0.68	0.32
		DT only		
	SNPPhena	0.53	0.23	0.14
	EU-ADR	0.25	0.24	0.10

Table 4: Cosine similarity score between target and source corpora for the three different pattern distributions. POS is part of speech pattern and DT is dependency type pattern.

Dictionary coverage. On the opposite, we observed that the efficiency of TL in our experiments can not be fully explained by the lexical similarity between source and target corpora. As shown in Table 5, the vocabulary overlap with the target corpora is almost equivalent whether we are considering DDI or ADE (53.4 *vs.* 51.2 and 58.9 *vs.* 60.5), whereas performances obtained with DDI were better than those obtained with ADE. Unsurprisingly, it is lower for reACE which is not a

biomedical corpus.

	DDI	ADE	reACE
SNPPhenA	53.4	51.2	39.8
EU-ADR	58.9	60.5	38.3

Table 5: Dictionary coverage. Percentage of words from the target copora present in the source corpora.

Lexical and semantic paradigms. We complete this analysis with few examples illustrating the lexical and semantic heterogeneity of sentences that may instantiate a same pattern. Table 6 provides 4 patterns and their instantiations in source and target corpora. One can observe that sentences instantiating a same pattern seems to have no particular similarity when considering lexical and semantic paradigms. A similar heterogeneity is observed when considering the lowest common ancestor term (or the *head*) of the patterns. Table 7 lists the most frequent lowest common ancestor in each corpus. Again, we observe no direct link with learning improvement.

7 Conclusion

In this paper, we empirically showed that a TL strategy can benefit biomedical RE tasks when using the TreeLSTM model, whereas it is mainly harmful with a model that does not consider syntax. This is of great interest for specific domains, such those of biomedicine, for which few annotated resources are available. Our TL approach led *(i)* to better performances than the state of the art for two biomedical RE tasks: SNP-phenotype and drug-disease RE; and *(ii)* to state-of-the-art results for two others focusing on target-disease and target-drug relationships. Interestingly, we showed that even a general domain corpus (reACE) may carry useful information and lead to improved performances. We proposed an analysis with syntax-based metrics and examples to provide elements of interpretation of this behavior and emphasize the key role of syntax in TL for RE. An exciting direction would be to explore this transfer strategy with Electronic Health Records of various origin.

Aknowledgement

This work is funded by the French National Research Agency (ANR) under the *PractiKPharma* project: ANR-15-CE23-0028, by the IDEX "Lorraine Université d'Excellence" (15-IDEX-0004) and by the *Snowball* Inria Associate Team.

---

[2]The stars mark the lowest common ancestor of the two entities in the dependency tree and are used to prevent similar pattern with different common ancestors to be considered the same. Note that the patterns are not directed, thus the two patterns *"NN nsubj *JJ* nmod NN nmod NN"* and *"NN nmod NN nmod *JJ* nsubj NN"* are equivalent.

Pattern	Corpus	Example of instantiation
NN conj NN	SNPPhenA	JJ NN VBD NN IN NN CC NN — conj — Multivariate analysis showed association between rs785677 and LOAD
	DDI	RB , JJ NN IN NNP CC NNP VBZ VBN . — conj — Therefore , concomitant use of TORADOL and probenecid is contraindicated .
ppmod *NN* ppmod NN ppmod NN	EU-ADR	DT NN IN JJ NNS IN DT NN IN JJ NNS VBZ RB VBN VBN . — ppmod — attr — ppmod — attr — The efficacy of antiepileptic agents in the prophylaxis of perioperative seizures has not been assessed .
	DDI	RB , NN IN NN IN NN IN NNP VBZ RB VBN . — ppmod — ppmod — Therefore , use of lamivudine in combination with zalcitabine is not recommended .
NN pmod *NN* ppmod NN	SNPPhenA	PRP VBD RB VB DT NN IN DT NN com NN IN NNS com NNS — ppmod — ppmod — We did not detect any association of the TCF7L2 HapA with adiposity measures
	reACE	DT JJ NN IN DT NNP com NNP com NNP IN DT NNP — ppmod — ppmod — the 45-year-old director of the Artificial Intelligence Laboratory at the MIT
NN dep *VBN* ppmod NN	SNPPhenA	NN com NN VBZ VBN VBN IN JJ attr NN — dep — ppmod — Apo genotype has been associated with systemic inflammation
	reACE	NNP com NNP V DT JJ JJ NN NN V VBZ VBN VBN IN NNP — dep — ppmod — Abu Talb , the last major prosecution witness , has been jailed in Sweden

Table 6: Examples of patterns and of their instantiation in corpora. Red words correspond to entities.

SNPPhenA	EU-ADR	DDI	ADE	reACE
associated (25.2)	analyzed (5.8)	*entity* (17.8)	*entity* (30.1)	*entity* (60.6)
entity (12.2)	associated (4.3)	administered (4.1)	developed (11.1)	is (2.2)
genotyped (5.4)	*entity* (2.9)	increase (3.0)	associated (4.1)	was (1.9)
association (4.4)	is (2.9)	administration (2.7)	is (2.7)	said (1.4)
showed (3.8)	polymorphisms (2.4)	reported (2.6)	induced (2.3)	
observed (3.3)	over-represented (2.4)	interact (2.6)	case (1.6)	
genes (2.6)	showed (2.4)	reduce (2.5)	following (1.4)	

Table 7: Terms corresponding to the lowest common ancestor in the POS + DT patterns. Their relative frequency in each corpus is provided in parenthesis. *Entity* means that the term is one of the two entities.

References

Shai Ben-David, John Blitzer, Koby Crammer, Alex Kulesza, Fernando Pereira, and Jennifer Wortman Vaughan. 2010. A theory of learning from different domains. *Machine Learning*, 79(1-2):151–175.

Christopher M. Bishop. 2007. *Pattern recognition and machine learning, 5th Edition*. Information science and statistics. Springer.

Behrouz Bokharaeian, Alberto Díaz Esteban, Nasrin Taghizadeh, Hamidreza Chitsaz, and Ramyar Chavoshinejad. 2017. SNPPhenA: a corpus for extracting ranked associations of single-nucleotide polymorphisms and phenotypes from literature. *J. Biomedical Semantics*, 8(1):14:1–14:13.

Àlex Bravo, Janet Piñero González, Núria Queralt-Rosinach, Michael Rautschka, and Laura Inés Furlong. 2015. Extraction of relations between genes and diseases from text and large-scale data analysis: implications for translational research. *BMC Bioinformatics*, 16:55:1–55:17.

Razvan C. Bunescu and Raymond J. Mooney. 2005. A shortest path dependency kernel for relation extraction. In *HLT/EMNLP 2005, Human Language Technology Conference and Conference on Empirical Methods in Natural Language Processing, Proceedings of the Conference, 6-8 October 2005, Vancouver, British Columbia, Canada*, pages 724–731.

Rich Caruana. 1997. Multitask learning. *Machine Learning*, 28(1):41–75.

Peggy Cellier, Thierry Charnois, and Marc Plantevit. 2010. Sequential patterns to discover and characterise biological relations. In *Computational Linguistics and Intelligent Text Processing, 11th International Conference, CICLing 2010, Iasi, Romania, March 21-27, 2010. Proceedings*, pages 537–548.

Ronan Collobert, Jason Weston, Léon Bottou, Michael Karlen, Koray Kavukcuoglu, and Pavel P. Kuksa. 2011. Natural language processing (almost) from scratch. *Journal of Machine Learning Research*, 12:2493–2537.

Nicolas Fiorini, David J. Lipman, and Zhiyong Lu. 2017. Towards PubMed 2.0. *Elife*, 6.

Lisheng Fu, Thien Huu Nguyen, Bonan Min, and Ralph Grishman. 2017. Domain adaptation for relation extraction with domain adversarial neural network. In *Proceedings of the Eighth International Joint Conference on Natural Language Processing (Volume 2: Short Papers)*, volume 2, pages 425–429.

Liang Ge, Jing Gao, Hung Q. Ngo, Kang Li, and Aidong Zhang. 2014. On handling negative transfer and imbalanced distributions in multiple source transfer learning. *Statistical Analysis and Data Mining*, 7(4):254–271.

Christoph Goller and Andreas Kuchler. 1996. Learning task-dependent distributed representations by back-propagation through structure. In *Neural Networks, 1996., IEEE International Conference on*, volume 1, pages 347–352. IEEE.

Harsha Gurulingappa, Abdul Mateen-Rajpu, and Luca Toldo. 2012. Extraction of potential adverse drug events from medical case reports. *Journal of biomedical semantics*, 3(1):15.

Ben Hachey, Claire Grover, and Richard Tobin. 2012. Datasets for generic relation extraction. *Natural Language Engineering*, 18(1):21–59.

María Herrero-Zazo, Isabel Segura-Bedmar, Paloma Martínez, and Thierry Declerck. 2013. The DDI corpus: An annotated corpus with pharmacological substances and drug-drug interactions. *Journal of Biomedical Informatics*, 46(5):914–920.

Fei Huang and Alexander Yates. 2009. Distributional representations for handling sparsity in supervised sequence-labeling. In *ACL 2009, Proceedings of the 47th Annual Meeting of the Association for Computational Linguistics and the 4th International Joint Conference on Natural Language Processing of the AFNLP, 2-7 August 2009, Singapore*, pages 495–503.

Yoon Kim. 2014. Convolutional neural networks for sentence classification. In *Proceedings of the 2014 Conference on Empirical Methods in Natural Language Processing, EMNLP 2014, October 25-29, 2014, Doha, Qatar, A meeting of SIGDAT, a Special Interest Group of the ACL*, pages 1746–1751.

Rémi Lebret and Ronan Collobert. 2014. Word embeddings through hellinger PCA. In *Proceedings of the 14th Conference of the European Chapter of the Association for Computational Linguistics, EACL 2014, April 26-30, 2014, Gothenburg, Sweden*, pages 482–490.

Joël Legrand and Ronan Collobert. 2014. Joint RNN-based greedy parsing and word composition. *CoRR*, abs/1412.7028.

Yang Liu, Furu Wei, Sujian Li, Heng Ji, Ming Zhou, and WANG Houfeng. 2015. A dependency-based neural network for relation classification. In *Proceedings of the 53rd Annual Meeting of the Association for Computational Linguistics and the 7th International Joint Conference on Natural Language Processing (Volume 2: Short Papers)*.

David McClosky and Eugene Charniak. 2008. Self-training for biomedical parsing. In *ACL 2008, Proceedings of the 46th Annual Meeting of the Association for Computational Linguistics*, pages 101–104.

Mike Mintz, Steven Bills, Rion Snow, and Dan Jurafsky. 2009. Distant supervision for relation extraction without labeled data. In *Proceedings of the Joint Conference of the 47th Annual Meeting of the ACL and the 4th International Joint Conference on Natural Language Processing of the AFNLP: Volume 2-Volume 2*, pages 1003–1011. Association for Computational Linguistics.

Makoto Miwa and Mohit Bansal. 2016. End-to-end relation extraction using LSTMs on sequences and tree structures. In *Proceedings of the 54th Annual Meeting of the Association for Computational Linguistics, ACL 2016*.

Erik M. van Mulligen, Annie Fourrier-Réglat, David Gurwitz, Mariam Molokhia, Ainhoa Nieto, Gianluca Trifirò, Jan A. Kors, and Laura Inés Furlong. 2012. The EU-ADR corpus: Annotated drugs, diseases, targets, and their relationships. *Journal of Biomedical Informatics*, 45(5):879–884.

Barbara Plank and Alessandro Moschitti. 2013. Embedding semantic similarity in tree kernels for domain adaptation of relation extraction. In *Proceedings of the 51st Annual Meeting of the Association for Computational Linguistics (Volume 1: Long Papers)*, volume 1, pages 1498–1507.

Chanqin Quan, Lei Hua, Xiao Sun, and Wenjun Bai. 2016. Multichannel Convolutional Neural Network for Biological Relation Extraction. *BioMed research international*, 2016:1850404.

Herbert Robbins and Sutton Monro. 1985. A stochastic approximation method. In *Herbert Robbins Selected Papers*, pages 102–109. Springer.

Chun-Wei Seah, Yew-Soon Ong, and Ivor W. Tsang. 2013. Combating negative transfer from predictive distribution differences. *IEEE Trans. Cybernetics*, 43(4):1153–1165.

Richard Socher, John Bauer, Christopher D. Manning, and Andrew Y. Ng. 2013. Parsing with compositional vector grammars. In *Proceedings of the 51st Annual Meeting of the Association for Computational Linguistics, ACL 2013*, pages 455–465.

Nitish Srivastava, Geoffrey E. Hinton, Alex Krizhevsky, Ilya Sutskever, and Ruslan Salakhutdinov. 2014. Dropout: a simple way to prevent neural networks from overfitting. *Journal of Machine Learning Research*, 15(1):1929–1958.

Kai Sheng Tai, Richard Socher, and Christopher D. Manning. 2015. Improved semantic representations from tree-structured long short-term memory networks. In *Proceedings of the 53rd Annual Meeting of the Association for Computational Linguistics, ACL 2015*, pages 1556–1566.

Karl R. Weiss, Taghi M. Khoshgoftaar, and Dingding Wang. 2016. A survey of transfer learning. *J. Big Data*, 3:9.

Kun Xu, Yansong Feng, Songfang Huang, and Dongyan Zhao. 2015. Semantic relation classification via convolutional neural networks with simple negative sampling. In *Proceedings of the 2015 Conference on Empirical Methods in Natural Language Processing, EMNLP 2015*, pages 536–540.

Yunlun Yang, Yunhai Tong, Shulei Ma, and Zhi-Hong Deng. 2016. A position encoding convolutional neural network based on dependency tree for relation classification. In *Proceedings of the 2016 Conference on Empirical Methods in Natural Language Processing, EMNLP 2016*, pages 65–74.

Daojian Zeng, Kang Liu, Siwei Lai, Guangyou Zhou, and Jun Zhao. 2014. Relation classification via convolutional deep neural network. In *COLING 2014, 25th International Conference on Computational Linguistics*, pages 2335–2344.

Han Zhao, Shanghang Zhang, Guanhang Wu, João P Costeira, José MF Moura, and Geoffrey J Gordon. 2017. Multiple source domain adaptation with adversarial training of neural networks. *arXiv preprint arXiv:1705.09684*.

In-domain Context-aware Token Embeddings Improve Biomedical Named Entity Recognition

Golnar Sheikhshab
Simon Fraser University
gsheikhs@sfu.ca

Inanc Birol
British Columbia Cancer Agency
ibirol@bcgsc.ca

Anoop Sarkar
Simon Fraser University
gsheikhs@sfu.ca

Abstract

Rapidly expanding volume of publications in the biomedical domain makes it increasingly difficult for a timely evaluation of the latest literature. That, along with a push for automated evaluation of clinical reports, present opportunities for effective natural language processing methods. In this study we target the problem of named entity recognition, where texts are processed to annotate terms that are relevant for biomedical studies. Terms of interest in the domain include gene and protein names, and cell lines and types. Here we report on a pipeline built on Embeddings from Language Models (ELMo) and a deep learning package for natural language processing (AllenNLP). We trained context-aware token embeddings on a dataset of biomedical papers using ELMo, and incorporated these embeddings in the LSTM-CRF model used by AllenNLP for named entity recognition. We show these representations improve named entity recognition for different types of biomedical named entities. We also achieve a new state of the art in gene mention detection on the BioCreative II gene mention shared task.

1 Introduction

Last decade witnessed substantial improvements in machine learning methods and their application to natural language processing tasks. Recently, Peters et al. (2018) introduced ELMo (Embeddings from Language Models), a system for deep contextualized word representation, and showed how it can be used in existing task-specific deep neural networks. The method improves the state of the art over a variety of NLP tasks such as question answering, word sense disambiguation, sentiment analysis, and named entity recognition. The developers of the tool also provide an ELMo model pre-trained on the Billion-word Language Model (LM) dataset (Chelba et al., 2014) as an off-the-shelf tool for use in a wide variety of NLP tasks and domains.

This begs the question of how the performance of downstream analysis would improve if the model were to be adapted to work with domain-specific texts. In this paper, we investigate the effect of an in-domain training set for ELMo in Named Entity Recognition (NER) applications. Our contributions are as follows:

1. Off-the-shelf ELMo has room for improvement in domain-specific applications

2. ELMo consistently improves biomedical named entity recognition when trained on in-domain data

3. Such improvement can be achieved even when the in-domain training dataset is smaller than the Billion-word LM data.

4. The resulting model achieves the highest precision/recall/F1 scores so far on BioCreative II Gene mention detection shared task (BC2GM).

We explain ELMo and AllenNER, the named entity recognizer we used, in sections 2 and section 3. Then, we describe our datasets in section 4, and we move on to report the results in section 5.

2 ELMo

ELMo (Peters et al., 2018) is a system that produces context-aware embeddings for word tokens. Similar to traditional context-independent word embeddings such as GloVe (Pennington et al., 2014) and Word2Vec (Mikolov et al., 2013), ELMo representations can be used as input to a neural network for downstream tasks. Though, ELMo is different from the traditional word embeddings in that it gives the representation of the

Proceedings of the 9th International Workshop on Health Text Mining and Information Analysis (LOUHI 2018), pages 160–164
Brussels, Belgium, October 31, 2018. ©2018 Association for Computational Linguistics

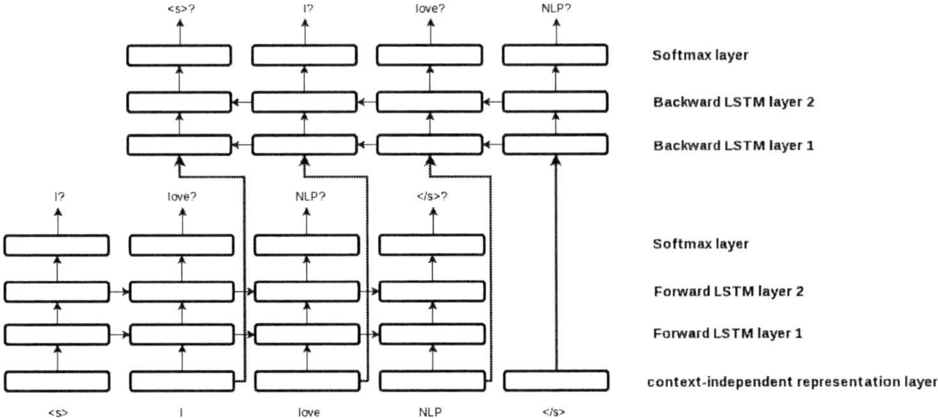

Figure 1: ELMo is a bidirectional LSTM for language modelling where the next or precedent tokens are predicted from the softmax layers over forward and backward LSTMs respectively.

word in the context of the specific given sentence; hence it is a context-aware word token representation as opposed to a word type representation.

It is trained using a language modeling objective function, where the objective is to predict the next word in the sequence; either sequentially left to right or right to left. As such, it can be viewed as learning a token level representation of words for a task that can be trained on unannotated data. These word representations can then be used for a task that is trained on labeled data. In our case, the task is biomedical named-entity recognition.

Figure 1 shows the architecture of ELMo as a recurrent language modelling network. The input to this system is a sequence of words $w_1 w_2 \ldots w_i \ldots w_n$. First, each word is converted to a context-independent embedding by a convolutional neural network (CNN) over its characters. These character-based representations are then fed into a two-layer bidirectional long short-term memory (LSTM) (Hochreiter and Schmidhuber, 1997) recurrent neural network. Output of the second layers of the forward and backward LSTMs are fed to a soft-max layer to predict w_{i+1} and w_{i-1}, respectively, at each position i.

Task-specific learned weights can be used later to combine all layers in ELMo model at position i and form the task-specific "ELMo representation of w_i".

Peters et al. (2018) showed that different layers in this deep recurrent model learn different aspects of a given token. The lower layers learn more syntactic features whereas higher layers learn the contextual aspects of the word. They linearly combined the layers using task-dependent weights, and their experiments show that for Named Entity Recognition tasks, the layers are combined with effectively the same weights.

3 Named Entity Recognition with AllenNLP

In our pipeline, we couple ELMo embeddings to AllenNLP (Gardner et al., 2017) for NER tasks.

AllenNLP uses a bidirectional two-layer LSTM-CRF (Lample et al., 2016) to perform NER as a sequence tagging task. Each word is tagged with an output that marks if it is at the beginning (B), in the middle (I), at the end (E or L), or outside (O) of an entity type. One-word entities are also marked (as S or U). For example B-Gene and I-Gene stand for beginning and inside of a Gene, whereas B-DNA and E-DNA stand for beginning and ending of a DNA entity type.

AllenNLP embeds the input words using a Convolutional Neural Network over characters. Rei et al. (2016) showed that word embeddings from character compositions outperform lookup embeddings such as word2vec, when used for named entity recognition.

AllenNLP combines the layers in ELMo Model using learned task-specific weights, concatenates the result for each token to context-independent word embeddings, and feed the concatenation into the LSTM-CRF as illustrated in Figure 2.

4 Datasets

We collected a focused domain-specific subset of PubMed Central (PMC) documents, and used

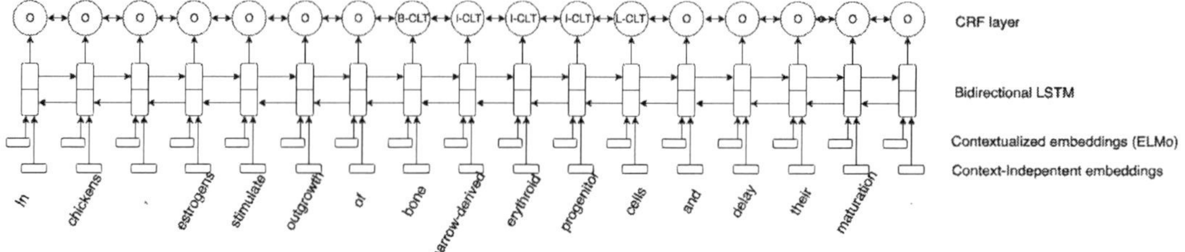

Figure 2: Architecture of LSTM-CRF (Lample et al., 2016) with ELMo. Traditional word embeddings and ELMo representations are concatenated and fed into a bidirectional LSTM. A CRF layer on top of bidirectional LSTM takes local label dependencies into account. At training time the log likelihood of gold label sequences is maximized. At test time, Viterbi (Viterbi, 1967) algorithm is used to decode the complete label sequence. -CLT in the labels of the example indicate cell_type entity.

them for training ELMo. This dataset is described in detail in section 4.1. We report results on two benchmark datasets, which we describe in sections 4.2 and 4.3.

4.1 ELMo Training Set

We downloaded the text files of a subset of PMC documents that are available at ftp://ftp.ncbi.nlm.nih.gov/pub/pmc in May 2018, and picked 3960 full-text documents that had a Medical Subject Heading (Mesh) term 'cancer'. We ran StanfordNLP/CoreNLP toolkit (Manning et al., 2014) on these documents for sentence splitting and tokenization. Tokens of each sentence were joined with space character in between to form the sentences in the training set. This dataset contains about 21 million tokens, and is substantially smaller than the One Billion Word Benchmark (Chelba et al., 2014) that Peters et al. (2018) used for training ELMo but contains in-domain text that is more likely to benefit the biomedical text analysis of interest in this paper.

4.2 BC2GM

BC2GM is the data set for BioCreative II Gene Mention detection shared task (Smith et al., 2008). This dataset contains 15000 training and 5000 test sentences, all from PubMed abstracts. Gold annotations give the gene mentions by providing the sentence ID, the start and end characters of the mention (ignoring all space characters), and the mention itself.

4.3 JNLPBA

JNLPBA (Kim et al., 2004) is the dataset for a shared task on biomedical entity detection. Its training set contains 2000 GENIA (Kim et al., 2003) abstracts, which the authors had collected

by searching MEDLINE abstracts for Mesh terms 'human', 'blood cells' and 'transcription factors'. The test set contain 404 abstracts, half of which are from the same domain and the other half are from a super-domain of 'blood cells' and 'transcription factors'. The documents are annotated for protein, DNA, RNA, cell_line, and cell_type entity classes.

5 Results

Table 1 shows the leading results in the literature (top four rows) in comparison with our results (bottom three rows) on BC2GM dataset.

In the year it was held, Ando (2007) had won the challenge with a semi-supervised system equipped with a lexicon and a combination of several classifiers. Gimli (Campos et al., 2013) is a supervised method based on conditional random fields (CRF) (Lafferty et al., 2001) with hand-engineered features that was the state of the art for gene mention detection before Graph-NER (Sheikhshab et al., 2018) obtained a higher F-score. GraphNER, obtained the distributions over labels from the CRF and propagated them on a graph of 3-grams similarities constructed over BC2GM.

Rei et al. (2016) set the previous state of the art on BC2GM by applying an LSTM-CRF based system with attention to characters. Our baseline, AllenNER (described in detail in section 3) is similar to their system, except AllenNER uses a convolutional neural network (CNN) over characters instead of using attention mechanism.

Our results, the lower part of Table 1, show that using the off-the-shelf ELMo, that is trained on the one Billion word language model benchmark (Chelba et al., 2014), improves the preci-

Model	Prec. (%)	Rec. (%)	F1 (%)
Ando (2007)	88.48	85.97	87.21
Gimli (2013)	90.22	84.32	87.17
GraphNER (2018)	89.18	85.57	87.34
Rei et al. (2016)	-	-	87.99
AllenNER with no ELMo (Baseline)	88.05	88.72	88.39
AllenNER + off-the-shelf ELMO	89.03	87.95	88.49
AllenNER + ELMO Trained In-Domain	**89.86**	**89.59**	**89.72***

Table 1: Leading results in the literature (up) in comparison with our results (down) on BC2GM dataset

Model	protein	DNA	RNA	cell_line	cell_type
AllenNER with no ELMo (Baseline)	70.47	70.87	63.94	57.17	73.55
AllenNER + off-the-shelf ELMO	69.96	70.56	**65.38**	59.70	73.21
AllenNER + ELMO Trained In-Domain	**75.08***	**73.13**	65.17	**61.15**	**75.87***

Table 2: F1-scores (%) for different entity types in JNLPBA dataset

sion on the expense of recall, modestly improving the F1 score. When ELMo is trained on approximately 21 million in-domain tokens both precision and recall are considerably improved resulting in a more than 1 percentage point improvement in the F1-score. A significance test using sigf (Padó, 2006) showed that this improvement is statistically significant ($p < 10^{-5}$), and the one from off-the-shelf ELMo is not ($p > 0.02$).

Table 2 shows our F1 scores on JNLPBA. It is evident from the table that using ELMo leads to salient improvements over the baseline if it is trained in-domain. The off-the-shelf ELMo has improved the performance for RNA and cell_line entity types but hurt the performance for protein, DNA, and cell_type. In-domain ELMo always obtains the best performance with the exception of RNA entity type where it is competitive with off-the-shelf ELMo and considerably better than the baseline.

Statistical significance tests using sigf (Padó, 2006) showed that most differences in Table 2 are not statistically significant after Bonferroni correction for multiple testing. The only statistically significant improvements are those of in-domain ELMo for protein and cell_type mention detections over both off-the-shelf ELMo and baseline. This could be due to the fact that proteins and cell_types are more frequent in JNLPBA when compared to other entities. Still, it is interesting to note that in-domain trained ELMo model is consistently performed better than the alternative ELMo models in all but one NER task. Table 3 shows the frequencies of different entity types in training and test sets of JNLPBA.

Our results on JNLPBA are not the state of the art. Habibi et al. (2017) report F1 scores as high as

Entity type	Training	Test
protein	30,269	5,067
DNA	9,533	1,056
RNA	951	118
cell_type	6,718	1,921
cell_line	3,830	500

Table 3: Frequencies of different entity types in training and test sets of JNLPBA

77.25% for protein and 63.31% for cell_line entity types when they use word embeddings trained on the union of (nearly 23 million) PubMed abstracts, (nearly 700,000) PMC full articles, and (approximately four million) English Wikipedia articles as input to an LSTM-CRF. Nevertheless, our results show the positive effect of using in-domain trained ELMo representations compared to a very strong baseline. We believe new state of the art will be achieved if in-domain ELMo representations are used to augment current state-of-the-art systems.

6 Conclusion

We show that token level context-aware embeddings trained on an auxiliary task of language modeling using the ELMo toolkit can be used to consistently improve biomedical named entity recognition tasks, but only when the pre-trained embeddings are trained on in-domain biomedical data. Using this technique we produce a new state of the art result on the BioCreative II dataset for gene mention detection.

Acknowledgments

The authors thank the funding organizations, Genome Canada, British Columbia Cancer Foundation, and Genome British Columbia for their partial support. The research was also partially supported by the Natural Sciences and Engineering Research Council of Canada grants NSERC RGPIN-2018-06437 and RGPAS-2018-522574 and a Department of National Defence (DND) and NSERC grant DGDND-2018-00025 to the third author.

References

Rie Kubota Ando. 2007. BioCreative II gene mention tagging system at IBM Watson. In *Proceedings of the Second BioCreative Challenge Evaluation Workshop*, volume 23, pages 101–103. Centro Nacional de Investigaciones Oncologicas (CNIO) Madrid, Spain.

David Campos, Sérgio Matos, and José Luís Oliveira. 2013. Gimli: open source and high-performance biomedical name recognition. *BMC bioinformatics*, 14(1):54.

Ciprian Chelba, Tomas Mikolov, Mike Schuster, Qi Ge, Thorsten Brants, Phillipp Koehn, and Tony Robinson. 2014. One billion word benchmark for measuring progress in statistical language modeling. *INTERSPEECH*.

Matt Gardner, Joel Grus, Mark Neumann, Oyvind Tafjord, Pradeep Dasigi, Nelson F. Liu, Matthew Peters, Michael Schmitz, and Luke S. Zettlemoyer. 2017. Allennlp: A deep semantic natural language processing platform.

Maryam Habibi, Leon Weber, Mariana Neves, David Luis Wiegandt, and Ulf Leser. 2017. Deep learning with word embeddings improves biomedical named entity recognition. *Bioinformatics*, 33(14):i37–i48.

Sepp Hochreiter and Jürgen Schmidhuber. 1997. Long short-term memory. *Neural computation*, 9(8):1735–1780.

J-D Kim, Tomoko Ohta, Yuka Tateisi, and Junichi Tsujii. 2003. Genia corpusa semantically annotated corpus for bio-textmining. *Bioinformatics*, 19(suppl_1):i180–i182.

Jin-Dong Kim, Tomoko Ohta, Yoshimasa Tsuruoka, Yuka Tateisi, and Nigel Collier. 2004. Introduction to the bio-entity recognition task at jnlpba. In *Proceedings of the international joint workshop on natural language processing in biomedicine and its applications*, pages 70–75. Association for Computational Linguistics.

John Lafferty, Andrew McCallum, and Fernando Pereira. 2001. Conditional random fields: Probabilistic models for segmenting and labeling sequence data. In *ICML*.

Guillaume Lample, Miguel Ballesteros, Sandeep Subramanian, Kazuya Kawakami, and Chris Dyer. 2016. Neural architectures for named entity recognition. In *Proceedings of the 2016 Conference of the North American Chapter of the Association for Computational Linguistics: Human Language Technologies*, pages 260–270, San Diego, California. Association for Computational Linguistics.

Christopher D. Manning, Mihai Surdeanu, John Bauer, Jenny Finkel, Steven J. Bethard, and David McClosky. 2014. The Stanford CoreNLP natural language processing toolkit. In *Association for Computational Linguistics (ACL) System Demonstrations*, pages 55–60.

Tomas Mikolov, Kai Chen, Greg Corrado, and Jeffrey Dean. 2013. Efficient estimation of word representations in vector space. *arXiv preprint arXiv:1301.3781*.

Sebastian Padó. 2006. *User's guide to* sigf: *Significance testing by approximate randomisation*.

Jeffrey Pennington, Richard Socher, and Christopher Manning. 2014. Glove: Global vectors for word representation. In *Proceedings of the 2014 conference on empirical methods in natural language processing (EMNLP)*, pages 1532–1543.

Matthew Peters, Mark Neumann, Mohit Iyyer, Matt Gardner, Christopher Clark, Kenton Lee, and Luke Zettlemoyer. 2018. Deep contextualized word representations. In *Proceedings of the 2018 Conference of the North American Chapter of the Association for Computational Linguistics: Human Language Technologies, Volume 1 (Long Papers)*, pages 2227–2237, New Orleans, Louisiana. Association for Computational Linguistics.

Marek Rei, Gamal KO Crichton, and Sampo Pyysalo. 2016. Attending to characters in neural sequence labeling models. *arXiv preprint arXiv:1611.04361*.

Golnar Sheikhshab, Elizabeth Starks, Readman Chiu, Aly Karsan, Anoop Sarkar, and Inanc Birol. 2018. Graphner: Using corpus level similarities and graph propagation for named entity recognition. In *Proceedings of the 2018 IEEE International Parallel and Distributed Processing Symposium Workshops*, pages 229–238.

Larry Smith, Lorraine K Tanabe, Rie Johnson nee Ando, Cheng-Ju Kuo, I-Fang Chung, Chun-Nan Hsu, Yu-Shi Lin, Roman Klinger, Christoph M Friedrich, Kuzman Ganchev, et al. 2008. Overview of biocreative ii gene mention recognition. *Genome biology*, 9(2):S2.

Andrew Viterbi. 1967. Error bounds for convolutional codes and an asymptotically optimum decoding algorithm. *IEEE transactions on Information Theory*, 13(2):260–269.

Self-training improves Recurrent Neural Networks performance for Temporal Relation Extraction

Chen Lin[a], Timothy A. Miller[a], Dmitriy Dligach[b], Hadi Amiri[a], Steven Bethard[c],Guergana Savova[a]

[a]Boston Children's Hospital Informatics Program, Harvard Medical School
`{firstname.lastname}@childrens.harvard.edu`
[b]Loyola University Chicago
`ddligach@luc.edu`
[c]University of Arizona
`bethard@email.arizona.edu`

Abstract

Neural network models are oftentimes restricted by limited labeled instances and resort to advanced architectures and features for cutting edge performance. We propose to build a recurrent neural network with multiple semantically heterogeneous embeddings within a self-training framework. Our framework makes use of labeled, unlabeled, and social media data, operates on basic features, and is scalable and generalizable. With this method, we establish the state-of-the-art result for both in- and cross-domain for a clinical temporal relation extraction task.

1 Introduction

Neural network methods have obtained spectacular successes in the fields of computer vision (He et al., 2016; Krizhevsky et al., 2012), speech recognition (Hinton et al., 2012; Graves and Jaitly, 2014), and machine translation (Sutskever et al., 2014), where large datasets are available for training. For extracting information from text, however, performance gains have been minimal or non-existent, with published work emphasizing that such performance parity is not obtainable without extensive feature engineering. Unlike other settings that have seen performance gains, information extraction tasks related to text typically have much smaller supervised training sets, and the neural network algorithms presumably do not see enough instances to optimally tune the large parameter space.

In this paper, we examine the important information extraction task of temporal relation extraction from clinical text. The state-of-the-art for this task is a machine learner with a heavily-engineered set of features (Sun et al., 2013; Lin et al., 2016a). The identification of temporal relations from the clinical text in the electronic medical records has been drawing growing attention because of its potential to provide accurate fine-grained analyses of many medical phenomena (e.g., disease progression, longitudinal effects of medications), with many clinical applications such as question answering (Das and Musen, 1995; Kahn et al., 1990), clinical outcomes prediction (Schmidt et al., 2005), and recognition of temporal patterns and timelines (Zhou and Hripcsak, 2007; Lin et al., 2014). Obtaining large supervised datasets for clinical tasks is expensive and difficult, so it has been challenging to show meaningful improvements from the recent explosion of sophisticated neural network methods.

Our hypothesis is that the range of interesting phenomena found in clinical data is much broader than what is covered by available gold standard datasets for temporal information extraction. The results of Clinical TempEval 2017 (Bethard et al., 2017) strongly support this latter point, as the performance of submitted systems drops severely when trained on gold instances in one domain and tested on a new domain. We are thus inspired to make use of unlabeled data in addition to gold standard data with a simple semi-supervised learning method–self-training and combine it with varieties of pre-trained word embeddings to overcome gaps in training data coverage. In self-training (Yarowsky, 1995; Riloff et al., 2003; Maeireizo et al., 2004), a classifier is first trained on existing labeled data, and then applied to unlabeled data (typically a much larger amount). The predicted instances above a confidence threshold are added to the training set and the classifier is re-trained. Self-training is especially attractive in a neural network setting because the primitive feature types used by these networks (i.e., tokens) are computationally more efficient to obtain than the sophisticated features typically used by feature engineering methods.

For pre-training, we investigate the use of multiple external data sources to train word embeddings that form the input layer of the model. Since our

Proceedings of the 9th International Workshop on Health Text Mining and Information Analysis (LOUHI 2018), pages 165–176
Brussels, Belgium, October 31, 2018. ©2018 Association for Computational Linguistics

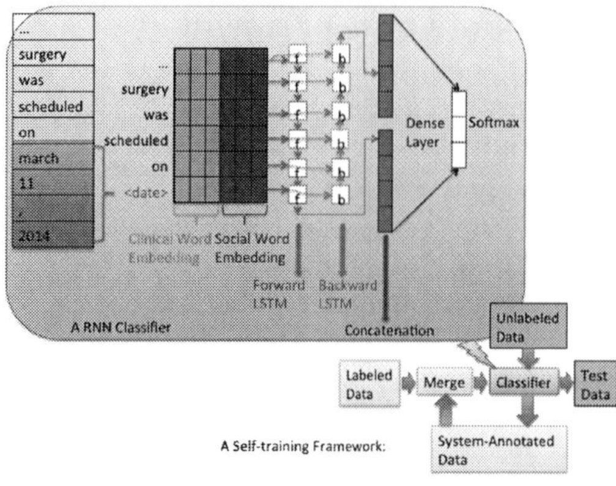

Figure 1: A RNN-based Self-training Framework

task is in the clinical setting, we use available clinical data sources, but also experiment with general domain sources trained on much larger datasets.

Besides showing that neural network approaches to information extraction can outperform feature-engineering approaches, we find that self-training works better in the neural network setting than with existing state-of-the-art feature-engineering approaches. Finally, we show that these methods generalize to new clinical domains better than the feature-engineering approaches we compare them to, obtaining state-of-the-art performance in an unsupervised domain adaptation setting.

2 Related Work

In recent years, several shared tasks on temporal relation extraction from clinical text have been organized. Among them, the i2b2 temporal challenge evaluates the i2b2 corpus (Sun et al., 2013), and Clinical TempEval series (Bethard et al., 2015, 2016, 2017) evaluate systems using the THYME corpus (Styler IV et al., 2014), which is annotated with time expressions (TIMEX3), events (EVENT), and temporal relations (TLINK) per an extension of the TimeML specifications (Pustejovsky et al., 2003; Pustejovsky and Stubbs, 2011). Challenge participants develop methods to extract EVENT and TIMEX3 entities, CONTAINS relations and document creation time relations. Herein, we focus on CONTAINS relation, which signals an EVENT occurs entirely within the temporal bounds of an *narrative container*. The *narrative container* is either another EVENT or TIMEX3.

Conventional learning methods, such as support vector machines (SVM) and conditional random fields (CRF) (Sun et al., 2013), have been developed for this task. Neural networks used in general relation extraction (Hashimoto et al., 2013; Socher et al., 2012), have also been adopted in clinical temporal relation extraction, such as structured perceptron (Leeuwenberg and Moens, 2017), convolutional neural networks (CNNs) (Dligach et al., 2017; Lin et al., 2017) and Long Short-Term memory (LSTM) networks (Tourille et al., 2017; Dligach et al., 2017). Classifiers are usually trained and tested in the same domain for the same medical condition, e.g. models are trained and tested on the colon cancer set of the THYME corpus for Clinical TempEval 2015 and 2016 (Bethard et al., 2015, 2016).

Clinical TempEval 2017 introduces the task of domain adaptation, as the most frequent use case would be the application of a model on a domain different from the domain it was trained on. The source domain of Clinical TempEval 2017 is colon cancer clinical text while the target domain is brain cancer clinical text. Few domain adaptation techniques are applied by the participants: 1) modeling unknown words to accommodate unseen vocabulary in the new domain; 2) using pretrained domain-independent word embeddings; 3) for supervised domain adaptation, assigning higher weights to samples from the new domain during model training. The performance on the domain adaptation task plummetted. Other domain adaptation methods used in general relation extraction include (Nguyen et al., 2014; Nguyen and Grishman, 2014; Plank and Moschitti, 2013).

Semi-supervised learning has been a popular approach for improving coverage and model generalizability for various information extraction tasks by exploring unlabeled data. Besides semi-supervised methods developed for feature-based learners (Le and Kim, 2015; Li and Zhou, 2010), there are such algorithms for deep neural network structures (DNN) (Laine and Aila, 2016; Kingma et al., 2014). Self-training or bootstrapping is a standard and straightforward semi-supervised learning method and widely used (Agichtein and Gravano, 2000; Pantel and Pennacchiotti, 2006; Greenwood and Stevenson, 2006; Rosenfeld and Feldman, 2007; Xu, 2008; Xu et al., 2007, 2010). To our best knowledge, we are the first to use self-training in a deep neural network setting for a clinical relation extraction task. Our motivation lies in two folds: 1) Self-training is computationally efficient as there is no other parallel learning goals such as minimizing the reconstruction errors in Generative Adversarial Networks-based semi-supervised learning. With primitive features, DNN-based self-training can effectively and efficiently evaluate a large amount of instances; 2) We hypothesize that not all unlabeled data are useful. Our goal is to use a straightforward method like self-training to study the unlabeled space and help to select the most informative instances.

3 Data

We collect a variety of external data sources, described below, to supplement the THYME dataset (Styler IV et al., 2014).

3.1 Labeled Clinical Data

Our labeled data is the THYME corpus (Styler IV et al., 2014) used for the Clinical TempEval tasks. The corpus contains internal medicine, oncology, pathology, and radiology reports for 200 *colon* cancer patients and 200 *brain* cancer patients for a total of 1200 notes. Following the unsupervised domain adaptation setting of Clinical TempEval 2017, we use colon cancer notes for model development, and brain cancer notes for cross-domain validation.

3.2 Unlabeled Clinical Data

We augment the labeled data with additional clinical notes for colon cancer patients for a total of $27,157$ notes (average length=135 words) from the same medical center as the THYME corpus from Section 3.1. On average, each patient has 125 notes of varied types – primary care, specialty care, pathology, radiology, etc. This set includes all electronic medical record notes at a single medical center for the 200 colon cancer THYME patients. We use it to automatically derive additional training instances, and refer to these generated instances as *silver* instances. We do not have access to additional unlabeled out-of-domain data (i.e. brain cancer clinical notes).

3.2.1 Clinical Word Embeddings

To train word embeddings with good vocabulary coverage and high representational power, we took advantage of the clinical notes from MIMIC-III (Medical Information Mart for Intensive Care) dataset (Johnson et al., 2016). The publicly available MIMIC III contains 879 million words from Beth Israel Deaconess Medical Center's Intensive Care Unit. We merged MIMIC-III data with the unlabeled colon cancer set above and trained 300-dimension embeddings with fastText (Joulin et al., 2016) and skip-gram (Guthrie et al., 2006) models.

3.2.2 Social Media Word Embeddings

While unlabeled clinical data provides a domain-matched source for training embeddings, additional data can be freely obtained from social media posts about colon cancer. To explore the benefits of extra coverage of such datasets versus the domain specificity of clinical embeddings, we obtain another set of embeddings using user-generated content about colon cancer from two social media platforms, namely Twitter and Reddit. For this purpose, we first generate a keyword list from two sources: a) the most frequent medical terms in the unlabeled colon cancer notes, these include any term that maps to the Unified Medical Language System concept unique identifiers (UMLS CUIs) (Bodenreider, 2004), b) the most frequent terms that map to ICD-9 billing codes related to colon cancer. These two lists results in a total number of 143 keywords. We use these keywords as a filter to collect 1.7 million publicly-available tweets about colon cancer. In addition, we collect 19K Reddit posts that contain at least one mention of colon cancer. We remove all occurrences of usernames, hash tags, URLs, and non-ASCII characters from the resulting data and employ fastText (Joulin et al., 2016) to obtain social media word embeddings.

In addition to the above embeddings, we utilize the Google News embeddings[1] trained by word2vec (Mikolov et al., 2013).

4 Methods

We develop a self-training framework to generate additional *(silver)* instances of CONTAINS relation (see Figure 1, lower-right). We focus on within-sentence CONTAINS relations and set aside all cross-sentence relations based on two motivations. First, the majority of the gold standard CONTAINS relations occur within a sentence.[2] Second, a sentence is a complete semantic and syntactic structure, which makes it an ideal unit for a sequence model, like RNN, to operate on. We therefore ignore cross-sentence CONTAINS links and focus on within-sentence CONTAINS relations. In addi-

---

[1] https://code.google.com/archive/p/word2vec/

[2] $4,3654$ within-sentence vs. 743 cross-sentence CONTAINS relations in colon cancer test set. We note that it is impractical to link all cross-sentence events and/or time expressions pairs due to the large number of potential links.

tion, since we use the official Clinical TempEval 2017 scoring tool, our models are penalized for the missed cross-sentence relations.

4.1 Preprocessing

We process the labeled and unlabeled clinical data through the sentence detection and tokenization modules of Apache cTAKES[3]. For the labeled clinical data, we use gold standard event and time expression annotations and their time classes (Styler IV et al., 2014) for both model development and final validation. For the unlabeled clinical text data, we use the cTAKES event annotator (Lin et al., 2016a) and time expression annotator (Miller et al., 2015) to automatically annotate event and time expressions along with their time classes (e.g., TIME, DATE, SET). Both labeled and unlabeled corpora are transformed to lower case as shown in Figure 2.

4.2 Instance Representation

We first create a dataset of within-sentence CONTAINS-relation candidates from the colon cancer text of the labeled clinical data. Given all gold standard events and time expressions within a sentence, we link every pair of events, and every event to a time expression (if present) to form CONTAINS candidates.

To mark the position of the relational arguments in a candidate pair, we adopt the same xml-tag marked-up token sequence representation as previous work (Dligach et al., 2017), and encode the time expression with its time class (Lin et al., 2017) for better generalizability. Figure 2 illustrates the marked-up token sequence representations for all three relational candidates, in which the event in an event-time relation pair is marked by ⟨e⟩ and ⟨/e⟩ and the time expression is marked by ⟨t⟩ and ⟨/t⟩. The time expression is further encoded by its time class, ⟨t⟩ ⟨date⟩ ⟨/t⟩, which is a gold standard attribute of a time expression annotation (Styler IV et al., 2014). Event-event instances are marked with additional indexes 1 and 2, e.g. *a ⟨e1⟩ surgery ⟨/e1⟩ is ⟨e2⟩ scheduled ⟨/e2⟩ on march* 11.

We also follow previous best practice in applying transitive closure to existing gold CONTAINS relations on the training data (Mani et al., 2006; Lin et al., 2016a). Depending on the order of the relational arguments, there are three types of gold standard relational labels, CONTAINS, CONTAINED-

---

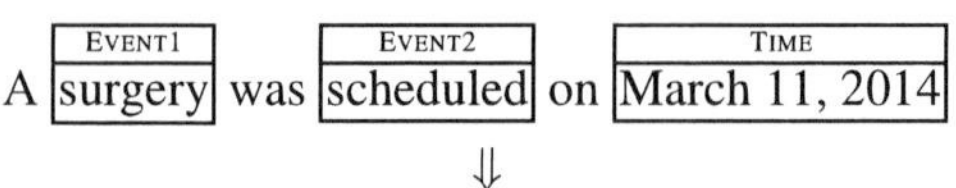

Candidate 1: a ⟨e⟩ surgery ⟨/e⟩ was scheduled on ⟨t⟩ ⟨date⟩ ⟨/t⟩;
Candidate 2: a surgery was ⟨e⟩ scheduled ⟨/e⟩ on ⟨t⟩ ⟨date⟩ ⟨/t⟩;
Candidate 3: a ⟨e1⟩ surgery ⟨/e1⟩ was ⟨e2⟩ scheduled ⟨/e2⟩ on march

Figure 2: Representations of event-event and event-time relational candidates in a sentence

BY, and NONE.

4.3 Bidirectional RNN Classifier

We use a bi-directional recurrent neural network to model the relational context similar to the state-of-the-art model (Tourille et al., 2017). As shown in Figure 1 (upper-left), each token in the token sequence input is represented by one set of clinical embeddings and one set of additional embeddings (either cancer-related social media embeddings or Google news embeddings) to capture the semantics exhibited by clinical and non-clinical terms.

As described in section 3.2.1, the clinical embeddings are derived from combining the MIMIC III and unlabeled colon cancer datasets. For the unlabeled colon cancer data, we use the extracted relational candidates as shown in Figure 2 to train embeddings, so that all xml-tag marked-up tokens and time-class tokens, e.g. ⟨/e⟩, ⟨/e1⟩, ⟨/t⟩, ⟨/date⟩, are represented. For each set of embeddings, an UNK token represents out-of-vocabulary words to accommodate unseen words in a new domain. Table 1 shows the coverage of each embedding set and their combinations over the labeled colon cancer training set. We will show the effect of the different embedding combinations in the experiments.

The two sets of embeddings for a given token are concatenated and fed into the two sequences of hidden states of RNN: forward states and backward states. The output of the two states is concatenated and fed into a dense layer and through a softmax layer to predict three relational labels as described in section 4.2. We evaluate two RNN models, Long Short-Term Memory (LSTM) (Hochreiter and Schmidhuber, 1997) and Gated recurrent units (GRUs) (Chung et al., 2014).

We implement the network in Keras (Chollet, 2015) with Theano (Theano Development Team, 2016) backend. We train our models with a batch

corpora	word#	coverage
(1) Clinical	136K	94.66%
(2) Cancer-related social media	60K	76.67%
(3) Google News	3M	83.69%
(1) + (2)	171K	95.69%
(1) + (3)	3M	95.70%

Table 1: Embedding word coverage (percentage of words in the THYME corpus covered by the vocabulary in each corpora); Clinical embeddings derived from the combination of MIMIC and unlabeled colon cancer datasets, see section 3.3; Cancer-related embedddings derived from the combination of relevant Reddit posts and tweets, see section 3.4

method	all silver	positive silver
joint bi-lstm	1.533M	19,441
SVM event-time	1.244M	57,462
SVM event-event	2.521M	36,960

Table 2: Number of generated silver training instances

size of 256, Stochastic Gradient Descent using Adam optimizer (Kingma and Ba, 2014), and a learning rate of 0.0001, on a GTX TitanX GPU. The hyper-parameters are optimized through a random search algorithm (Li et al., 2016) and the size of the hidden states of the forward and backward recurrent neural networks are set 512. We keep 10% of the training samples as a validation split, and applied a 0.5 dropout ratio and 0.0001 L2-regularized penalties to the embedding layers. For the high-precision model, we increased the weight of the L2-regularizer from 0.0001 to 0.001.

4.4 Self-Training

We apply the high-precision bi-directional RNN model trained on the labeled data to generate CONTAINS predictions on the unlabeled colon cancer data for *silver* annotations. We retain instances with a confidence score as generated by the `softmax` function of greater than 0.9 (a higher threshold will result in too few positive instances, a lower threshold will result in disproportionately many negative instances). We find that a lower threshold leads to low quality predictions and a higher threshold generates too few CONTAINS relations. The retained silver instances are merged with the gold ones and input into the bi-directional-RNN for a second-round of training.

As a comparison, we use self-training with the state-of-the-art SVM model (Lin et al., 2016a,b) to generate *silver* relations. The SVM-based THYME system is the latest release of Apache cTAKES v4 temporal module. For a comparison with the best setting of RNN-based self-training, we add all positive (CONTAINS, CONTAINED-BY) *silver* relations with the confidence threshold of greater than 0.9 to the gold training data of THYME corpus

and then retrain the SVM model.

Table 2 shows the number of silver instances generated by each learning algorithm. The high-precision bi-directional RNN model (joint-bi-lstm) is built upon LSTM networks with clinical and social media embeddings, and trained on the training split of the colon cancer set of THYME corpus.

5 Experiments

We experimented with several combinations of clinical and cancer-related social media and Google news embeddings. We tested three modes of merging silver instances with gold annotations (Figure 1, lower right): 1) Posi-Merge: merging the positive predictions (i.e. CONTAINS and CONTAINED-BY relations) with the gold relations; 2) sub-Merge: merging a subset of the silver data (a random sample of 45K silver samples including CONTAINS, CONTAINED-BY, and NONE relations) with the gold relations; and 3) all-Merge: merging all silver data with the gold relations. After merging, we shuffled gold and silver instances together to balance the batch-wise computation.

Models utilizing self-training were trained on the gold colon cancer training set of the THYME corpus and silver instances predicted from the unlabeled colon cancer data. Models were tested on the gold colon cancer and gold brain cancer development sets of the THYME corpus, comparing in-domain and cross-domain performance to select the best models for testing. The best models were tested on the gold colon cancer and brain cancer test sets (Clinical TempEval 2017 test sets).

All models were evaluated with the metrics precision (P), recall (R) and F1-score (F), using the standard Clinical TempEval evaluation script, where the P and R definitions are enhanced through temporal closure (UzZaman and Allen, 2011; UzZaman et al., 2012): when calculating precision, we run temporal closure on the gold relations but not on the system-generated ones; when calculating recall, we run temporal closure on the system-generated relations but not on the gold ones.

6 Results

Table 3 shows performance of the THYME system and various bi-directional RNN methods on the colon cancer and brain cancer development sets. For RNNs, we evaluated both LSTM and GRU models. For embedding combinations, we tested using the clinical embedding alone (C), using both clinical and cancer-related social media embeddings (CS), using both clinical and Google News embeddings (CG), and using Google News Embeddings alone (G). For ways to merge silver samples with gold instances we tested *no-self-training* in which no silver instances were used, *all-Merge* in which all silver instances were used, *sub-Merge* in which a subset of silver samples were used, and *Posi-Merge* in which only the positive silver instances were used. Among all settings, *bi-LSTM CG Posi-Merge* and *bi-LSTM CS Posi-Merge* achieved the best F1-score (F1b) on the brain development set; *bi-LSTM CS Posi-Merge* had the best F1-score (F1c) on the colon development set. These two best performing neural models along with the *THYME no-self-training* system were tested on the Clinical TempEval test splits.

Table 4 shows that the bi-LSTM models outperform the SVM-based THYME system and the Clinical TempEval 2017 top system, especially on the cross-domain experiments. The THYME system performance on the colon test set is 0.621 F1 which is an improvement over previously reported results (Lin et al., 2016b). The THYME system result on the brain cancer test is reported here for the first time. Note that the THYME system was trained on all gold colon cancer annotations (training, development and test), while the bi-LSTM models were trained on gold training colon cancer data and positive silver colon cancer samples. The best Clinical TempEval result on the gold colon cancer test set – 0.613 F1-score – is reported by the LIMSI-COT system which makes use of cTAKES-generated features (Tourille et al., 2017). The best Clinical TempEval result on the gold brain cancer test set – 0.34 F1-score – is achieved by the GUIR system (MacAvaney et al., 2017), while LIMSI-COT obtains 0.33 cross-domain F1-score.

7 Discussion

7.1 Comparison with SVM Self-Training

The top two rows of Table 3 show that the self-training technique did not improve the SVM-based THYME system. While recall reached its peak with the self-trained SVM, the precision trade-off was disastrous and F1 suffers dramatically. Our interpretation of this result is that the SVM is simply adjusting its class priors, labeling more instances as positive, but its fixed feature set and linear model constrain it from learning anything of interest from the silver data. The SVMs we use have extensively-engineered representations that were implicitly fit to the training and development sets of the colon cancer data. These feature sets may not have the representational power to find useful new patterns in the silver data. In contrast, the neural network models learn to extract features in their lower layers, and when given new data (e.g., silver data from self-training), the representation learning parts of the model are able to adapt and potentially find new patterns. This suggests that self-training for neural networks has higher potential than for SVMs, and that in the SVM setting, self-training should be accompanied by additional feature engineering.

Another difference between the models is that the SVM model relies on sophisticated linguistic features (parse trees, event and time expression attributes) that cannot be as reliably extracted from silver data. A token-sequence neural model, in contrast, makes use of very basic features and maintains a relatively accurate performance on the unlabeled data. It is possible that SVM performance is actually hurt by the lower quality features available from the silver training instances it encounters.

It is also worth noting that extracting additional silver instances for the SVM model is slower as it takes longer to generate the complex features that the SVM models use, while the token-based features of the neural model are extremely fast.

For all these reasons, we believe that neural networks are a more practical solution and better suited for a semi-supervised learning framework such as self-training.

7.2 Impact of Embeddings

Adding a broader range of embeddings as input to the bi-LSTM self-trained models improved the performance for the cross-domain task (rows 6-8 of table 3). It is possible that the clinical embeddings, even though trained on the mixture of MIMIC III and unlabeled colon cancer corpora, still do not provide semantic representation for the brain cancer notes. The diseases, symptoms, procedures, linguistic choices, etc. may vary substantially be-

Model	F1 drop ratio: (F1c-F1b)/F1c	colon cancer relations			brain cancer relations		
		P	R	F1c	P	R	F1b
1. THYME no-self-training	15.46%	0.661	0.587	0.621	0.533	0.518	0.525
2. THYME Posi-Merge	27.11%	0.185	**0.608**	0.284	0.123	**0.664**	0.207
3. bi-lstm CS no-self-training	16.59%	0.711	0.541	0.615	0.514	0.511	0.513
4. bi-lstm CS all-Merge	**8.87%**	**0.727**	0.431	0.541	**0.582**	0.428	0.493
5. bi-lstm CS sub-Merge	10.48%	0.712	0.549	0.620	0.567	0.543	0.555
6. bi-lstm C Posi-Merge	13.50%	0.712	0.551	0.622	0.528	0.549	0.538
7. bi-lstm CS Posi-Merge	9.63%	0.690	0.567	0.623	0.523	0.609	**0.563**
8. bi-lstm CG Posi-Merge	10.63%	0.684	0.584	**0.630**	0.513	0.624	**0.563**
9. bi-gru CS Posi-Merge	10.43%	0.702	0.559	0.623	0.522	0.600	0.558
10. bi-lstm G Posi-Merge	14.33%	0.673	0.530	0.593	0.475	0.545	0.508

Table 3: Model performance of *CONTAINS* relation on colon cancer and brain cancer development sets. C: clinical embeddings representation; CS: clinical and social media embeddings representation; CG: clinical and Google News embeddings representations; G: Google News embeddings. all-Merge: all silver instances added to gold training data; Posi-Merge: positive silver instances added to gold training data; sub-Merge: a subset of silver data added to gold training data.

Model	F1 drop ratio (F1c-F1b)/F1c	colon cancer relations			brain cancer relations		
		P	R	F1	P	R	F1
best Clinical TempEval	44.54%	0.657	0.575	0.613	0.52	0.25	0.34
THYME no-self-training	15.46%	0.661	**0.587**	0.621	**0.533**	0.518	0.525
bi-lstm CS Posi-Merge	13.14%	**0.700**	0.563	**0.624**	0.520	0.566	**0.542**
bi-lstm CG Posi-Merge	**13.04%**	0.692	0.576	**0.629**	0.514	**0.585**	**0.547**

Table 4: *CONTAINS* relations on colon cancer and brain cancer test set

tween these two cancer populations. Cancer-related social media and Google News embeddings come in with additional word coverage and more general semantic representations and thus help with the cross-domain performance. Word coverage increments are shown in Table 1. However, using non-clinical (Google News) embeddings on its own (row 10 of table 3) decreased both in-domain and cross-domain performance, even worse than the THYME system (row 1). It's possible that even though Google News embedding have good word coverage, general senses dominate clinical-specific senses, demonstrating the need for some clinical-specific data.

One interesting fact is that the cancer-related social media embedding has a much smaller vocabulary size than the Google News embeddings. Still, the CS option achieves the same F1-score as the CG option on the gold development brain set. Because of its better coverage and general semantic representation, CG option performs the best on the colon development set and the test sets of both colon and brain cancer data as shown in Table 4. We experimented with concatenating all three embeddings (clinical, cancer-related social media, and Google News), but did not observe any performance improvements.

7.3 Sampling of Silver Instances

Adding all high-confidence silver data to the gold training data clearly hurts performance (row 4). One possible explanation is the negative-to-positive instance ratio which is much higher in the silver data (80:1) than in the gold data (8:1). Adding the highly unbalanced silver samples may weight the system towards predicting the negative class, thus row 4 has higher precision but lower recall. Adding a random subset of silver samples to the gold samples provides additional information without skewing the class distribution, and we observe that in this setting the bi-LSTM model outperforms the THYME system, row 5 of Table 3. However, this setup may provide unpredictable performance due to the randomness of sampling the silver data.

The best merging option is the Posi-Merge. The models in rows 6-9 of Table 3 all outperform the THYME system, even for a single clinical embedding setting in row 6 of Table 3. Posi-Merge provides a stable sample of the silver data, strengthens the positive signals and achieves good cross-domain performance.

7.4 Analysis of Improvements

We are interested in understanding the different contributions of self-training and pre-trained embeddings. Embeddings can provide a kind of adaptation for words in a new domain that are similar to words in the training data (e.g., *brain* in the brain cancer corpus may behave similarly to *colon* in a colon cancer corpus). However, self-training may still provide benefit if there are words in the test set that do not have correlates in the training data, but that can be found in the silver data. In these cases, confident silver instances provide information to the neural network about how these words should be integrated into the learned representations for predicting the relation category.

To investigate this possibility, we visualized the embeddings for gold training data, silver data, and development set data, all for colon cancer patients. We hope to find a sub-space in the embedding space where there is overlap between words in the silver data and development, but no nearby words from the training data. Figure 3 shows a visualized scatter-plot (Maaten and Hinton, 2008) of one such space, showing words from gold training set (blue), silver data (red), and the gold development set (yellow), given the clinical embedding. The upper-left cluster of silver words encloses several words occurring in the development set which are not represented or even close to the nearest words from the training set visualized in the lower right corner. Figure 3 shows that through self-training the vocabulary coverage is extended to less represented areas thus the model variance error is reduced which makes the model more generalizable.

7.5 LSTM vs. GRU

Given the same settings (rows 7 and 9 of Table 3), a GRU model performs similarly to a LSTM model for the in-domain task, but differently for the cross-domain task. GRUs are related to LSTMs, both utilize gating mechanisms to manage the vanishing gradient problem, though GRUs have fewer parameters. The performance difference may not be meaningful; we selected the LSTM for the test set evaluation due to its nominally better performance. However, given the small magnitude of these differences, future work may investigate whether GRUs may have advantages in reducing overfitting.

Figure 3: A part of T-SNE-visualized space

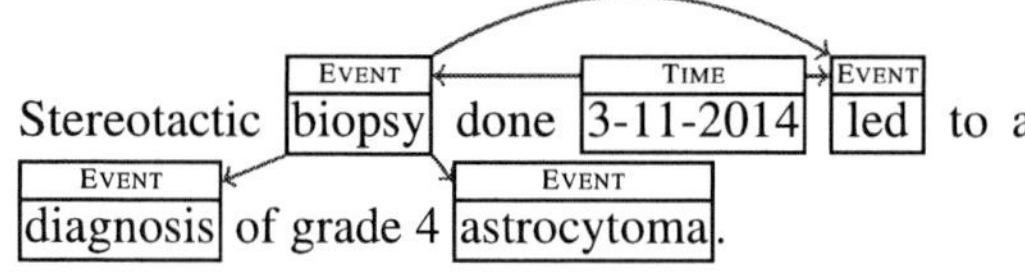

Figure 4: System annotations for a brain cancer sentence. Each arrow represents a CONTAINS relation.

7.6 Error Analysis

By comparing the error outputs of the THYME system and the best self-trained bi-LSTM system of Table 3 (rows 1 and 8) on the gold brain cancer development set, we find that the THYME system tends to pick up short-distance relation pairs, while the bi-LSTM model performs well on both short- and long- distance relations. One such example is shown in Figure 4. It represents a complex set of relations between four events and one time expression. All marked entities are participating in at least one CONTAINS relation, e.g. CONTAINS (3-11-2014, biopsy), CONTAINS (3-11-2014, led), CONTAINS (biopsy, led), CONTAINS (biopsy, diagnosis), CONTAINS (biopsy, astrocytoma). The link between two of the events, *biopsy* and *astrocytoma*, spans almost across the entire sentence. The bi-LSTM model predicts all relations correctly even without the assistance of transitive closure. We hypothesize that the benefit is due to the bidirectional setting of the LSTM model, which models the sentence structure very well. With the additional silver instances, two sets of embedding representations, and the memory capabilities, the self-trained bi-LSTM model adapts to a new domain to cover both short- and long-distance relations.

8 Conclusion

We show that neural models for temporal information extraction are able to take advantage of self-training. Compared with SVM models that leverage sophisticated features, our RNN-based self-training framework for temporal relation extraction operates on primitive features, models the sentence structure well, and is highly scalable and generalizable. Our RNN framework establishes a new state-of-the-art result for Clinical TempEval 2017 domain adaptation task. Experiments with externally-trained embeddings suggest that health-related social media or large scale general-domain text data can complement domain-specific text for a domain adaptation task. We will open source our learning framework in the near future.

Acknowledgments

The project is supported by 1U24CA184407-01 from the National Cancer Institute and R01LM010090 from the National Library Of Medicine at the US National Institutes of Health. The content is solely the responsibility of the authors and does not necessarily represent the official views of the National Institutes of Health. We thank James H. Martin and the anonymous reviewers for their valuable suggestions and constructive criticism. The Titan Xp GPU used for this research was donated by the NVIDIA Corporation.

References

Eugene Agichtein and Luis Gravano. 2000. Snowball: Extracting relations from large plain-text collections. In *Proceedings of the fifth ACM conference on Digital libraries*, pages 85–94. ACM.

Steven Bethard, Leon Derczynski, Guergana Savova, Guergana Savova, James Pustejovsky, and Marc Verhagen. 2015. Semeval-2015 task 6: Clinical tempeval. In *Proceedings of the 9th International Workshop on Semantic Evaluation (SemEval 2015)*, pages 806–814.

Steven Bethard, Guergana Savova, Wei-Te Chen, Leon Derczynski, James Pustejovsky, and Marc Verhagen. 2016. Semeval-2016 task 12: Clinical tempeval. *Proceedings of SemEval*, pages 1052–1062.

Steven Bethard, Guergana Savova, Martha Palmer, James Pustejovsky, and Marc Verhagen. 2017. Semeval-2017 task 12: Clinical tempeval. *Proceedings of the 11th International Workshop on Semantic Evaluation (SemEval-2017)*, pages 563–570.

Olivier Bodenreider. 2004. The unified medical language system (umls): integrating biomedical terminology. *Nucleic acids research*, 32(suppl 1):D267–D270.

François Chollet. 2015. Keras. https://github.com/fchollet/keras.

Junyoung Chung, Caglar Gulcehre, KyungHyun Cho, and Yoshua Bengio. 2014. Empirical evaluation of gated recurrent neural networks on sequence modeling. *arXiv preprint arXiv:1412.3555*.

Amar K Das and Mark A Musen. 1995. A comparison of the temporal expressiveness of three database query methods. In *Proceedings of the Annual Symposium on Computer Application in Medical Care*, page 331. American Medical Informatics Association.

Dmitriy Dligach, Timothy Miller, Chen Lin, Steven Bethard, and Guergana Savova. 2017. Neural temporal relation extraction. *EACL 2017*, page 746.

Alex Graves and Navdeep Jaitly. 2014. Towards end-to-end speech recognition with recurrent neural networks. In *Proceedings of the 31st International Conference on Machine Learning (ICML-14)*, pages 1764–1772.

Mark A Greenwood and Mark Stevenson. 2006. Improving semi-supervised acquisition of relation extraction patterns. In *Proceedings of the Workshop on Information Extraction beyond the Document*, pages 29–35. Association for Computational Linguistics.

David Guthrie, Ben Allison, Wei Liu, Louise Guthrie, and Yorick Wilks. 2006. A closer look at skip-gram modelling. In *Proceedings of the 5th international Conference on Language Resources and Evaluation (LREC-2006)*, pages 1–4. sn.

Kazuma Hashimoto, Makoto Miwa, Yoshimasa Tsuruoka, and Takashi Chikayama. 2013. Simple customization of recursive neural networks for semantic relation classification. In *Proceedings of the 2013 Conference on Empirical Methods in Natural Language Processing*, pages 1372–1376.

Kaiming He, Xiangyu Zhang, Shaoqing Ren, and Jian Sun. 2016. Deep residual learning for image recognition. In *Proceedings of the IEEE conference on computer vision and pattern recognition*, pages 770–778.

Geoffrey Hinton, Li Deng, Dong Yu, George E Dahl, Abdel-rahman Mohamed, Navdeep Jaitly, Andrew Senior, Vincent Vanhoucke, Patrick Nguyen, Tara N Sainath, et al. 2012. Deep neural networks for acoustic modeling in speech recognition: The shared views of four research groups. *IEEE Signal Processing Magazine*, 29(6):82–97.

Sepp Hochreiter and Jürgen Schmidhuber. 1997. Long short-term memory. *Neural computation*, 9(8):1735–1780.

Alistair EW Johnson, Tom J Pollard, Lu Shen, Liwei H Lehman, Mengling Feng, Mohammad Ghassemi, Benjamin Moody, Peter Szolovits, Leo Anthony Celi, and Roger G Mark. 2016. Mimic-iii, a freely accessible critical care database. *Scientific data*, 3.

Armand Joulin, Edouard Grave, Piotr Bojanowski, and Tomas Mikolov. 2016. Bag of tricks for efficient text classification. *arXiv preprint arXiv:1607.01759*.

Michael G Kahn, Larry M Fagan, and Samson Tu. 1990. Extensions to the time-oriented database model to support temporal reasoning in medical expert systems. *Methods of information in medicine*, 30(1):4–14.

Diederik Kingma and Jimmy Ba. 2014. Adam: A method for stochastic optimization. *arXiv preprint arXiv:1412.6980*.

Diederik P Kingma, Shakir Mohamed, Danilo Jimenez Rezende, and Max Welling. 2014. Semi-supervised learning with deep generative models. In *Advances in Neural Information Processing Systems*, pages 3581–3589.

Alex Krizhevsky, Ilya Sutskever, and Geoffrey E Hinton. 2012. Imagenet classification with deep convolutional neural networks. In *Advances in neural information processing systems*, pages 1097–1105.

Samuli Laine and Timo Aila. 2016. Temporal ensembling for semi-supervised learning. *arXiv preprint arXiv:1610.02242*.

Thanh-Binh Le and Sang-Woon Kim. 2015. Modified criterion to select useful unlabeled data for improving semi-supervised support vector machines. *Pattern Recognition Letters*, 60:48–56.

Tuur Leeuwenberg and Marie-Francine Moens. 2017. Structured learning for temporal relation extraction from clinical records. In *Proceedings of the 15th Conference of the European Chapter of the Association for Computational Linguistics*.

Lisha Li, Kevin Jamieson, Giulia DeSalvo, Afshin Rostamizadeh, and Ameet Talwalkar. 2016. Hyperband: A novel bandit-based approach to hyperparameter optimization. *arXiv preprint arXiv:1603.06560*.

Yu-Feng Li and Zhi-Hua Zhou. 2010. Improving semi-supervised support vector machines through unlabeled instances selection. *arXiv preprint arXiv:1005.1545*.

Chen Lin, Dmitriy Dligach, Timothy A Miller, Steven Bethard, and Guergana K Savova. 2016a. Multi-layered temporal modeling for the clinical domain. *Journal of the American Medical Informatics Association*, 23(2):387–395.

Chen Lin, Elizabeth W Karlson, Dmitriy Dligach, Monica P Ramirez, Timothy A Miller, Huan Mo, Natalie S Braggs, Andrew Cagan, Vivian Gainer, Joshua C Denny, and Guergana K Savova. 2014. Automatic identification of methotrexate-induced liver toxicity in patients with rheumatoid arthritis from the electronic medical record. *Journal of the American Medical Informatics Association*.

Chen Lin, Timothy Miller, Dmitriy Dligach, Steven Bethard, and Guergana Savova. 2016b. Improving temporal relation extraction with training instance augmentation. In *Proceedings of the 15th Workshop on Biomedical Natural Language Processing*, pages 108–113. Association for Computational Linguistics.

Chen Lin, Timothy Miller, Dmitriy Dligach, Steven Bethard, and Guergana Savova. 2017. Representations of time expressions for temporal relation extraction with convolutional neural networks. *BioNLP 2017*, pages 322–327.

Laurens van der Maaten and Geoffrey Hinton. 2008. Visualizing data using t-sne. *Journal of Machine Learning Research*, 9(Nov):2579–2605.

Sean MacAvaney, Arman Cohan, and Nazli Goharian. 2017. Guir at semeval-2017 task 12: A framework for cross-domain clinical temporal information extraction. In *Proceedings of the 11th International Workshop on Semantic Evaluation (SemEval-2017)*, pages 1024–1029, Vancouver, Canada. Association for Computational Linguistics.

Beatriz Maeireizo, Diane Litman, and Rebecca Hwa. 2004. Co-training for predicting emotions with spoken dialogue data. In *Proceedings of the ACL 2004 on Interactive poster and demonstration sessions*, page 28. Association for Computational Linguistics.

Inderjeet Mani, Marc Verhagen, Ben Wellner, Chong Min Lee, and James Pustejovsky. 2006. Machine learning of temporal relations. In *Proceedings of the 21st International Conference on Computational Linguistics and the 44th annual meeting of the Association for Computational Linguistics*, pages 753–760. Association for Computational Linguistics.

Tomas Mikolov, Ilya Sutskever, Kai Chen, Greg S Corrado, and Jeff Dean. 2013. Distributed representations of words and phrases and their compositionality. In *Advances in neural information processing systems*, pages 3111–3119.

Timothy A Miller, Steven Bethard, Dmitriy Dligach, Chen Lin, and Guergana K Savova. 2015. Extracting time expressions from clinical text. In *Proceedings of the 2015 Workshop on Biomedical Natural Language Processing (BioNLP 2015)*, pages 81–91. Association for Computational Linguistics.

Minh Luan Nguyen, Ivor W Tsang, Kian Ming A Chai, and Hai Leong Chieu. 2014. Robust domain adaptation for relation extraction via clustering consistency. In *Proceedings of the 52nd Annual Meeting of the Association for Computational Linguistics (Volume 1: Long Papers)*, volume 1, pages 807–817.

Thien Huu Nguyen and Ralph Grishman. 2014. Employing word representations and regularization for domain adaptation of relation extraction. In *Proceedings of the 52nd Annual Meeting of the Association for Computational Linguistics (Volume 2: Short Papers)*, volume 2, pages 68–74.

Patrick Pantel and Marco Pennacchiotti. 2006. Espresso: Leveraging generic patterns for automatically harvesting semantic relations. In *Proceedings of the 21st International Conference on Computational Linguistics and the 44th annual meeting of the Association for Computational Linguistics*, pages 113–120. Association for Computational Linguistics.

Barbara Plank and Alessandro Moschitti. 2013. Embedding semantic similarity in tree kernels for domain adaptation of relation extraction. In *Proceedings of the 51st Annual Meeting of the Association for Computational Linguistics (Volume 1: Long Papers)*, volume 1, pages 1498–1507.

James Pustejovsky, José M Castano, Robert Ingria, Roser Sauri, Robert J Gaizauskas, Andrea Setzer, Graham Katz, and Dragomir R Radev. 2003. Timeml: Robust specification of event and temporal expressions in text. *New directions in question answering*, 3:28–34.

James Pustejovsky and Amber Stubbs. 2011. Increasing informativeness in temporal annotation. In *Proceedings of the 5th Linguistic Annotation Workshop*, pages 152–160. Association for Computational Linguistics.

Ellen Riloff, Janyce Wiebe, and Theresa Wilson. 2003. Learning subjective nouns using extraction pattern bootstrapping. In *Proceedings of the seventh conference on Natural language learning at HLT-NAACL 2003-Volume 4*, pages 25–32. Association for Computational Linguistics.

Benjamin Rosenfeld and Ronen Feldman. 2007. Using corpus statistics on entities to improve semi-supervised relation extraction from the web. In *Proceedings of the 45th Annual Meeting of the Association of Computational Linguistics*, pages 600–607.

Reinhold Schmidt, Stefan Ropele, Christian Enzinger, Katja Petrovic, Stephen Smith, Helena Schmidt, Paul M Matthews, and Franz Fazekas. 2005. White matter lesion progression, brain atrophy, and cognitive decline: the austrian stroke prevention study. *Annals of neurology*, 58(4):610–616.

Richard Socher, Brody Huval, Christopher D Manning, and Andrew Y Ng. 2012. Semantic compositionality through recursive matrix-vector spaces. In *Proceedings of the 2012 joint conference on empirical methods in natural language processing and computational natural language learning*, pages 1201–1211. Association for Computational Linguistics.

William F Styler IV, Steven Bethard, Sean Finan, Martha Palmer, Sameer Pradhan, Piet C de Groen, Brad Erickson, Timothy Miller, Chen Lin, Guergana Savova, et al. 2014. Temporal annotation in the clinical domain. *Transactions of the Association for Computational Linguistics*, 2:143–154.

Weiyi Sun, Anna Rumshisky, and Ozlem Uzuner. 2013. Evaluating temporal relations in clinical text: 2012 i2b2 challenge. *Journal of the American Medical Informatics Association*, 20(5):806–813.

Ilya Sutskever, Oriol Vinyals, and Quoc V Le. 2014. Sequence to sequence learning with neural networks. In *Advances in neural information processing systems*, pages 3104–3112.

Theano Development Team. 2016. Theano: A Python framework for fast computation of mathematical expressions. *arXiv e-prints*, abs/1605.02688.

Julien Tourille, Olivier Ferret, Aurelie Neveol, and Xavier Tannier. 2017. Neural architecture for temporal relation extraction: A bi-lstm approach for detecting narrative containers. In *Proceedings of the 55th Annual Meeting of the Association for Computational Linguistics (Volume 2: Short Papers)*, volume 2, pages 224–230.

Naushad UzZaman and James F Allen. 2011. Temporal evaluation. In *Proceedings of the 49th Annual Meeting of the Association for Computational Linguistics: Human Language Technologies: short papers-Volume 2*, pages 351–356. Association for Computational Linguistics.

Naushad UzZaman, Hector Llorens, James Allen, Leon Derczynski, Marc Verhagen, and James Pustejovsky. 2012. Tempeval-3: Evaluating events, time expressions, and temporal relations. *arXiv preprint arXiv:1206.5333*.

Fei-Yu Xu. 2008. *Bootstrapping relation extraction from semantic seeds*. Saarland Univ., Department of Computational Linguistics and Phonetics.

Feiyu Xu, Hans Uszkoreit, Sebastian Krause, and Hong Li. 2010. Boosting relation extraction with limited closed-world knowledge. In *Proceedings of the 23rd International Conference on Computational Linguistics: Posters*, pages 1354–1362. Association for Computational Linguistics.

Feiyu Xu, Hans Uszkoreit, and Hong Li. 2007. A seed-driven bottom-up machine learning framework for extracting relations of various complexity. In *Proceedings of the 45th annual meeting of the Association of Computational Linguistics*, pages 584–591.

David Yarowsky. 1995. Unsupervised word sense disambiguation rivaling supervised methods. In *Proceedings of the 33rd annual meeting on Association for Computational Linguistics*, pages 189–196. Association for Computational Linguistics.

Li Zhou and George Hripcsak. 2007. Temporal reasoning with medical dataa review with emphasis on medical natural language processing. *Journal of biomedical informatics*, 40(2):183–202.

Listwise temporal ordering of events in clinical notes

Serena Jeblee
Department of Computer Science
University of Toronto
Toronto, Ontario, Canada
sjeblee@cs.toronto.edu

Graeme Hirst
Department of Computer Science
University of Toronto
Toronto, Ontario, Canada
gh@cs.toronto.edu

Abstract

We present metrics for listwise temporal ordering of events in clinical notes, as well as a baseline listwise temporal ranking model that generates a timeline of events that can be used in downstream medical natural language processing tasks.

1 Introduction

For medical narratives such as clinical notes, event and time information can be useful in automated classification and prediction tasks. For example, the timeline of a patient's medical history can be used to predict whether they will be readmitted to the hospital within a certain time window. A medical timeline can also be used for other tasks such as disease classification, and for summarizing a patient history for physicians.

Because events are not necessarily mentioned in chronological order in such documents, once the individual events are identified, the model needs to determine the temporal relationships between them. Temporal relations are categorical labels that describe how two events are related. These relations can be binary (related or not) or simple (BEFORE, AFTER, OVERLAP), or they can capture more complex relationships such as partial overlap or adjacency. A popular temporal relation scheme for clinical notes is the CONTAINS relation, which specifies whether a time phrase or event subsumes another event.

However, most temporal relation methods use pairwise classification, which can result in inconsistent relationships and which requires classifying n^2 pairs of events, many of which have no defined relation. What is needed is an overall timeline of medically relevant events that ideally can capture event duration and overlap. A listwise ordering of events inherently captures all pairwise relationships between events and prevents inconsistencies that can arise in pairwise ordering.

While ranking methods have generally been applied to information retrieval tasks such as searching, we can view temporal ordering as a ranking task. In this work, we examine a baseline listwise ranking method for events in clinical notes and we establish a set of metrics for evaluating listwise temporal ordering of these events.

1.1 Listwise vs. pairwise ordering

Temporal relation extraction is typically framed as a pairwise classification problem: generate all pairs of events in a document, and then determine what type of temporal relation exists between them, if any. The major problem with this approach is that the vast majority of event pairs have no relationship, or the relation between them is unknown. This results in an unbalanced classification problem, and there is no guarantee that the predicted pairwise relations are consistent with one another. Because of the sparsity of annotated long-distance relations, many pairwise classification models have been limited to events mentioned within the same sentence or within some small window of the text. It is often difficult for humans to analyze the relations from an entire document quickly, especially when they are inconsistent.

In contrast, a document-level list inherently captures pairwise relations between all events in the document, regardless of whether or not they appear in the same sentence. Thus, we choose to represent the events as a temporally ordered list instead of as pairs of temporal relations.

However, since pairwise relations often capture relationships that are more complex than just BEFORE, AFTER, and OVERLAP, we add time information to the events in the reference list when available. This information includes event start, end, and overlap times, based on the annotated re-

Proceedings of the 9th International Workshop on Health Text Mining and Information Analysis (LOUHI 2018), pages 177–182
Brussels, Belgium, October 31, 2018. ©2018 Association for Computational Linguistics

lationships to time phrases in the text. For this work, we sort the list by event start time, but in principle we could sort by end time or examine event overlaps. All time information can be either exact, relative (before or after a certain time), or unknown.

2 Related work

Most existing work on temporal relation extraction for clinical text relies on human-labeled spans and relations. The input to these models is usually pairs of events (or events and times) that a human has identified as being related, and all the model has to do is decide the type of relation. However, given an unlabeled dataset the task is much more difficult – the system must first identify the events and time phrases, decide which pairs are related, and then determine the type of each relation. Derczynski (2017) covers the general topic of temporal ordering of events in text.

For the medical domain, the Clinical TempEval task at SemEval (Bethard et al., 2017) has multiple tasks that involve identification of events, time expressions, and attributes in clinical notes, as well as relation classification. SemEval 2015 also had a task on cross-document event ordering, although the data was in the news domain (Minard et al., 2015).

Additionally, most recent work has focused on small relation sets, such as narrative container relations (CONTAINS, NO-RELATION), which were originally introduced by Pustejovsky and Stubbs (2011), or simple relations (BEFORE, AFTER, OVERLAP, NONE), although some work has attempted to classify with Allen's complete set of 13 temporal relations (Allen, 1984).

Dligach et al. (2017) and Lin et al. (2017) achieved state-of-the-art performance on identifying container relations in the THYME corpus (Styler et al., 2014); however, they considered only relations in which both entities appear in the same sentence. This is a limitation in many temporal relation systems. Since clinical notes are often long and may refer to distant entities such as the admission or discharge date, cross-sentence relations should not be ignored. Tourille et al. (2017) identified cross-sentence container relations in the THYME corpus, in addition to intra-sentence relations, using a bi-directional LSTM. They used word and character embeddings of gold-standard event attributes and attributes generated by cTAKES (Savova et al., 2010).

Tannier and Muller (2011) addressed relation closure in temporal graphs with all 13 Allen relations. In our current work we deal only with simple relations, but this is something we would like to expand in the future.

For our ranking model, we build upon List-Net (Cao et al., 2007), which describes a listwise approach to ranking. The ranking function is a linear neural network which assigns a relevance score for each document in a set related to a query (such as in a document retrieval task). The loss function is typically based on top-k probability, i.e., the probability of a given document being ranked among the top k documents with respect to a query. More recent work such as IntervalRank (Moon et al., 2010) used isotonic regression with a maximum margin criteria to optimize for correct relative rankings.

3 Data

3.1 Dataset

We use the THYME corpus (Styler et al., 2014), which contains de-identified clinical notes with human-annotated times, events, and temporal relations, using the TimeML schema (Pustejovsky et al., 2003). This dataset is publicly available with a data use agreement. We use the provided train/dev/test split and the gold-standard EVENT, TIMEX3, and temporal relation annotations, including document creation time (DCT) relations.

For now, our listwise ordering method can represent only simple relations (BEFORE, AFTER, OVERLAP), so we map the BEFORE/OVERLAP relation and ENDS-ON relation to BEFORE (since we are ranking by start time), we transform AFTER relations to BEFORE, and all other relations in the THYME dataset to OVERLAP (including the CONTAINS relation).

One of the limitations of the annotations in the THYME dataset is that event annotations are always applied to just a single word, even though there are many instances where the event would be better represented by a phrase. Unfortunately, this is common in temporally annotated datasets.

3.2 Converting gold-standard pairwise relations to list representations

In order to evaluate listwise ordering methods, we need a reference list to compare against. To our knowledge, all temporally annotated clinical

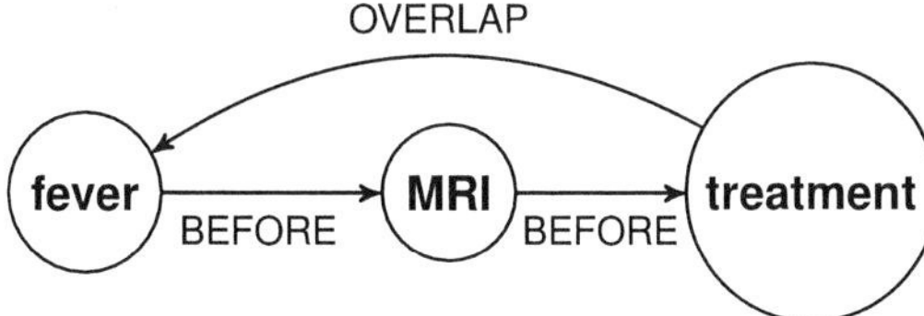

Figure 1: Example of the type of cycle in the temporal graph where the OVERLAP link would be dropped.

datasets have pairwise annotations, so we convert these pairwise relations into reference lists for training our model. We use the graph of gold-standard pairwise event relations to extract a grouped listwise ordering. This is not a straightforward process, since not all event pairs are annotated. The vast majority of relations are between event and the document creation time (DCT), which makes it difficult to determine how events are related to each other when there is no explicit annotation for that pair.

First, we take only the gold-standard event–event relations and create a directed graph representation[1]. Unfortunately, we find that some of these graphs have cycles, which indicate inconsistent orderings in the gold-standard annotation (such as A BEFORE B, B BEFORE C, C OVERLAP A). We have no choice but to drop some relation links in order to resolve these cycles. We choose to drop OVERLAP links, since these are the least specific, as the relation it does not specify *how* the events overlap. Since we are ordering by start time, for two events to have different ranks means only that one starts before the other; it doesn't mean that they don't overlap. Therefore we favor preserving the BEFORE relations. In total, 30 OVERLAP links were removed from the test set. See Figure 1 for a fabricated example of the type of inconsistency where the OVERLAP link is dropped.

We then augment the graph with transitive and time-based relations. For annotated event–time relations, we add the associated time information to the event, along with the part of the interval that the time specifies (start, end, or overlap). We use the Python *dateparser* module to convert the string representation to an ISO date–time format. We then compare the time intervals of every pair of events to discover more BEFORE and AFTER re-

---

[1] AFTER relations are inverted to become BEFORE relations, and INITIALIZES and FINISHES are converted to BEGINS-ON and ENDS-ON respectively.

lations. We compare the start and end times of events first, and if that information is not available, we compare the overlap times of the two events.

Lastly, we group the events that all have the same incoming and outgoing relations and have either overlap relations or no specified relations with each other. This results in a number of 'bins', which can each contain one or more events, and all of the relations from the individual events. We then order these bins according to the BEFORE and AFTER relations between bins, which are preserved from the individual events. All events in the same bin are assigned the same rank. The final list of events, including associated time information, can be easily viewed and understood by humans.

We verify that the output list preserves the pairwise relations by checking that for each event–event relation in the original set, the events are ordered correctly in the list. For time–event relations, we check that the associated interval information is consistent with the time relation. As discussed above, we are forced to ignore some of the OVERLAP links and leave the events in separate bins because combining them would create conflicts between the merged edges. We also note that there may be many variants of the listwise ordering that are consistent with the pairwise gold-standard relations.

The list conversion code is available at `https://github.com/sjeblee/chrononet`.

4 Listwise evaluation metrics

Traditional ranking models are usually evaluated according to normalized discounted cumulative gain (NDCG) and mean average precision (MAP). However, both of these metrics are focused on the top k ranked documents, which makes sense for a document retrieval task but is not an appropriate metric for temporal ranking, where we care about the ordering of all events.

Here we present two listwise ranking metrics, in addition to the standard pairwise recall:

Mean squared error (MSE)

$$MSE = \frac{\sum_{y \in Y}(rank_t(y) - rank_p(y))^2}{|Y|} \quad (1)$$

where Y is the set of events, $rank_t$ is the correct rank, and $rank_p$ is the predicted rank. This is an absolute metric that measures how correct the rank score is for each individual event. However, this does not measure how correct the relative rankings are, so we introduce a second metric:

Pairwise ordering accuracy (POA)

$$POA = \frac{p_O + p_E}{|L_O| + |L_E|} \qquad (2)$$

$$p_O = \sum_{u,v \in L_O} \begin{cases} 1, & \text{if } rank_p(u) < rank_p(v) \\ 0, & \text{otherwise} \end{cases} \qquad (3)$$

$$p_E = \sum_{u,v \in L_E} \begin{cases} 1, & \text{if } |rank_p(u) - rank_p(v)| \leq \varepsilon \\ 0, & \text{otherwise} \end{cases} \qquad (4)$$

where L_O is the set of ordered pairs (u, v) in the reference list such that $rank_t(u) < rank_t(v)$ (i.e., event u is ranked before event v), and L_E is the set of pairs (u, v) where $rank_t(u) = rank_t(v)$.

Since the ranking model may output slightly different rank values for events that are close together, we consider rank values to be the same if they are within ε of each other. For this paper we set $\varepsilon = 0.01$, and rank values are between 0 and 1 inclusive, with 0 being the earliest start time.

Although this metric looks at pairs of events, it measures the overall accuracy of the whole list. If two events are swapped but are otherwise in roughly the correct position in the list, POA will penalize the model less than it would for an event that is placed far away from its correct position.

Gold-standard pairwise relation recall (GPR)

From the list output, it is easy to extract all event–event relation pairs. From these we can compute the pairwise classification accuracy. Since many event–event relations are not present in the gold-standard annotations, we report recall only.

5 Models

For our ranking model we use an open-source implementation of ListNet[2], substituting rank MSE as the loss function.

The model input is the concatenation of an embedding vector and a normalized vector of numerical features. The embedding vector contains the word embedding of the event text concatenated with the word embeddings of the previous and next 3 words, and the second feature vector contains the gold-standard event attributes and the

---

[2] https://github.com/shiba24/learning2rank

Model	MSE	POA	GPR$_B$	GPR$_O$
Reference list	.000	1.000	.844	.246
ListNet	**.072**	**.517**	.420	.254
Text order	.148	.413	**.640**	.000
Random ordering	.170	.366	.485	.000
Pairwise NN	–	–	.186	**.624**

Table 1: Listwise ordering on the THYME test set. MSE: mean squared error, POA: list pairwise ordering accuracy, GPR$_B$: gold-standard pairwise relation recall of BEFORE relations, GPR$_O$: overlap relations

Model	P	R	F$_1$
Pairwise NN	.006	.540	.011
Pairwise NN (no NONE)	.841	.851	.825

Table 2: Pairwise relation classification on the THYME test set. The first pairwise neural network (NN) model includes all possible pairs, including NONE relations. The second model is restricted to only pairs that are known to have a relation. P: precision, R: recall

span start of the event. We use publicly available word embeddings trained on Wikipedia, PubMed, and PMC (Pyysalo et al., 2013). The target rank of each event is the position in the reference list, scaled to $[0, 1]$. Any number of events can share the same rank.

The pairwise classification model is a feedforward neural network implemented in PyTorch (Paszke et al., 2017), with one hidden layer with 256 nodes and ReLU activations, trained for 10 epochs. Each event pair is represented with the same features as the ranking model, and the scaled character distance between the two events in the text. The goal of this classification model is not to beat the state of the art, but rather to compare the listwise method to a simple pairwise model.

6 Results

Table 1 shows the accuracy of the ListNet ranking model according to the listwise metrics (MSE and POA), as well as gold-standard pair relation recall (GPR). As a baseline, we include the results from random ranking (every event is randomly assigned a ranking value between 0 and 1), and ranking by the order of mention in the text (since many events are indeed mentioned in chronological order). Scores from random ranking are averaged over 10 runs. We also include GPR results from the pairwise classification model for comparison.

While the ListNet ranking model has plenty of room for improvement in terms relative ordering, it outperforms both the random ordering and the order of mention in the text.

Table 2 shows the accuracy of the pairwise classification with respect to the gold-standard annotations. We cannot extract a listwise ordering from the pairwise model results because the predicted relations have cycles. In addition, most temporal relation models using THYME data have used the full set of relations or only container relations, and thus are not comparable to this model.

7 Discussion and future work

For many health-related NLP tasks, listwise ordering offers several benefits over pairwise ordering. The list avoids cycles and inconsistent pair relations, and is also a more compact representation – all pairwise relations can be inferred from the list. Moreover, the list of events and associated time information is easy for humans to review.

Although simple listwise ordering does not capture more-finely grained interval temporal relations such as partial event overlap and endpoint relations, the inclusion of interval time information for each event allows us to choose how to order them. For example, we could choose to order the list by end time instead of start time. In the future we hope to represent more-complex event relations and handle relative time phrases.

8 Conclusion

We have shown that events in clinical text can be ordered in a listwise fashion, which prevents many of the issues that occur in pairwise classification. The metrics presented here are an alternative to pairwise-only metrics, which we hope will serve as a foundation for further listwise temporal ordering work in the medical domain.

Acknowledgments

The authors thank Prabhat Jha and Mireille Gomes for their advice, and the Mayo Clinic for providing the THYME dataset through the Health Natural Language Processing Center. This work was supported by a Google Faculty Award, the U.S. National Institutes of Health, and the University of Toronto.

References

James F Allen. 1984. Towards a general theory of action and time. *Artificial Intelligence*, 23(2):123–154.

Steven Bethard, Guergana Savova, Martha Palmer, and James Pustejovksy. 2017. SemEval-2017 Task 12: Clinical TempEval. In *Proceedings of the 11th Workshop on Semantic Evaluations (SemEval 2017)*, pages 565–572. Association for Computational Linguistics.

Zhe Cao, Tao Qin, Tie-Yan Liu, Ming-Feng Tsai, and Hang Li. 2007. Learning to rank: From pairwise approach to listwise approach. In *Proceedings of the 24th International Conference on Machine Learning*, ICML '07, pages 129–136, New York, NY, USA. ACM.

Leon RA Derczynski. 2017. *Automatically Ordering Events and Times in Text*, volume 677 of *Studies in Computational Intelligence*. Springer International Publishing.

Dmitriy Dligach, Timothy Miller, Chen Lin, Steven Bethard, and Guergana Savova. 2017. Neural temporal relation extraction. In *Proceedings of the 15th Conference of the European Chapter of the Association for Computational Linguistics: Volume 2, Short Papers*, pages 746–751. Association for Computational Linguistics.

Chen Lin, Timothy Miller, Dmitriy Dligach, Steven Bethard, and Guergana Savova. 2017. Representations of time expressions for temporal relation extraction with convolutional neural networks. In *Proceedings of the 16th Workshop on Biomedical Natural Language Processing*, pages 322–327. Association for Computational Linguistics.

Anne-Lyse Minard, Manuela Speranza, Eneko Agirre, Itziar Aldabe, Marieke Van Erp, Bernardo Magnini, German Rigau, and Fondazione Bruno Kessler. 2015. SemEval-2015 Task 4: TimeLine: Cross-document event ordering. *Proceedings of the 9th International Workshop on Semantic Evaluation (SemEval 2015)*, (SemEval):778–786.

Taesup Moon, Alex Smola, Yi Chang, and Zhaohui Zheng. 2010. IntervalRank: Isotonic regression with listwise and pairwise constraints. In *Proceedings of the Third ACM International Conference on Web Search and Data Mining*, WSDM '10, pages 151–160, New York, NY, USA. ACM.

Adam Paszke, Sam Gross, Soumith Chintala, Gregory Chanan, Edward Yang, Zachary DeVito, Zeming Lin, Alban Desmaison, Luca Antiga, and Adam Lerer. 2017. Automatic differentiation in PyTorch. In *NIPS Workshop*.

James Pustejovsky, José Castaño, Robert Ingria, Roser Saurí, Robert Gaizauskas, Andrea Setzer, and Graham Katz. 2003. TimeML: Robust specification of event and temporal expressions in text. In *IWCS-5,*

Fifth International Workshop on Computational Semantics, pages 1–11.

James Pustejovsky and Amber Stubbs. 2011. Increasing informativeness in temporal annotation. In *Proceedings of the 5th Linguistic Annotation Workshop*, pages 152–160.

Sampo Pyysalo, Filip Ginter, Hans Moen, Tapio Salakoski, and Sophia Ananiadou. 2013. Distributional semantics resources for biomedical text processing. In *Proceedings of the Fifth International Symposium on Languages in Biology and Medicine*, pages 39–44.

Guergana K Savova, James J Masanz, Philip V Ogren, Jiaping Zheng, Sunghwan Sohn, Karin C Kipper-Schuler, and Christopher G Chute1. 2010. Mayo clinical text analysis and knowledge extraction system (cTAKES): Architecture, component evaluation and applications. *Journal of the American Medical Informatics Association: JAMIA*, 17(5):507–513.

William F Styler, IV, Steven Bethard, Sean Finan, Martha Palmer, Sameer Pradhan, Piet C de Groen, Brad Erickson, Timothy Miller, Chen Lin, Guergana Savova, and James Pustejovsky. 2014. Temporal annotation in the clinical domain. *Transactions of the Association for Computational Linguistics*, 2014(2):143–154.

Xavier Tannier and Philippe Muller. 2011. Evaluating temporal graphs built from texts via transitive reduction. *Journal of Artificial Intelligence Research*, 40:375–413.

Julien Tourille, Olivier Ferret, Aurélie Névéol, and Xavier Tannier. 2017. Neural architecture for temporal relation extraction: A bi-LSTM approach for detecting narrative containers. In *Proceedings of the 55th Annual Meeting of the Association for Computational Linguistics*, pages 224–230.

Time Expressions in Mental Health Records for Symptom Onset Extraction

Natalia Viani, Lucia Yin,
Joyce Kam, André Bittar
IoPPN, King's College London
London, UK

Ayunni Alawi
University of Sheffield
Sheffield, UK

Rina Dutta, Rashmi Patel,
Robert Stewart
IoPPN, King's College London
SLaM NHS Foundation Trust
London, UK

Sumithra Velupillai
IoPPN, King's College London
London, UK;
KTH Royal Institute of Technology
Stockholm, Sweden

Abstract

For psychiatric disorders such as schizophrenia, longer durations of untreated psychosis are associated with worse intervention outcomes. Data included in electronic health records (EHRs) can be useful for retrospective clinical studies, but much of this is stored as unstructured text which cannot be directly used in computation. Natural Language Processing (NLP) methods can be used to extract this data, in order to identify symptoms and treatments from mental health records, and temporally anchor the first emergence of these. We are developing an EHR corpus annotated with time expressions, clinical entities and their relations, to be used for NLP development. In this study, we focus on the first step, identifying time expressions in EHRs for patients with schizophrenia. We developed a gold standard corpus, compared this corpus to other related corpora in terms of content and time expression prevalence, and adapted two NLP systems for extracting time expressions. To the best of our knowledge, this is the first resource annotated for temporal entities in the mental health domain.

1 Introduction and Background

For psychiatric disorders such as schizophrenia, prolonged periods of time without treatment are associated with worse intervention outcomes (Kisely et al., 2006). The number of days between first symptom onset and initiation of adequate treatment is defined as duration of untreated psychosis (DUP). For patients with schizophrenia, a longer DUP has been linked to poorer cognitive function at the time of first presentation (Lappin et al., 2007). In addition, it has been shown to predict more severe symptoms and greater social and functional impairment (Hill et al., 2012). Therefore, identifying and reducing the DUP could significantly improve both clinical and functional outcomes. Starting from this observation, there is an increasing interest in measuring the DUP across large clinical samples, to provide a quality measure for mental health services, and in developing international guidelines aimed at reducing this value, thus improving outcomes at different levels (Connor et al., 2013).

Routinely collected data from health services, such as electronic health records (EHRs), can be useful for large-scale retrospective clinical studies. In mental health services, a large proportion of clinically relevant information is recorded only in open text fields. To make this information available for computational analysis, Natural Language Processing (NLP) methods can be used (Meystre et al., 2008; Wang et al., 2018). When applying NLP techniques to the clinical domain, one crucial task involves the identification of *temporal information*. In general, for temporal information modeling, three different steps are typically outlined: (i) the identification of relevant concepts, such as symptoms (*hallucinations*) and treatments (*Clozapine*), (ii) the identification of time expressions (*May 1st*), and (iii) the identification of temporal relations between entity pairs ({hallucinations} BEFORE {Clozapine}).

Over the past years, methods for temporal information extraction have been developed with promising results, mainly based on the ISO-TimeML specification language that was devel-

Proceedings of the 9th International Workshop on Health Text Mining and Information Analysis (LOUHI 2018), pages 183–192
Brussels, Belgium, October 31, 2018. ©2018 Association for Computational Linguistics

oped for the general NLP domain (Pustejovsky et al., 2010). In the clinical domain, a few manually annotated corpora (*gold standards*) have been created. As part of the Informatics for Integrating Biology and the Bedside (i2b2) project, a set of 310 de-identified discharge summaries from an Intensive Care Unit (ICU) were annotated with events, time expressions, and temporal relations (Sun et al., 2013a). This corpus was then used for organizing the 2012 i2b2 Challenge on temporal relation extraction (Sun et al., 2013b). In the oncology field, Styler IV et al. developed a corpus of 1,254 de-identified EHR notes, annotated for both clinical and temporal information (THYME corpus) (Styler IV et al., 2014). This corpus has been used in different NLP challenges, among which Clinical TempEval 2015 and 2016 focused on temporal information extraction (440 and 591 documents, respectively) (Bethard et al., 2015, 2016). In both these corpora, four main TimeML types of time expressions (TIMEXes) are defined: Date, Duration, Frequency (or Set), and Time. The THYME corpus also includes two additional TIMEX types specific to the oncology domain: PrePostExp (expressions indicating Pre- and Post-operational concepts) and Quantifier (expressions like *twice* or *four times*).

Compared to other clinical domains, mental health records are characterized by a greater extent of narrative portions, describing symptomology and health progression without relying on structured fields. In this framework, relevant temporal information (e.g., associated to symptom onset or treatment initiation) is not always well represented by current temporal models. For example, identifying expressions like *at age 8* or *in 3rd year of secondary school* is not straightforward, especially as regards the normalization phase (e.g., converting *6th May 2018* to "2018-05-06").

Our long-term goal is to accurately identify symptoms and treatments from mental health records, and anchoring these on a timeline, to be able to calculate DUP and other clinically relevant temporal constructs on a large patient cohort. To address this long-term goal, we are developing a corpus with annotations that cover all necessary elements (time expressions, clinical entities and their relations).

In this study, we focus on one subgoal: addressing the problem of accurately identifying time expressions in mental health records related to pa-

tients who have been diagnosed with schizophrenia. Our aim was (i) to develop a gold standard corpus with time expression annotations, (ii) to analyze and compare typical time expressions in this corpus with other clinical corpora that have been annotated with time information (i2b2 2012, Clinical TempEval 2016), and (iii) to perform a feasibility study on adapting existing NLP systems for extracting time expressions.

2 Materials and Methods

2.1 Data

In this study, we used anonymized[1] mental health records from the Clinical Record Interactive Search (CRIS) database (Perera et al., 2016)[2]. This database was derived in 2008 from the EHR system adopted by a large mental healthcare provider in southeast London: the South London and Maudsley National Health Service (NHS) Foundation Trust (SLaM).

Mental health records for patients who had received a diagnosis related to schizophrenia were extracted. To identify these patients, we queried the CRIS database for patients who had been documented with an ICD-10 code for this disease (F20*) or, if not documented with a structured code, we relied on the output of an NLP tool which extracts diagnoses from free text (based on the keyword "schizophrenia") (Perera et al., 2016), resulting in 8,483 documents for 1,691 patients[3]. To make the task feasible for manual annotation and relevant to the clinical use-case, two main document sample steps were taken:

1. Only documents that were written within 3 months of first presentation to mental health services were extracted, on the assumption that these early documents would most likely contain the richest description of the patient's clinical history and presenting complaints related to relevant symptoms;

2. From these documents, only the longest document (in terms of total number of characters) for each patient was extracted to be used for annotation, on the assumption that this

---

[1]Textual portions are automatically anonymized by removing patient and relative identifiers, such as names and postal codes.

[2]This database has ethical approval for secondary analysis (Oxford REC C, reference 08 H0606 71+5).

[3]Data were extracted on March 31st 2016 for patients accepted in services after January 1st 2012.

document most likely would contain most information about the patient history;

From the extracted set, a random sample of 52 documents (one document per patient) was used in the time expression annotation task for creating our corpus.

2.2 Time Expression Annotations

As a first step for the extraction of psychosis symptom onset, it is necessary to identify all the time expressions occurring inside the text (e.g., *May 2012, a year ago*). These expressions can then be used at a later stage, to link each mentioned symptom to the corresponding date or time.

To enable the development of an accurate temporal extraction system, we manually annotated the available corpus with occurrences of time expressions, marking both TIMEX spans and types (e.g., Date). To facilitate this task, we prepared domain-specific annotation guidelines, inspired by the guidelines used in the 2012 i2b2 challenge (Sun et al., 2013a) and the THYME project (Styler IV et al., 2014).

In addition, we performed a comparative analysis with existing corpora (i2b2 2012 and Clinical TempEval 2016), to highlight similarities and differences, and to gain deeper knowledge in domain-specific characteristics related to how time information is documented in clinical text.

Comparison to Related Corpora and Guidelines Adaptation

Both the i2b2 2012 and the Clinical TempEval 2016 corpora are characterized by relatively short notes, with content organized in semi-structured sections (e.g., "History of present illness", "Hospital course"). To develop guidelines tailored to the mental health domain, we manually reviewed a few example documents to identify initial domain-specific requirements. In our corpus, most documents have few or no systematic section, with clinical and temporal information scattered across many different paragraphs. Moreover, symptoms and their onset are frequently associated to vague dates, as opposed to most events documented within the ICU and the oncology domain (e.g., problems, exams, operations). As a consequence, we found that the examples included in the i2b2 2012 and THYME guidelines did not capture all the time expressions that are typical of the mental health domain, and we de-

cided to adapt them in order to simplify and clarify the annotation task. First, we only kept the TIMEX types that were relevant to the considered clinical use-case[4]: Date (e.g., *in May 2012, yesterday*), Time (e.g., *in the morning, 3 pm*), Duration (e.g., *for three years, over the past two weeks*), and Frequency (e.g., *daily, twice a week*). Within Dates, we explicitly included generic expressions such as *past* and *current*, to enable temporal contextualization of events that cannot be anchored to specific TIMEXes. As for Frequencies, we put a particular focus on medication-related TIMEXes and domain-specific expressions (e.g., *OD* for "once daily"). We also defined an additional TIMEX type for "age-related" expressions, to capture clinically relevant temporal patient information. Although this type is not included in common TimeML models, it has been previously investigated as it can encompass relevant temporal information in a clinical setting (Wang et al., 2016). In this study, besides looking at the patient's current age (e.g., *28-year-old* man), we included all the expressions that rely on the date of birth in order to be correctly normalized (see Section 3.1). The final guidelines, which were written and revised by two NLP researchers, describe: the annotation task and goal, the TimeML TIMEX types (with sentences taken from the reference guidelines), and the domain-specific TIMEX types and examples.

Annotation Process

Annotations were carried out by three medical students, using the eHOST annotation tool (South et al., 2012). The students were all native English speakers and in their 1st-3rd year of medical studies. The corpus was randomly divided into five batches of documents (9-13 documents in each batch), and each batch was independently annotated by two different annotators. After the completion of the first batch (*development set*, 10 documents), we jointly discussed issues that had arisen during the annotation process, to refine and reach a consensus on improvements and edits in the guidelines. As a result, we added specific rules for the time expressions that had caused disagreements, and removed ambiguous sentences and examples. For instance, we found that "dates" and "durations" were sometimes hard to distinguish, and created specific rules to disambiguate those

---

[4]PrePostExp and Quantifier TIMEX types were not considered.

(e.g., *over the last week* should be annotated as a Duration, and not a Date). The updated guidelines were then applied to annotate the remaining documents. When all batches had been double-annotated, we carried out an adjudication phase in order to create a gold standard corpus. The adjudicator decided which annotations to include in the gold standard in case of disagreement between annotators, added missing annotations and omitted or corrected erroneous ones.

2.3 Automated Time Expression Extraction

In this study, we explored two well-known time expression taggers, SUTime (Chang and Manning, 2012) and HeidelTime (Strötgen and Gertz, 2010), which were developed and evaluated on general domain corpora. When applied to the TempEval-2 newspaper data, both systems achieved state-of-the-art performance (F1 scores of 92% and 86%, respectively, for time expression identification) (Verhagen et al., 2010). Moreover, they have previously been used for the automatic processing of clinical narratives (Jindal and Roth, 2013; Wang et al., 2016).

Both SUTime and HeidelTime use a list of pattern matching rules, built on regular expressions, to recognize and normalize time expressions inside the text. As a main difference, while SUTime links relative TIMEXes (such as *yesterday*) to the document creation date, HeidelTime uses different normalization strategies depending on documents' types (e.g., *news, narratives*).

To adapt the systems to the mental health domain, we first evaluated their original versions on the development set[5], to see what the increase in performance over non-domain-specific rules would be. Then, we manually reviewed the TIMEXes present in the development set, and modified and added rules as needed. The performance of the updated systems was then evaluated on a *validation set*, consisting of two batches (23 documents in total). To allow for future development and evaluation, we did not use the remaining batches (*test set*, 19 documents) in this study. The documents we used to adapt and evaluate the temporal taggers were from the adjudicated gold standard corpus.

2.4 Evaluation Metrics

To assess the quality of the developed corpus, we calculated inter-annotator agreement (IAA) for each annotated batch, using the metrics that were used for i2b2 2012 (average of precision and recall) and Clinical TempEval 2016 (F1 score). First, we computed the average of precision and recall: the entities marked by one annotator were used as the gold reference, while the entities identified by the second annotator were considered as the system's output (switching these two roles would not change the final result). Moreover, we measured the F1 score (i.e., the harmonic mean of precision and recall), which provides a good way to quantify agreement for entity extraction tasks (Hripcsak and Rothschild, 2005).

To evaluate the performance of SUTime and HeidelTime, we defined: (i) true positives (TP), as the gold TIMEXes that were found in the system's output; (ii) false negatives (FN), as the gold TIMEXes that were not found in the system's output; and (iii) false positives (FP), as the system TIMEXes that were not found among gold annotations. In this case, we assessed the system's performance in terms of precision, recall, and F1 score.

3 Results

3.1 Time Expression Annotations

The total number of gold TIMEXes in our corpus is 3,413, with an average of 65.6 annotations per document[6]. Table 1 reports the prevalence of TIMEX types in the corpus, divided into development, validation, and test sets. Overall, the majority of TIMEXes are represented by Dates (55.8%). Durations, Times, and Frequencies account for 16.5%, 10.7%, and 8.1%, respectively.

As mentioned, we defined a new TIMEX type referring to the patient's age: "Age-related". This type was assigned to 8.9% of all TIMEXes. Some examples include:

- *at the age of 8*: requires adding 8 years to the date of birth for normalization;

- *when he was a child*: requires the date of birth and a shared definition of "child years" for normalization;

---

[5]To compute the performance of the original systems, we used: the SUTime grammar included in the Stanford CoreNLP (*https://stanfordnlp.github.io/CoreNLP/*) distribution dated 2017-06-09, and the HeidelTime resources included in the GATE (*https://gate.ac.uk/*) distribution 8.3.

[6]Annotators worked 20-24 hours, and annotated 2/3 of the corpus each (33-39 docs): the average time required for corpus development was around 35-40 minutes per document.

	development set	validation set	test set	total
# documents	10	23	19	52
# TIMEXes	964 (96.4/doc)	1,401 (60.9/doc)	1,048 (55.2/doc)	3,413 (65.6/doc)
Date	593 (61.5%)	803 (57.3%)	507 (48.4%)	1,903 (55.8%)
Duration	148 (15.3%)	215 (15.3%)	200 (19.1%)	563 (16.5%)
Time	94 (9.8%)	129 (9.2%)	143 (13.6%)	366 (10.7%)
Frequency	60 (6.2%)	127 (9.1%)	89 (8.5%)	276 (8.1%)
Age-related	69 (7.2%)	127 (9.1%)	109 (10.4%)	305 (8.9%)

Table 1: TIMEX annotation results: prevalence of types in our corpus.

- *since his teens*: requires the date of birth and a shared definition of "teens years" for normalization;

- *during his first year* (implicitly referring to the first year of university): requires the date of birth and the usual timing of university for normalization;

With respect to IAA, we computed results on TIMEX spans (without considering the different TIMEX types, as this was not calculated for the corpora used for comparison), for both partial and exact matches. In the first case, the average of precision and recall was 78%, and the F1 score was 77%. In the second case, both metrics resulted in 60%.

For partial matches, the IAA per batch was in the range of 73.6%-83.7% (average of precision and recall), and 73.5%-83.3% (F1 score). We also computed the percentage of TIMEX type matches for those time expressions that the annotators agreed on with respect to overlapping spans, resulting in 91% percentage match.

3.2 Comparison to Related Corpora

In Table 2, our corpus is compared to the i2b2 2012 and the Clinical TempEval 2016 corpora. Specifically, the table reports the size, the number of TIMEXes, the type prevalence[7], and the IAA values for the three considered corpora. To allow comparing TIMEX types among these corpora, we merged Clinical TempEval annotations as follows: PrePostExp time expressions were considered among Dates, while Quantifier time expressions were considered as Frequencies. No modifications were required in order to compare the i2b2 2012 corpus. Also, since we added the

new TIMEX type Age_related, we were not able to compare these annotations in either corpus.

3.3 Temporal Expression Extraction System Adaptation

In this work, we used SUTime and HeidelTime to identify TIMEX spans in the developed corpus[8]. The results of this domain adaptation are shown in Table 3. First, we ran the original versions of the two systems on the development set, obtaining an F1 score of 72.5% for SUTime and 63.6% for HeidelTime (allowing partial matches). As expected, these scores are much lower than those obtained on general domain corpora (92% and 86% F1 scores on TempEval-2 newspaper data). After tuning the systems' rules on the development set, we achieved scores of 79.7% and 77.3%, respectively. By running the updated systems on the validation set, we obtained a final result of 79.5% and 75.8%, respectively.

It is important to mention that, although the original version of SUTime included rules to capture some "age" expressions (e.g., *28-year-old*), these were considered as Durations. In the original version of HeidelTime, instead, these expressions were explicitly excluded, as they were probably not considered as proper time expressions. This is one of the reasons why the original version of HeidelTime had much lower recall than SUTime (Table 3, "HeidelTime original" row).

4 Discussion

Extracting temporal information from mental health records is particularly challenging, as this domain is characterized by a large proportion of free-text and heterogeneity in self-reported experiences (i.e., mental health symptoms), circum-

---

[7]These numbers were computed on released data, for i2b2 2012, and on publicly available annotations, for Clinical TempEval 2016.

[8]For determining Age-related TIMEXes, we applied a set of post-processing rules to the output of the two temporal taggers.

	Our corpus	**i2b2 2012**	**Clinical TempEval 2016**
Domain	Mental health	Intensive care	Oncology
# documents	52	310	591
# tokens	206,661 (3,974/doc)	178,000 (574/doc)	550,487 (931/doc)
# TIMEXes	3,413 (1.65/100tok)	4,184 (2.35/100tok)	7,863 (1.43/100tok)
Prevalence	Date: 55.8% Duration: 16.5% Time: 10.7% Frequency: 8.1% Age_related: 8.9%	Date: 68.4% Duration: 17.8% Time: 3.1% Frequency: 10.7% Age_related: NA	Date: 76.1% Duration: 10.6% Time: 3.4% Frequency: 9.9% Age_related: NA
IAA (Avg P-R)	Partial: 78% Strict: 60%	Partial: 89% Strict: 73%	NA
IAA (F1 score)	Partial : 77% Strict : 60%	NA	Partial: NA Strict: 73%

Table 2: Comparison between our corpus, i2b2 2012, and Clinical TempEval 2016. IAA: inter-annotator agreement; Avg P-R: average of precision and recall; NA: not applicable (TIMEX type not annotated or IAA metric not calculated in these corpora).

Set	System	P	R	F1
dev	SUTime original	71.4%	73.6%	72.5%
	HeidelTime original	71.7%	57.2%	63.6%
dev	SUTime updated	72.9%	87.8%	79.7%
	HeidelTime updated	73.6%	81.3%	77.3%
valid	SUTime updated	72.8%	87.7%	79.5%
	HeidelTime updated	70.5%	81.9%	75.8%

Table 3: SUTime and HeidelTime results. P: precision; R: recall.

stances (e.g., social support networks, recent or past stressful experiences, psychoactive substance use), and treatment and outcomes. In this study, we annotated time expressions related to patients with schizophrenia in EHRs. The documents in our corpus are long when compared to similar corpora (3,974 tokens/doc), and include a large proportion of relevant time expressions (65.6 TIMEXes/doc). In addition, they might contain information taken from structured forms (e.g., questions, references to health care legislation), which are not relevant to the patient's clinical history, but could still include references to time.

4.1 Comparison to Related Corpora

When comparing our corpus to other related corpora, there are differences in the documentation types that can have an impact on the development of temporal information extraction systems. For instance, the discharge summaries in the 2012 i2b2 corpus each start with the admission and discharge date, which are annotated as TIMEXes.

Similarly, the Clinical TempEval 2016 documents are organized in sections with semi-structured date information, that can be useful to then link and anchor clinically relevant events in the documents. The documents in our corpus, on the contrary, include various paragraphs describing both past and current events related to the patient, without necessarily following a predefined structure.

Regarding TIMEX types, there was a greater prevalence of Date expressions in the i2b2 2012 (ICU domain) and Clinical TempEval 2016 (oncology domain) corpora, as compared to our corpus (Table 2). This might relate to the fact that, in the ICU and oncology clinical settings, treatment episodes are likely to be shorter and changes in physical health parameters and onset/duration of treatment occur over shorter periods of time. As another interesting observation, our corpus is characterized by a higher prevalence of the Time type, which is probably due to the fact that many events are described as happening at a specific time of day (*this morning*, *at night*). It is important to point out that in both i2b2 2012 and Clinical TempEval 2016, age-related information was not marked. One reason for this might be that these types of constructs were not considered useful for the use-cases that these corpora were developed for.

As regards the IAA, we obtained a value of 60%/78% (strict/partial) for the average of precision and recall, and a value of 60%/77% (strict/partial) for the F1 score. Although these results are lower in comparison to those of i2b2 2012 and Clinical TempEval 2016 (Table 2), this can be considered a promising result, given the intrinsic complexity of the mental health domain. As an important remark, the difference between partial and strict IAA measures indicates that identifying the spans of time expressions is not straightforward. This is also reflected in the results obtained on the i2b2 2012 corpus. In our case, the main reason for disagreement was the inclusion/exclusion of prepositions or determiners in TIMEX spans (e.g., *for three years* instead of *three years*). We also analyzed disagreements in TIMEX type assignments. Differentiating between Date and Duration was one of the main disagreements (accounting for 42% of disagreements). For instance, an expression like *this week* was assigned Date (interpreted as a point in time) by one annotator, and Duration (interpreted as a period) by another.

4.2 Domain-specific Time Expressions

As an interesting result of the annotation task, we identified a set of time expressions which were not present in the other corpora, but which are essential to allow capturing symptom onset. As previously mentioned, these expressions are those related to the age of the patient, which account for 8.9% of all TIMEXes (Table 1). Despite not being particularly frequent, Age-related TIMEXes can be crucial to determine the first onset of symptoms, which is often reported by patients or their relatives in a vague way. For example, extracting these kinds of expressions is essential for sentences like[9]:

- *she first experienced hallucinations **at the age of 18***

- *he started hearing voices **when he was 15***

- *he has been experiencing these symptoms **since his teens***

---

[9]The reported sentences have been paraphrased.

Besides defining a new TIMEX type, we also found some example TIMEXes that are specific to the analyzed domain and were not present in the compared corpora. As a first example, we identified a few expressions that are related to drug prescriptions, such as *OD* (i.e., once daily) and *NOCTE* (i.e., every night). Moreover, we noticed that the expression *15 minute* is often used as a Frequency, rather than a Duration, as this is the usual interval of time used to observe patients with schizophrenia (e.g., "*he was placed on **15 minute** visual observations*"). Determining the correct interpretation is not straightforward, as this relies on domain knowledge (in the sentence *I went for a **15-minute** walk*, the same TIMEX represents a Duration). As another interesting example, we realized that, in the field of mental health, the expressions */7*, */12*, and */52* can be used to refer to days (*3/7 ago* = three days ago), months (*for 3/12* = for three months), and weeks (*in 2/52* = in two weeks), respectively. In our corpus, we found a total of 12 expressions of this kind (4 Dates, 6 Durations, and 2 Age_related). To normalize them, it is possible to create specific rules that map the different patterns to the corresponding temporal values.

4.3 Time Expression System Adaptation

We applied SUTime and HeidelTime on the development set through an iterative process (Table 3). By running the two original versions of the systems, we found that SUTime performed better than HeidelTime, especially in terms of recall (73.6% versus 57.2%): this is probably due to the fact that, in our annotation schema, we included a few expressions which were already taken into account by the first system, but not by the second (e.g., *28 years old, past*). In the adaptation process, we identified false negatives (FNs) for both systems, and then refined rules to capture them. It is important to point out that, in this preliminary experiment, we focused on improving the systems' performance for partial matches, rather than identifying exact TIMEX spans. While a few of the performed adaptations involved general domain rules (e.g., dates in the form "dd/MM" were not recognized by SUTime), we mostly needed to address TIMEXes specific to the mental health domain. By adding extraction rules for these expressions, we were able to reduce the number of FNs, thus obtaining an improvement in recall from 73.6%

to 87.8%, for SUTime, and from 57.2% to 81.3% for HeidelTime (development set). As for precision, lowering the number of FPs was not trivial, as these rule-based systems cannot distinguish between time expressions that are patient-related (thus relevant to our goal) and those that are not (e.g., form-related).

After an error analysis, we found a few non-trivial TIMEXes that were not correctly captured by the system and will require further adaptation/improvement. For instance, SUTime was not able to identify age-related expressions like *between the ages of 10 and 12* and Time intervals like *9.30-10*. On the other hand, ambiguous words such as *present* (as in *present at the appointment*) and *minutes* (as in *minutes of the meeting*) were erroneously considered as TIMEXes. Also, all the time expressions that were included in form-like paragraphs (e.g., *The Activities of **Daily** Living include...*) were counted among false positives, as we were not interested in extracting these.

In this study, the best final F1 score was obtained with the adapted version of SUTime (79.5% on the validation set), which represents a promising result if compared to the IAA of 77% (F1 score). This could reflect the fact that time expressions often follow specific patterns: by adequately tuning extraction rules, it is possible to obtain a good extraction performance, which can be even higher than that of a human annotator (this is particularly true for recall).

4.4 Limitations and Future Work

As a main limitation of this work, we only addressed the extraction of TIMEX spans and types, without dealing with TIMEX value normalization, which would require assigning a standardized value to each TIMEX. For instance, the expression *for three years* should be normalized as *P3Y*, while the expression *at the age of 8* would require the date of birth in order to be correctly normalized. We are in the process of extending our TIMEX annotations with normalized values. In future work, we will use these annotations to develop suitable rules for time expression extraction and value normalization. Moreover, while adapting TIMEX extraction systems, we did not write contextual rules to disambiguate expressions that can belong to different types depending on their context, although we did disambiguate these during manual annotation (e.g., *at night* was marked as a Time, when

referring to a single episode, or as a Frequency, when referring to a drug prescription). As a future improvement, we will address this task by dealing with the context in which time expressions appear, for example, by using word embeddings to represent each word with automatically derived contextual features (Mikolov et al., 2013). Finally, since we are interested in assessing the usability of SUTime and HeidelTime in other clinical domains, we plan to extend the adaptability study presented in this work to other clinical corpora, such as the i2b2 and Clinical TempEval corpora used for comparison here.

As previously mentioned, the extraction of time expressions represents a first step towards our final goal, i.e., the identification of symptom onset and DUP in free text. The next step will involve the annotation of clinically relevant entities (symptoms primarily), to be linked to the available temporal information. Extracting entities such as symptoms could be done by knowledge-based approaches based on keyword lists, or using word embedding models (or a combination of both). We are currently exploring different alternatives. As for temporal linking, we will need to refer each clinical entity to a specific time expression. For example, given the sentence *he first experienced hallucinations in 2008*, the following link should be identified: "{2008} CONTAINS {hallucinations}". To reach this goal, we are currently experimenting with the annotation of a set of documents, where relevant events and time expressions have been pre-annotated by using automatic approaches.

5 Conclusion

In this paper, we described the annotation of time expressions in mental health records related to schizophrenia, thus creating an annotated corpus. To the best of our knowledge, this is the first gold standard developed in this domain for a specific mental health use-case: onset and DUP extraction. In addition, this is the first study explicitly incorporating age-related information, which is not captured by current temporal models. As an important aspect, we also assessed the adaptability of two existing rule-based TIMEX extraction systems to the new analyzed use-case, obtaining promising results.

Due to governance regulations, the corpus annotated in this study cannot be made publicly available. However, there are procedures in place to provide researchers with controlled access to the CRIS database. Moreover, the developed guidelines and the adapted versions of SUTime and HeidelTime have been made publicly available[10], and could be easily reused or adapted for other temporal information extraction tasks.

Acknowledgments

NV and SV are supported by the Swedish Research Council (2015-00359), Marie Skłodowska Curie Actions, Cofund, Project INCA 600398.

RS, RD and RP are funded by the NIHR Specialist Biomedical Research Centre for Mental Health at the South London and Maudsley NHS Foundation Trust and Institute of Psychiatry, King's College London.

RP has received support from a Medical Research Council (MRC) Health Data Research UK Fellowship (MR/S003118/1) and a Starter Grant for Clinical Lecturers (SGL015/1020) supported by the Academy of Medical Sciences, The Wellcome Trust, MRC, British Heart Foundation, Arthritis Research UK, the Royal College of Physicians and Diabetes UK.

References

Steven Bethard, Leon Derczynski, Guergana Savova, James Pustejovsky, and Marc Verhagen. 2015. Semeval-2015 task 6: Clinical tempeval. In *Proceedings of the 9th International Workshop on Semantic Evaluation (SemEval 2015)*, pages 806–814.

Steven Bethard, Guergana Savova, Wei-Te Chen, Leon Derczynski, James Pustejovsky, and Marc Verhagen. 2016. Semeval-2016 task 12: Clinical tempeval. In *Proceedings of the 10th International Workshop on Semantic Evaluation (SemEval-2016)*, pages 1052–1062.

Angel X Chang and Christopher D Manning. 2012. Sutime: A library for recognizing and normalizing time expressions. In *LREC*.

Charlotte Connor, Max Birchwood, Colin Palmer, Sunita Channa, Nick Freemantle, Helen Lester, Paul Patterson, and Swaran Singh. 2013. Don't turn your back on the symptoms of psychosis: a proof-of-principle, quasi-experimental public health trial to reduce the duration of untreated psychosis in Birmingham, UK. *BMC psychiatry*, 13:67.

Michele Hill, Niall Crumlish, Mary Clarke, Peter Whitty, Elizabeth Owens, Laoise Renwick, Stephen Browne, Eric A. Macklin, Anthony Kinsella, Conall Larkin, John L. Waddington, and Eadbhard O'Callaghan. 2012. Prospective relationship of duration of untreated psychosis to psychopathology and functional outcome over 12 years. *Schizophrenia research*, 141(2-3):215–221.

George Hripcsak and Adam S Rothschild. 2005. Agreement, the f-measure, and reliability in information retrieval. *Journal of the American Medical Informatics Association*, 12(3):296–298.

Prateek Jindal and Dan Roth. 2013. Extraction of events and temporal expressions from clinical narratives. *Journal of biomedical informatics*, 46:S13–S19.

Stephen Kisely, Anita Scott, Jennifer Denney, and Gregory Simon. 2006. Duration of untreated symptoms in common mental disorders: association with outcomes. *The British Journal of Psychiatry*, 189(1):79–80.

Julia M. Lappin, Kevin D. Morgan, Craig Morgan, Paola Dazzan, Abraham Reichenberg, Jolanta W. Zanelli, Paul Fearon, Peter B. Jones, Tuhina Lloyd, Jane Tarrant, Annette Farrant, Julian Leff, and Robin M. Murray. 2007. Duration of untreated psychosis and neuropsychological function in first episode psychosis. *Schizophrenia research*, 95(1-3):103–110.

Stéphane M Meystre, Guergana K Savova, Karin C Kipper-Schuler, and John F Hurdle. 2008. Extracting information from textual documents in the electronic health record: a review of recent research. *Yearbook of medical informatics*, pages 128–144.

Tomas Mikolov, Kai Chen, Greg Corrado, and Jeffrey Dean. 2013. Efficient estimation of word representations in vector space. *arXiv preprint arXiv:1301.3781*.

Gayan Perera, Matthew Broadbent, Felicity Callard, Chin-Kuo Chang, Johnny Downs, Rina Dutta, Andrea Fernandes, Richard D Hayes, Max Henderson, Richard Jackson, Amelia Jewell, Giouliana Kadra, Ryan Little, Megan Pritchard, Hitesh Shetty, Alex Tulloch, and Robert Stewart. 2016. Cohort profile of the South London and Maudsley NHS Foundation Trust Biomedical Research Centre (SLaM BRC) Case Register: current status and recent enhancement of an Electronic Mental Health Record-derived data resource. *BMJ Open*, 6(3).

James Pustejovsky, Kiyong Lee, Harry Bunt, and Laurent Romary. 2010. Iso-timeml: An international standard for semantic annotation. In *LREC*.

Brett R South, Shuying Shen, Jianwei Leng, Tyler B Forbush, Scott L DuVall, and Wendy W Chapman. 2012. A prototype tool set to support machine-assisted annotation. In *Proceedings of the 2012*

---

[10]The guidelines developed in this study are available at: *https://github.com/medesto/annotation-guidelines*. The adapted versions of SUTime and HeidelTime are available at: *https://github.com/medesto/systems-adaptation*.

Workshop on Biomedical Natural Language Processing, pages 130–139.

Jannik Strötgen and Michael Gertz. 2010. Heideltime: High quality rule-based extraction and normalization of temporal expressions. In *Proceedings of the 5th International Workshop on Semantic Evaluation*, pages 321–324.

William F Styler IV, Steven Bethard, Sean Finan, Martha Palmer, Sameer Pradhan, Piet C de Groen, Brad Erickson, Timothy Miller, Chen Lin, Guergana Savova, and James Pustejovsky. 2014. Temporal annotation in the clinical domain. *Transactions of the Association for Computational Linguistics*, 2:143–154.

Weiyi Sun, Anna Rumshisky, and Ozlem Uzuner. 2013a. Annotating temporal information in clinical narratives. *Journal of biomedical informatics*, 46:S5–S12.

Weiyi Sun, Anna Rumshisky, and Ozlem Uzuner. 2013b. Evaluating temporal relations in clinical text: 2012 i2b2 challenge. *Journal of the American Medical Informatics Association*, 20(5):806–813.

Marc Verhagen, Roser Sauri, Tommaso Caselli, and James Pustejovsky. 2010. Semeval-2010 task 13: Tempeval-2. In *Proceedings of the 5th international workshop on semantic evaluation*, pages 57–62.

Wei Wang, Kory Kreimeyer, Emily Jane Woo, Robert Ball, Matthew Foster, Abhishek Pandey, John Scott, and Taxiarchis Botsis. 2016. A new algorithmic approach for the extraction of temporal associations from clinical narratives with an application to medical product safety surveillance reports. *Journal of biomedical informatics*, 62:78–89.

Yanshan Wang, Liwei Wang, Majid Rastegar-Mojarad, Sungrim Moon, Feichen Shen, Naveed Afzal, Sijia Liu, Yuqun Zeng, Saeed Mehrabi, Sunghwan Sohn, and Hongfang Liu. 2018. Clinical information extraction applications: A literature review. *Journal of biomedical informatics*, 77:34–49.

Evaluation of a Sequence Tagging Tool for Biomedical Texts

Julien Tourille[1,6], Matthieu Doutreligne[1,2], Olivier Ferret[3],
Nicolas Paris[1,2,4], Aurélie Névéol[1], Xavier Tannier[4,5]

[1]LIMSI, CNRS, Université Paris-Saclay, [2]WIND-DSI, AP-HP, [3]CEA, LIST,
[4]Sorbonne Université, [5]Inserm, LIMICS, [6]Univ. Paris-Sud
`{julien.tourille,matthieu.doutreligne,aurelie.neveol}@limsi.fr,`
`olivier.ferret@cea.fr, nicolas.paris@aphp.fr,`
`xavier.tannier@sorbonne-universite.fr`

Abstract

Many applications in biomedical natural language processing rely on sequence tagging as an initial step to perform more complex analysis. To support text analysis in the biomedical domain, we introduce Yet Another SEquence Tagger (YASET), an open-source multi purpose sequence tagger that implements state-of-the-art deep learning algorithms for sequence tagging. Herein, we evaluate YASET on part-of-speech tagging and named entity recognition in a variety of text genres including articles from the biomedical literature in English and clinical narratives in French. To further characterize performance, we report distributions over 30 runs and different sizes of training datasets. YASET provides state-of-the-art performance on the CoNLL 2003 NER dataset (F1=0.87), MEDPOST corpus (F1=0.97), MERLoT corpus (F1=0.99) and NCBI disease corpus (F1=0.81). We believe that YASET is a versatile and efficient tool that can be used for sequence tagging in biomedical and clinical texts.

1 Introduction

Many applications in biomedical Natural Language Processing (NLP), including relation extraction or text classification, rely on sequence tagging as an initial step to perform more complex analysis. To support text analysis in the biomedical domain, we present Yet Another SEquence Tagger (YASET), an open-source multi-purpose sequence tagger written in Python. YASET aims at providing NLP researchers with fast and accurate implementations of cutting-edge deep learning sequence tagging models. YASET is built using TensorFlow (Abadi et al., 2015), an open source software library for numerical computation. The code is licensed under the version 3 of the *GNU Gen-* *eral Public License*[1] and is freely available online[2]. The main contributions of this work are:

- a fast and accurate implementation of a state-of-the-art sequence tagging model based on Long Short-Term Memory Networks (LSTMs) (Hochreiter and Schmidhuber, 1997). The architecture is similar to the one described in Lample et al. (2016) and is able to process mini-batches for faster training. Furthermore, YASET supports the use of handcrafted features in combination with word and character embeddings;

- an easy-to-use interface based on a central configuration file that is used to setup experiments, with default parameters that are suitable for most sequence tagging tasks;

- an evaluation on various biomedical corpora and on the CoNLL 2003 corpus, studying the stability of our model and the effect of training data size. We compare YASET performance with state-of-the-art results published in the literature.

2 Related Work

Several open-source implementations for sequence tagging based on neural network architectures have become available over the last years. Lample et al. (2016) provide a Python implementation[3] for the two models presented in their paper. They are implemented with Theano (Al-Rfou et al., 2016), a Python library for deep learning.

---

[1]`https://www.gnu.org/licenses/gpl-3.0.`
`en.html`
[2]`https://github.com/jtourille/yaset`
[3]`https://github.com/glample/tagger`

193

Proceedings of the 9th International Workshop on Health Text Mining and Information Analysis (LOUHI 2018), pages 193–203
Brussels, Belgium, October 31, 2018. ©2018 Association for Computational Linguistics

NeuroNER[4] (Dernoncourt et al., 2017) targets non-expert users and is based on Lample's Bi-LSTM model. The authors intended to make the tool easy to use by providing automatic format conversion from the brat format (Stenetorp et al., 2012) to the input format and from the output format to the brat format. The tool produces several plots during training for performance analysis. It is implemented in Python and makes use of the TensorFlow library. Both implementations of the Bi-LSTM model suffer from a very long training time which makes them cumbersome to use. YASET offers a faster implementation of the model by allowing mini-batch training and by using the pipeline API of TensorFlow.

Rei and Yannakoudakis (2016) released a Python implementation[5] of different models presented in their works (Rei and Yannakoudakis, 2016; Rei et al., 2016; Rei, 2017). One major difference with YASET resides in the possibility to use a language modeling objective during training.

Recently, Yang and Zhang (2018) introduced NCRF++[6], a tool presented as *the neural version of CRF++*[7] and implemented with Py-Torch (Paszke et al., 2017). The tool is very close to YASET, with the possibility to define hand-crafted word features and to perform *nbest* decoding.

Another type of neural network model, based on Convolutional Neural Networks (CNNs) for character level representation, is presented in Ma and Hovy (2016). The authors implemented the architecture with PyTorch. The code is freely available online[8]. The same type of architecture is implemented in the tool released by Reimers and Gurevych (2017). It is implemented with Keras (Chollet et al., 2015) and is freely available online[9].

Outside the peer-reviewed scientific environment, many other implementations are freely available online. However, we will not review them in this paper.

---

[4]https://github.com/
Franck-Dernoncourt/NeuroNER

[5]https://github.com/marekrei/
sequence-labeler

[6]https://github.com/jiesutd/NCRFpp

[7]https://taku910.github.io/crfpp

[8]https://github.com/XuezheMax/
NeuroNLP2

[9]https://github.com/UKPLab/
emnlp2017-bilstm-cnn-crf

3 Neural Network Model

There is currently one neural network model implemented in YASET. This model is mostly based on Lample et al. (2016). However, similar architectures are presented in other work (Collobert et al., 2011; Ma and Hovy, 2016; Rei and Yannakoudakis, 2016; Rei et al., 2016). Other network architectures will be implemented in the future.

3.1 Main Component

Our approach relies on LSTMs. The architecture of our model is presented in Figure 1. For a given sequence of tokens, represented as vectors, we compute representations of left and right contexts of the sequence at every token. These representations are computed using two LSTMs (forward and backward LSTM in Figure 1).

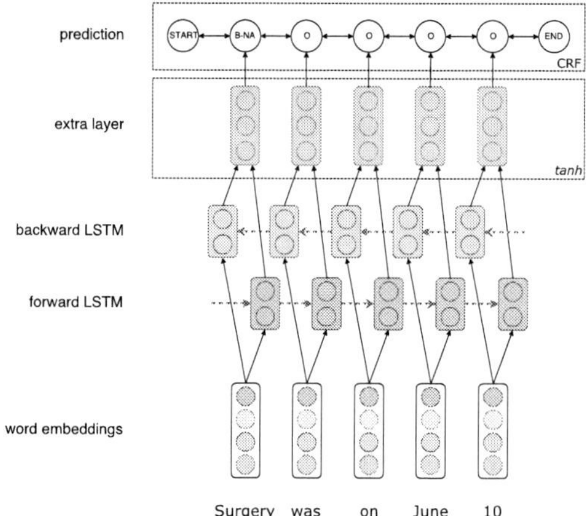

Figure 1: YASET neural network architecture overview. The example is extracted from the THYME corpus (Styler IV et al., 2014) used during the shared tasks Clinical TempEval (Bethard et al., 2015, 2016, 2017). One of the objectives was to extract medical events. In the example, the event *Surgery* is marked as a medical event with the type *N/A*.

These representations are concatenated and passed through a $tanh$ activation layer whose size is equal to the size of the concatenated vector.

Finally, the output of the last layer is linearly projected to a n-dimensional vector representing the number of categories. Following previous work (Dernoncourt et al., 2017; Lample et al., 2016; Ma and Hovy, 2016), we add a final Conditional Random Field (CRF) layer to take into account the previous label during training and prediction.

3.2 Input Embeddings

Vectors representing tokens are built by concatenating a character-based embedding, a word embedding and one embedding per feature.

An overview of the embedding computation is presented in Figure 2. Following Lample et al. (2016), the character-based representation is computed with a Bi-LSTM whose parameters are defined by users. First, a random embedding is generated for every character in the training corpus. Token characters are then processed with a forward and backward LSTM architecture. The final character-based representation results from the concatenation of the forward and backward representations. The character-based representation is then concatenated to a pre-trained word-embedding and one embedding per feature. We apply dropout on the final input embedding.

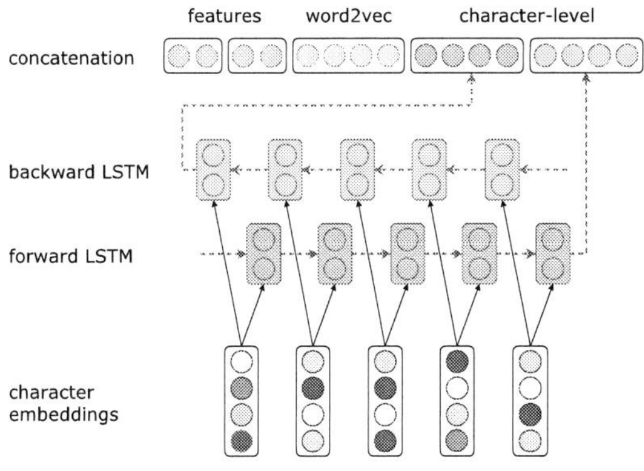

Figure 2: YASET input embedding computation overview. Embeddings result from the concatenation of a pre-trained word embedding, a character-level representation of the token and one embedding per categorical feature.

4 Tool Overview

In this section, we present a general overview of YASET. First we describe input data formats (sequences and pre-trained word embeddings). Then we present the input pipeline, the network training phase, the implemented evaluation metrics and the management of its parameters.

4.1 Input and Output Data

YASET takes CoNLL-like[10] formatted files as input. Sequences are separated by empty lines and there must be one token per line. For each token,

the first and last columns are reserved namely for the token itself and the token label. Each token must have a label.

Users can add several categorical features. Feature columns must be specified in the configuration file by providing their indexes. Besides the first and last columns, and the feature columns, users can add other columns. They will be ignored by the system.

YASET can take training and development files as input but can also create development instances if there is no development file provided by users. In this case, users can specify the train/dev split ratio and may specify a random seed for experiment reproducibility.

Train and development instances consistency is checked upon startup. Each label and feature values from the development instances must appear in the train instances. An example of YASET input file format is presented in Figure 3.

In prediction mode, test data is supplied in the same format, without the token label column, which will be added with the predicted labels.

```
...
Lien NNP I-NP I-PER
. . O O

China NNP I-NP I-LOC
says VBZ I-VP O
time NN I-NP O
right RB I-ADVP O
for IN I-PP O
Taiwan NNP I-NP I-LOC
talks NNS I-NP O
. . O O

BEIJING VBG I-VP I-LOC
1996-08-22 CD I-NP O
...
```

Figure 3: Example extracted from the CoNLL 2003 shared task corpus (Sang and Meulder, 2003). For each token (col. #1), there is a label (col. #4, IOB format) and two categorical features (cols. #2 and #3), the *part-of-speech tag* and a *chunk label* (IOB format).

4.2 Word Embeddings

YASET supports embeddings in the word2vec (Mikolov et al., 2013b) or Gensim (Řehůřek and Sojka, 2010) formats. Other formats of embeddings, for instance FastText (Bojanowski et al., 2017) or Glove (Pennington et al., 2014), must first be converted to either accepted format.

---

[10] http://universaldependencies.org/docs/format.html

Out Of Vocabulary (OOV) tokens can be addressed by two strategies. In the first one, users provide a vector for OOV tokens. In this case, the vector is expected to be part of the provided word embedding matrix and users must specify the vector ID. In the second strategy, we follow the methodology described in Lample et al. (2016). We insert a randomly initialized vector in the embedding matrix that will be used for OOV tokens. Then we replace singletons in the training corpus by the OOV token with a probability of p, which is defined by users. Word embeddings can be fine-tuned during training in both cases by setting the appropriate flag to *true* in the configuration file.

4.3 Input Pipeline

Minimizing the memory footprint was one of the core objectives during the development of the tool. YASET leverages the pipeline API of TensorFlow to build a lightweight and fast input pipeline. Input instances are stored in the binary TensorFlow format. This avoids the need to store the whole dataset in memory. Mini-batches are extracted randomly and fed to the network.

Training instances can be bucketized according to their lengths. This possibility can speedup training with large corpora. Buckets boundaries are automatically computed. To assert effective randomization during training, buckets will contain enough instances for several mini-batches.

4.4 Neural Network Training

YASET alternates between two phases during training. In the *iteration phase*, mini-batches are extracted from the input pipeline and fed to the neural network model. Optimization is performed according to the parameters set by users in the configuration file.

At the end of each iteration, YASET enters the *evaluation phase* where the current model state is used to make predictions on the development instances. The performance score is logged for further analysis and a snapshot of the model is kept if the performance score is the best obtained so far on the development instances. Since model snapshots can take a lot of memory, we only keep the best one, i.e. the one that performs best on the development instances.

Network training is performed via back-propagation. Users can select the neural network model architecture (only one available for now),

the maximum number of iterations n and the patience criterion p. Training will stop if the maximum number n is reached or if there are p iterations without performance improvement on the development instances.

Users can also set several parameters related to the learning algorithm such as the initial learning rate, and gradient clipping and exponential decay factors. Finally, users can select the mini-batch size, the number of CPU cores to use and the evaluation metric.

Another set of parameters are related to the neural network model. These parameters allow to select the dropout rate and the different hidden layer and embedding sizes.

4.5 Metrics

Model performance is measured on the development instances at the end of each iteration. Two metrics are currently implemented in YASET: *accuracy* and *CoNLL*. The former measures the fraction of correctly predicted labels among all the predicted labels. The latter is an entity-based metric which outputs precision, recall and F1-measure scores. Precision measures the fraction of correctly predicted entities among all predicted entities. Recall assesses the fraction of correctly predicted entities among all the entities that should have been detected. F1-measure is the harmonic mean of precision and recall. The implementation of the CoNLL metric is inspired by the official evaluation script used during the CoNLL-2003 shared task (Sang and Meulder, 2003) and by the Python portage of the script by Sampo Pyysalo[11]. In the case of performance evaluation with the CoNLL metric, token labels must follow either the IOB, IOBE or IOBES tagging scheme[12].

4.6 Parameters

YASET parameters are fully customizable and centralized in one configuration file which is used to setup experiments. YASET also targets end-users from a broader community by providing hyperparameter value suggestions and insights on how to choose them for various sequence-tagging situations.

---

[11]`https://github.com/spyysalo/conlleval.py`

[12]Inside, Outside, Begin, End, Single

5 Experiments

We demonstrate the performance of YASET by applying the model to four different corpora selected to cover a variety of languages, text genres, sequence types and annotation densities. We focus our effort on biomedical texts, using MedPost (Smith et al., 2004), a corpus of biomedical abstracts annotated with Part-of-Speech (PoS) tags, the NCBI Disease corpus (Islamaj Dogan and Lu, 2012), a dataset of biomedical abstracts annotated with disease related entities and MERLoT (Campillos et al., 2018), a corpus of clinical documents written in French which were annotated with two different Named Entity Recognition (NER) tag sets (Protected Health Information (PHI) and biomedical entities).

We also apply YASET on the CoNLL 2003 English NER corpus (Sang and Meulder, 2003), a classic benchmark corpus for NER in the general domain.

5.1 Presentation of the Corpora

We use the corpus partition provided with the dataset distributions when available. For NCBI and CoNLL 2003, models are trained on the train sets and evaluated on the test sets while the development sets are used to determine early stopping. For MERLoT and MedPost, partition details are outlined below. The code used to preprocess and analyze the corpora is available online[13].

The MedPost corpus is a collection of tokenized MEDLINE abstracts annotated with PoS tags. It contains 6,701 sentences ($\approx$182,000 tokens). The corpus is divided in 13 files of different sizes, from which we extracted one file to serve as development set (*tag_mb.ioc*). The tag set contains originally 63 unique entities that we grouped in a coarser set of 51 entities. Grouping affected punctuation signs, which were assigned a unique PUNCT tag.

The NCBI corpus contains 793 MEDLINE abstracts with 11,350 annotations of diseases and disease modifiers. There are 4 different entities. Train and development sets contain 6,594 sentences ($\approx$151,000 tokens).

MERLoT is a French clinical corpus with several levels of annotations. It contains 500 medical reports with a total of 25,087 sentences ($\approx$177,000 tokens). We used it for two tasks. The first one is de-identification with a tag set comprising 11 types of PHI (e.g. *names*, *dates*). The second one is medical NER with 19 entity classes (e.g. *anatomy*, *disorder*).

The medical entity annotations include nested entities. Because this study aims to experiment on a variety of datasets using the same model, we did not attempt to extract several levels of nested entities. When nested entities occur, our experiments only addressed the outer layer corresponding to the largest entity. For example, the mention "cancer du sein" (*breast cancer*) was originally annotated with one DISORDER entity (cancer du sein, *breast cancer*) and one ANATOMY entity (sein, *breast*). Nested entity removal reduced the annotations to only one tag per token and in this case, the DISORDER annotation was kept while the ANATOMY annotation was removed. In total, 5,218 entities (8.4% of the complete set) were pruned. For our experiments, we also removed the headers and footers of the documents, which were not available to annotators working towards the gold standard and could cause ambiguities with some entity classes (e.g. *person* or *hospital*). This results in a smaller corpus, compared to the corpus used for de-identification experiments ($\approx$123,000 tokens).

The CoNLL 2003 corpus is a common dataset for evaluating NER algorithms. It is based on the Reuters corpus (Lewis et al., 2004) and is the only dataset outside the biomedical domain used in our experiments. The training and development set together contain 18,451 sentences ($\approx$256,000 tokens) annotated with 4 unique entities.

A detailed overview of the corpus characteristics is available in Table 1.

5.2 Experimental Setup

In this section, we present the experimental setup used for our experiments. First, we describe how we selected the hyper-parameters. Then, we detail the pre-trained word embeddings that were used in this work. Finally, we present the neural network training parameters.

5.2.1 Hyper-parameter Selection

We used Hyperopt[14] (Bergstra et al., 2011, 2013) to select a set of hyper-parameters that performed well on the MERLoT de-identification dataset. Due to heavy computation times, we re-used this

---

[13]https://github.com/strayMat/
bio-medical_ner

[14]http://hyperopt.github.io/hyperopt/

Corpus	# sent.	# tok.	# ann.	Entities
Name: MedPost **Task**: POS **Domain**: MEDLINE abstracts	6,701	182,319	182,319 (100%)	51 POS tagging detailed in Smith et al. (2004)
Name: NCBI disease **Task**: NER **Domain**: MEDLINE abstracts	7,279	151,005	11,350 (7.5%)	DiseaseClass, SpecificDisease, Composite-Mention, Modifier
Name: MERLoT medical **Task**: NER **Domain**: Medical reports from the Hepato-gastro-enterology and Nutrition ward	5,137	123,942	56,680 (46%)	Concept_Idea, MedicalProcedure, Hospital, Persons, Temporal, BiologicalProcessOrFunction, Devices, Measurement, Disorder, Aspect, Chemicals_Drugs, Dosage, SignOrSymptom, Anatomy, Localization, Livingbeings, Strength, AdministrationRoute, Drugform
Name: MERLoT de-identification **Task**: NER **Domain**: Same as MERLoT medical	25,599	177,158	31,723 (18%)	first name, last name, initials, address, zip code, town, date, hospital, identifier, phone number, email
Name: CoNLL 2003 **Task**: NER **Domain**: News articles from the Reuters corpus	18,451	256,145	42,646 (17%)	PER, ORG, LOC, MISC

Table 1: Overview of the corpora. The number of annotations corresponds to the number of annotated tokens to be comparable to the size of the corpus measured in number of tokens

set of hyper-parameters on the other datasets. Parameter search space includes the number of units of the character Bi-LSTM, the number of units of the main Bi-LSTM, the dimension of the character embeddings and the dropout rate. The retained setup uses character embeddings of size 24, 32 units for the character-level Bi-LSTM, 64 units for the main Bi-LSTM and a dropout rate of 0.5.

5.2.2 Embeddings

Pre-trained word embeddings were shown to boost the performance in various NLP tasks (Collobert et al., 2011; Mikolov et al., 2013a) and specifically in NER (Lample et al., 2016; Dernoncourt et al., 2017).

For each corpus, we used pre-trained word embeddings created with datasets that were consistent in genre and domain with the corpora used in this work. All word embedding models were computed using word2vec. For CoNLL 2003, we trained a model on Google News. For NCBI disease and MedPost, which both comprise abstracts from biomedical articles, we trained a model on the PubMed corpus. For MERLoT, we trained a model on the corpus from which MERLoT is extracted. It contains about 138,000 clinical notes written in French (Campillos et al., 2018).

For each of these word vector models, we computed low-dimensional representations by applying Principal Component Analysis (PCA) on the original high-dimensional word vectors (300 di-

mensions). During hyper-parameter search, we varied the dimension of the word embeddings and observed important impacts on the performance (up to 2.5 points of F1). We finally picked word vectors of dimension 25 (reduced by PCA), which showed to be the most efficient. This choice improved the scores as well as diminished drastically the computation times.

We also ran a few experiments using Fast-Text (Bojanowski et al., 2017) embeddings trained on Wikipedia. Preliminary results on CoNLL 2003 show a major performance improvement. An analysis of the impact of the word embedding model on performance will be conducted in following work.

5.2.3 Training

In all our experiments, the network was trained using the Adam optimizer (Kingma and Ba, 2015) with an initial learning rate of 0.001. Early-stopping was set to 10 epochs.

6 Results

In this section, we present the performance obtained on the corpora using the experimental setup described in the previous section. First, we report corpus-specific results. Then, we study the impact of the corpus size on performance.

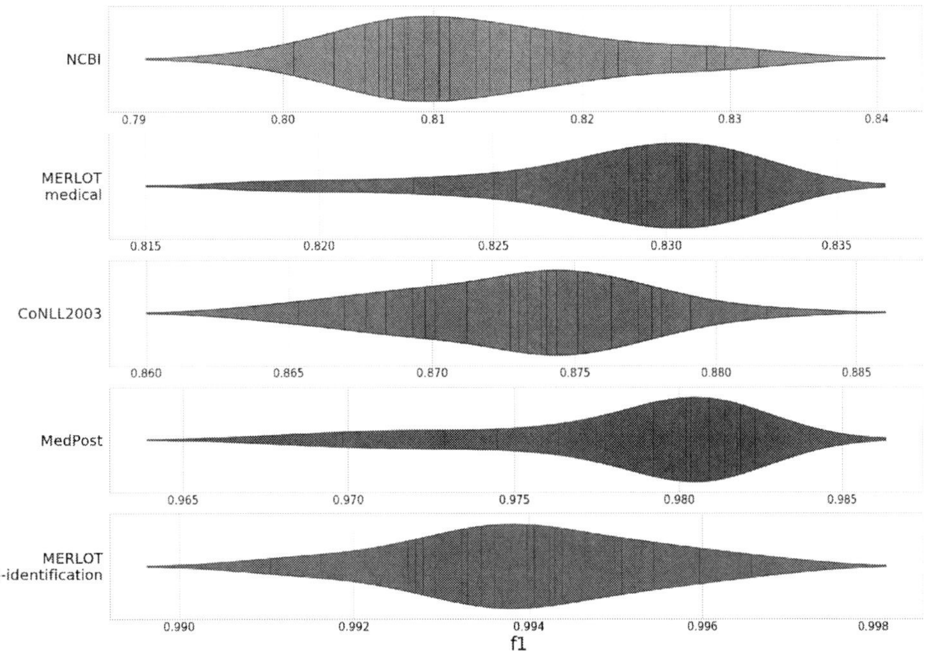

Figure 4: Distributions of F1 scores for the four corpora over 30 trainings for each corpus. Vertical lines refer to each sample. The scales are specific to each dataset for a proper visualization of each distribution.

6.1 Corpus Specific Results

Reimers and Gurevych (2017) report that neural network model training is highly non-deterministic and is subject to the random seed choice. Because of this variability during the training phase, it is crucial to report results on numerous trainings. Therefore we ran 30 experiments for each task presented in this work. We report F1-score statistics in Table 2. We also plot F1-score distributions for each task in Figure 4.

Our experiments show that performance varies across datasets, reflecting the heterogeneous difficulties of the tasks, inherent to the nature of the labels and the quality of the annotations. We notice that the standard deviations are similar for all NER tasks. It suggests that the variability of the score is independent from the task and mainly due to the model architecture. Previous work performances are reported in Table 3 for every dataset.

Although our model does not show state-of-the-art performances for all of them, it obtains competitive results, demonstrating the generalization ability of the selected shared set of hyper-parameters and embeddings.

Dataset	Model	F1
NCBI	Dang et al. (2018)	**84.41**
	Islamaj Dogan and Lu (2012)	81.80
	This paper	81.33
MERLoT med.[a]	Campillos et al. (2018)	81.40
	This paper	**82.87**
CoNLL 2003	Lample et al. (2016)	90.94
	Ma and Hovy (2016)	91.21
	Yang and Zhang (2018)	91.35
	Peters et al. (2018)	**92.22**
	This paper	87.31
MedPost[a]	Smith et al. (2004)	97.43
	This paper	**97.83**
MERLoT PHI	Grouin and Névéol (2014)	94.00
	This paper	**99.40**

[a] Corpora where the experimental set-up differed between our experiments and that of prior work. For MEDPOST, we used 51 categories instead of 63; for MERLoT med. we removed nested entities.

Table 3: State of the art results obtained in prior work on the datasets.

Task	Dataset	Mean F1	σ	Max. Diff.
NER	NCBI disease	81.33	0.83	3.27 (4.1%)
NER	MERLoT medical	82.87	0.36	1.40 (1.7%)
NER	CoNLL 2003	87.31	0.41	1.76 (2.0%)
POS	MedPost	97.83	0.39	1.44 (1.5%)
NER	MERLoT de-identification	99.40	0.14	0.568 (0.57%)

Table 2: Performance (%) of YASET on the 4 datasets.

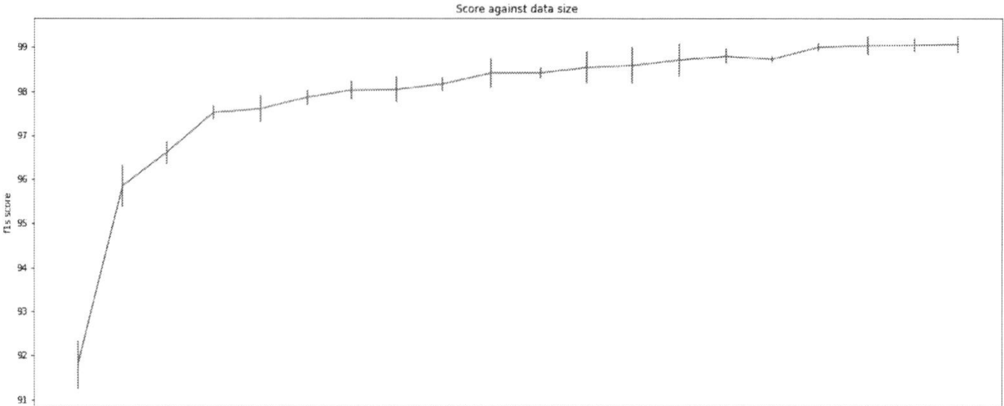

Figure 5: Effect of the training set size on the F1 score for the MERLoT dataset with de-identification annotations. These scores are computed over 6 training iterations per chunk. Vertical bars show the standard deviation σ.

6.2 Performance According to Training Data Size

The development of a labeled dataset for training annotation models is often a heavy investment in time and resources. Having some insight on the performance of the model for different training data sizes is crucial. Thus, we investigate the impact of this parameter on model performance. Focusing on the de-identification task of MERLoT, we split the training set into 20 subsets of equal sizes. Then, we built sub-training sets of growing size, that we refer as *chunks*. This resulted in 20 chunks where small chunks are subsets of the bigger one. Finally, we train YASET 6 times on each chunk, measuring how the performance improves with the size of the chunks. Figure 5 shows the progression of the F1-score according to the number of tokens. Table 4 presents the model performance according to the dataset size.

Dataset	5%	10%	25%	50%	100%
MERLoT PHI	91.79 (0.55)	95.86 (0.47)	97.60 (0.30)	98.41 (0.33)	99.06 (0.19)

Table 4: Performance (F1) against train set size (as percentage of the total training set indicated on the first row). Standard errors appear between parentheses.

We observe that the performance improves logarithmically with every chunk added in the training data as shown in Table 4. This finding is similar to the observation of Sun et al. (2017) for vision tasks. Further addition of data will slightly improve the performance as the maximum performance plateau is almost reached.

7 Conclusion and Future work

We propose an easy-to-use annotation tool implementing a state-of-the-art Bi-LSTM-CRF neural architecture. We apply the tool to PoS tagging and NER on clinical, biomedical and general domain texts. By running multiple experiment for each dataset, we confirm previous observations that neural network training is highly non-deterministic and dependent on the random seed choice.

Although we did not search for optimal hyper-parameters for each task, we obtain high performance in all our experiments, suggesting that tailoring hyper-parameters for a specific task improves only lightly the performance of the model and the neural network architecture implemented in YASET is robust with regards to the performance obtained.

Concerning the impact of the training dataset size on model performance, we confirm the intuition that adding more data allows to improve the performance of neural network models. However, result analysis suggests that the growth is logarithmic with the training data size.

One limitation of our model is its inability to directly handle nested entities such as those found in the MERLoT medical dataset and commonly used in the biomedical and clinical domains. When filtering these nested entities, specific classes are heavily impacted, including anatomy, disorder, localization, and medical procedure entities. Several strategies have been proposed to handle such cases. Campillos et al. (2018) present a 3-layer CRF model that annotates different non-overlapping clinical entities at each layer. Ju et al. (2018) present a dynamic end-to-end neural network model capable of handling an undetermined number of nesting levels. Katiyar and Cardie (2018) model the task as an hypergraph whose structure is learned with an LSTM network.

Future research will focus on the influence of word embedding models which were shown to significantly impact on performance. Specifically, models taking into account sub-token information (Bojanowski et al., 2017) or emphasizing context (Peters et al., 2018) should be further explored. Moreover, other neural network models for NER such as the ones proposed by Rei et al. (2016) and Ma and Hovy (2016) will be investigated and implemented in YASET. Having a centralized implementation of different NER models will allow us to compare their performances on various corpora.

Acknowledgements

The authors thank the Biomedical Informatics Department at the Rouen University Hospital for providing access to the LERUDI corpus for this work. This work was supported in part by the French National Agency for Research under grant CABeRneT ANR-13-JS02-0009-01 and by Labex Digicosme, operated by the Foundation for Scientific Cooperation (FSC) Paris-Saclay, under grant CÔT.

References

Martín Abadi, Ashish Agarwal, Paul Barham, et al. 2015. TensorFlow: Large-Scale Machine Learning on Heterogeneous Systems.

Rami Al-Rfou, Guillaume Alain, Amjad Almahairi, et al. 2016. Theano: A Python framework for fast computation of mathematical expressions. *CoRR*.

James Bergstra, Daniel Yamins, and David Cox. 2013. Making a Science of Model Search: Hyperparameter Optimization in Hundreds of Dimensions for Vision Architectures. In *Proceedings of the 30th International Conference on Machine Learning*, pages 115–123.

James S. Bergstra, Rémi Bardenet, Yoshua Bengio, and Balázs Kégl. 2011. Algorithms for Hyper-Parameter Optimization. In *Advances in Neural Information Processing Systems 24*, pages 2546–2554.

Steven Bethard, Leon Derczynski, Guergana Savova, James Pustejovsky, and Marc Verhagen. 2015. SemEval-2015 Task 6: Clinical TempEval. In *Proceedings of the 9th International Workshop on Semantic Evaluation*, pages 806–814. Association for Computational Linguistics.

Steven Bethard, Guergana Savova, Wei-Te Chen, et al. 2016. SemEval-2016 Task 12: Clinical TempEval. In *Proceedings of the 10th International Workshop on Semantic Evaluation*, pages 1052–1062. Association for Computational Linguistics.

Steven Bethard, Guergana Savova, Martha Palmer, and James Pustejovsky. 2017. SemEval-2017 Task 12: Clinical TempEval. In *Proceedings of the 11th International Workshop on Semantic Evaluation*, pages 565–572. Association for Computational Linguistics.

Piotr Bojanowski, Edouard Grave, Armand Joulin, and Tomas Mikolov. 2017. Enriching Word Vectors with Subword Information. *Transactions of the Association of Computational Linguistics*, 5:135–146.

Leonardo Campillos, Louise Deléger, Cyril Grouin, et al. 2018. A French Clinical Corpus with Comprehensive Semantic Annotations: Development of the Medical Entity and Relation LIMSI annOtated Text corpus (MERLOT). *Language Resources and Evaluation*, 52(2):571–601.

François Chollet et al. 2015. Keras. https://github.com/fchollet/keras.

Ronan Collobert, Jason Weston, Léon Bottou, et al. 2011. Natural Language Processing (Almost) from Scratch. *Journal of Machine Learning Research*, 12:2493–2537.

Thanh Hai Dang, Hoang-Quynh Le, Trang M Nguyen, and Sinh T Vu. 2018. D3NER: biomedical named entity recognition using CRF-biLSTM improved with fine-tuned embeddings of various linguistic information. *Bioinformatics*.

Franck Dernoncourt, Ji Young Lee, Özlem Uzuner, and Peter Szolovits. 2017. De-identification of Patient Notes with Recurrent Neural Networks. *Journal of the American Medical Informatics Association*, 24(3):596–606.

Cyril Grouin and Aurélie Névéol. 2014. De-identification of clinical notes in French: towards a protocol for reference corpus development. *Journal of Biomedical Informatics*, 50:151–161.

Sepp Hochreiter and Jürgen Schmidhuber. 1997. Long Short-Term Memory. *Neural Computation*, 9(8):1735–1780.

Rezarta Islamaj Dogan and Zhiyong Lu. 2012. An improved corpus of disease mentions in PubMed citations. *Proceedings of the 2012 Workshop on Biomedical Natural Language Processing*, pages 91–99.

Meizhi Ju, Makoto Miwa, and Sophia Ananiadou. 2018. A Neural Layered Model for Nested Named Entity Recognition. In *Proceedings of the 2018 Conference of the North American Chapter of the Association for Computational Linguistics: Human Language Technologies*, pages 1446–1459. Association for Computational Linguistics.

Arzoo Katiyar and Claire Cardie. 2018. Nested Named Entity Recognition Revisited. In *Proceedings of the 2018 Conference of the North American Chapter of the Association for Computational Linguistics: Human Language Technologies*, pages 861–871. Association for Computational Linguistics.

Diederik P. Kingma and Jimmy Ba. 2015. Adam: A Method for Stochastic Optimization. In *Proceedings of the 3rd International Conference on Learning Representations*.

Guillaume Lample, Miguel Ballesteros, Sandeep Subramanian, Kazuya Kawakami, and Chris Dyer. 2016. Neural Architectures for Named Entity Recognition. In *Proceedings of the 2016 Conference of the North American Chapter of the Association for Computational Linguistics: Human Language Technologies*, pages 260–270. Association for Computational Linguistics.

David D. Lewis, Yiming Yang, Tony G. Rose, and Fan Li. 2004. RCV1: A New Benchmark Collection for Text Categorization Research. *Journal of Machine Learning Research*, 5:361–397.

Xuezhe Ma and Eduard Hovy. 2016. End-to-end Sequence Labeling via Bi-directional LSTM-CNNs-CRF. In *Proceedings of the 54th Annual Meeting of the Association for Computational Linguistics*, pages 1064–1074. Association for Computational Linguistics.

Tomas Mikolov, Kai Chen, Greg Corrado, and Jeffrey Dean. 2013a. Efficient Estimation of Word Representations in Vector Space. *CoRR*, abs/1301.3781.

Tomas Mikolov, Ilya Sutskever, Kai Chen, Greg S. Corrado, and Jeff Dean. 2013b. Distributed Representations of Words and Phrases and their Compositionality. In *Advances in Neural Information Processing Systems 26*, pages 3111–3119.

Adam Paszke, Sam Gross, Soumith Chintala, et al. 2017. Automatic differentiation in PyTorch. In *NIPS-W*.

Jeffrey Pennington, Richard Socher, and Christopher D. Manning. 2014. GloVe: Global Vectors for Word Representation. In *Proceedings of the 14th Conference of the European Chapter of the Association for Computational Linguistics*, pages 1532–1543. Association for Computational Linguistics.

Matthew Peters, Mark Neumann, Mohit Iyyer, et al. 2018. Deep Contextualized Word Representations. In *Proceedings of the 2018 Conference of the North American Chapter of the Association for Computational Linguistics: Human Language Technologies*, pages 2227–2237. Association for Computational Linguistics.

Radim Řehůřek and Petr Sojka. 2010. Software Framework for Topic Modelling with Large Corpora. In *Proceedings of the LREC 2010 Workshop on New Challenges for NLP Frameworks*, pages 45–50. European Language Resources Association.

Marek Rei. 2017. Semi-supervised Multitask Learning for Sequence Labeling. In *Proceedings of the 55th Conference of the Association for Computational Linguistics*, pages 2121–2130. Association for Computational Linguistics.

Marek Rei, Gamal Crichton, and Sampo Pyysalo. 2016. Attending to Characters in Neural Sequence Labeling Models. In *Proceedings of the 26th International Conference on Computational Linguistics*, pages 309–318.

Marek Rei and Helen Yannakoudakis. 2016. Compositional Sequence Labeling Models for Error Detection in Learner Writing. In *Proceedings of the 54th Annual Meeting of the Association for Computational Linguistics*, pages 1181–1191. Association for Computational Linguistics.

Nils Reimers and Iryna Gurevych. 2017. Reporting Score Distributions Makes a Difference: Performance Study of LSTM-networks for Sequence Tagging. In *Proceedings of the 2017 Conference on Empirical Methods in Natural Language Processing*, pages 338–348.

Erik F. Tjong Kim Sang and Fien De Meulder. 2003. Introduction to the CoNLL-2003 Shared Task: Language-Independent Named Entity Recognition. In *Proceedings of the 7th Conference on Natural Language Learning*. Association for Computational Linguistics.

Laurence H. Smith, Thomas C. Rindflesch, and W. John Wilbur. 2004. MedPost: a Part-of-speech Tagger for BioMedical Text. *Bioinformatics*, 20(14):2320–2321.

Pontus Stenetorp, Sampo Pyysalo, Goran Topić, et al. 2012. brat: a Web-based Tool for NLP-Assisted Text Annotation. In *Proceedings of the 13th Conference of the European Chapter of the Association for Computational Linguistics*, volume Demonstration Papers, pages 102–107. Association for Computational Linguistics.

William F. Styler IV, Steven Bethard, Sean Finan, et al. 2014. Temporal Annotation in the Clinical Domain. *Transactions of the Association for Computational Linguistics*, 2:143–154.

Chen Sun, Abhinav Shrivastava, Saurabh Singh, and Abhinav Gupta. 2017. Revisiting Unreasonable Effectiveness of Data in Deep Learning Era. *CoRR*, abs/1707.02968.

Jie Yang and Yue Zhang. 2018. NCRF++: An Open-source Neural Sequence Labeling Toolkit. In *Proceedings of ACL 2018, System Demonstrations*, pages 74–79. Association for Computational Linguistics.

Learning to Summarize Radiology Findings

Yuhao Zhang, Daisy Yi Ding, Tianpei Qian,
Christopher D. Manning, Curtis P. Langlotz
Stanford University
Stanford, CA 94305
{yuhaozhang, dingd, tianpei}@stanford.edu
{manning, langlotz}@stanford.edu

Abstract

The Impression section of a radiology report summarizes crucial radiology findings in natural language and plays a central role in communicating these findings to physicians. However, the process of generating impressions by summarizing findings is time-consuming for radiologists and prone to errors. We propose to automate the generation of radiology impressions with neural sequence-to-sequence learning. We further propose a customized neural model for this task which learns to encode the study background information and use this information to guide the decoding process. On a large dataset of radiology reports collected from actual hospital studies, our model outperforms existing non-neural and neural baselines under the ROUGE metrics. In a blind experiment, a board-certified radiologist indicated that 67% of sampled system summaries are at least as good as the corresponding human-written summaries, suggesting significant clinical validity. To our knowledge our work represents the first attempt in this direction.

1 Introduction

The radiology report documents and communicates crucial findings in a radiology study. As shown in Figure 1, a standard radiology report usually consists of a Background section that describes the exam and patient information, a Findings section, and an Impression section (Kahn Jr et al., 2009). In a typical workflow, a radiologist first dictates the detailed findings into the report, and then summarizes the salient findings into the more concise Impression section based also on the condition of the patient.

The impressions are the most significant part of a radiology report that communicate the findings. Previous studies have shown that over 50% of referring physicians read only the impression statements in a report (Lafortune et al., 1988;

Background: history: swelling; pain. technique: 3 views of the left ankle were acquired. comparison: no prior study available.
Findings: there is normal mineralization and alignment. no fracture or osseous lesion is identified. the ankle mortise and hindfoot joint spaces are maintained. there is no joint effusion. the soft tissues are normal.
Human Impression: normal left ankle radiographs.
Extractive Baseline: there is no joint effusion.
Pointer-Generator: normal right ankle.
Our model: normal radiographs of the left ankle.

Figure 1: An example radiology report with study background information organized into a **Background** Section, and radiology findings in a **Findings** Section. The human-written summary (or impression) and predicted summaries from different models are also shown. The extractive baseline does not summarize well, the baseline pointer-generator model generates spurious sequence, while our model gives correct summary by incorporating the background information.

Bosmans et al., 2011). Despite its importance, the generation of the impression statements is error-prone. For example, crucial findings may be forgotten, which would cause significant miscommunications (Gershanik et al., 2011). Additionally, the process of writing the impression statements is time-consuming and highly repetitive with the dictation of the findings. This suggests a crucial need to automate the radiology impression generation process.

In this work, we propose to automate the generation of radiology impressions with neural sequence-to-sequence learning. In particular, we argue that this task could be viewed as a text summarization problem, where the source sequence is the radiology findings and the target sequence the

Proceedings of the 9th International Workshop on Health Text Mining and Information Analysis (LOUHI 2018), pages 204–213
Brussels, Belgium, October 31, 2018. ©2018 Association for Computational Linguistics

impression statements. We collect a dataset of radiology reports from actual hospital radiographic studies, and find that this task involves both *extractive summarization* where descriptions of radiology observations can be taken directly from the findings, and *abstractive summarization* where new words and phrases, such as conclusions of the study, need to be generated from scratch. We empirically evaluate existing popular summarization systems on this task and find that, while existing neural models such as the pointer-generater network can generate plausible summaries, they sometimes fail to model the study background information and thus generate spurious results. To solve this problem, we propose a customized summarization model that properly encodes the study background information and uses the encoded information to guide the decoding process.

We show that our model outperforms existing non-neural and neural baselines on our dataset measured by the standard ROUGE metrics. Moreover, in a blind experiment, a board-certified radiologist indicated that 67% of sampled system summaries are at least as good as the reference summaries written by well-trained radiologists, suggesting significant clinical validity of the resulting system. We further show through detailed analysis that our model could be reliably transferred to radiology reports from another organization, and that the model can sometimes summarize radiology studies for body parts unseen during training.

To review, our main contributions in this paper include: (i) we propose to summarize radiology findings into impression statements with neural sequence-to-sequence learning, and to our knowledge our work represents the first attempt in this direction; (ii) we propose a new customized summarization model to this task that improves over existing methods by better leveraging study background information; (iii) we further show via a radiologist evaluation that the summaries generated by our model have significant clinical validity.

2 Related Work

Early Summarization Systems. Early work on summarization systems mainly focused on extractive approaches, where the summaries are generated by scoring and selecting sentences from the input. Luhn (1958) proposed to represent the input by topic words and score each sentence by the amount of topic words it contains. Kupiec et al. (1995) scored sentences with a feature-based statistical classifier. Steinberger and Jezek (2004) applied latent semantic analysis to cluster the topics and then select sentences that cover the most topics. Meanwhile, various graph-based methods, such as the LexRank (Mihalcea and Tarau, 2004) and the TextRank algorithm (Erkan and Radev, 2004), were applied to model sentence dependency by representing sentences as vertices and similarities as edges. Sentences are then scored and selected via modeling of the graph properties.

Neural Summarization Systems. Summarization systems based on neural network models enable abstractive summarization, where new words and phrases are generated to form the summaries. Rush et al. (2015) first applied an attention-based neural encoder and a neural language model decoder to this task. Nallapati et al. (2016) used recurrent neural networks for both the encoder and the decoder. To address the limitation that neural models with a fixed vocabulary cannot handle out-of-vocabulary words, a pointer-generator model was proposed which uses an attention mechanism that copies elements directly from the input (Nallapati et al., 2016; Merity et al., 2017; See et al., 2017). See et al. (2017) further proposed a coverage mechanism to address the repetition problem in the generated summaries. Paulus et al. (2018) applied reinforcement learning to summarization and more recently, Chen and Bansal (2018) obtained improved result with a model that first selects sentences and then rewrites them.

Summarization of Radiology Reports. Most prior work that attempts to "summarize" radiology reports focused on classifying and extracting information from the report text (Friedman et al., 1995; Hripcsak et al., 1998; Elkins et al., 2000; Hripcsak et al., 2002). More recently, Hassanpour and Langlotz (2016) studied extracting named entities from multi-institutional radiology reports using traditional feature-based classifiers. Goff and Loehfelm (2018) built an NLP pipeline to identify asserted and negated disease entities in the impression section of radiology reports as a step towards report summarization. Cornegruta et al. (2016) proposed to use a recurrent neural network architecture to model radiological language in solving the medical named entity recognition and negation detection tasks on radiology reports. To our knowledge, our work represents the first attempt at automatic summarization of radiology findings into natural language impression statements.

3 Task Definition

We now give a formal definition of the task of summarizing radiology findings. Given a passage of findings represented as a sequence of tokens $\mathbf{x} = \{x_1, x_2, \ldots, x_N\}$, with N being the length of the findings, our goal is to find a sequence of tokens $\mathbf{y} = \{y_1, y_2, \ldots, y_L\}$ that best summarizes the salient and clinically significant findings in $\mathbf{x}$, with L being an arbitrary length of the summary.[1] Note that the mapping between $\mathbf{x}$ and $\mathbf{y}$ can either be modeled in an unsupervised way (as done in unsupervised summarization systems), or be learned from a dataset of findings-summary pairs.

4 Models

In this section we introduce our model for the task of summarizing radiology findings. As our model builds on top of existing work on neural sequence-to-sequence learning and the pointer-generator model, we start by introducing them.

4.1 Neural Sequence-to-Sequence Model

At a high-level, our model implements the summarization task with an encoder-decoder architecture, where the encoder learns hidden state representations of the input, and the decoder decodes the input representations into an output sequence.

For the encoder, we use a Bi-directional Long Short-Term Memory (Bi-LSTM) network. Given the findings sequence $\mathbf{x} = \{x_1, x_2, \ldots, x_N\}$, we encode $\mathbf{x}$ into hidden state vectors with:

$$\mathbf{h} = \text{Bi-LSTM}(\mathbf{x}), \qquad (1)$$

where $\mathbf{h} = \{h_1, h_2, \ldots, h_N\}$. Here h_N combines the last hidden states from both directions in the encoder.

After the entire input sequence is encoded, we generate the output sequence step by step with a separate LSTM decoder. Formally, at the t-th step, given the previously generated token y_{t-1} and the previous decoder state s_{t-1}, the decoder calculates the current state s_t with:

$$s_t = \text{LSTM}(s_{t-1}, y_{t-1}). \qquad (2)$$

We then use s_t to predict the output word. For the initial decoder state we set $s_0 = h_N$.

The vanilla sequence-to-sequence model that uses only s_t to predict the output word has a major limitation: it generates the entire output sequence based solely on a vector representation of the input (i.e., h_N), which may result in significant information loss. For better decoding we therefore employ the attention mechanism (Bahdanau et al., 2015; Luong et al., 2015), which uses a weighted sum of all input states at every decoding step.

Given the decoder state s_t and an input hidden state h_i, we calculate an input distribution a^t as:

$$e_i^t = v^\top \tanh(W_h h_i + W_s s_t), \qquad (3)$$
$$a^t = \text{softmax}(e^t), \qquad (4)$$

where W_h, W_s and v are learnable parameters.[2] We then calculate a weighted input vector as:

$$h_t^* = \sum_i a_i^t h_i. \qquad (5)$$

h_t^* encodes the salient input information that is useful at decoding step t. Lastly, we obtain the output vocabulary distribution at step t as:

$$P(y_t|\mathbf{x}, y_{<t}) = \text{softmax}(V' \tanh(V[s_t; h_t^*])), \qquad (6)$$

where V' and V are learnable parameters.

4.2 Pointer-Generator Network

While the encoder-decoder framework described above can generate impressions from a fixed vocabulary, the model can clearly benefit from being able to "copy" salient observations directly from the input findings. To add such "copying" capacity into the model, we use a pointer-generator network similar to the one described in See et al. (2017).

The main idea is that at each decoding step t, we allow the model to either generate a word from the vocabulary with a generation probability p_{gen}, or copy a word directly from the input sequence with probability $1 - p_{\text{gen}}$. We model p_{gen} as:

$$p_{\text{gen}} = \sigma(w_{h^*}^\top h_t^* + w_s^\top s_t + w_y y_{t-1}), \qquad (7)$$

where y_{t-1} denotes the previous decoder output, w_{h^*}, w_s and w_y learnable parameters and σ a sigmoid function. For the copy distribution, we reuse the attention distribution a^t calculated in (4). Therefore, the overall output distribution in the pointer-generator network is:

$$P(y_t|\mathbf{x}, y_{<t}) = p_{\text{gen}} P_{\text{vocab}}(y_t) + (1 - p_{\text{gen}}) \sum_{i:x_i = y_t} a_i^t, \qquad (8)$$

---

[1] While the name "impression" is often used in clinical settings, we use "summary" and "impression" interchangably.

[2] For clarity we leave out the bias terms in all linear layers.

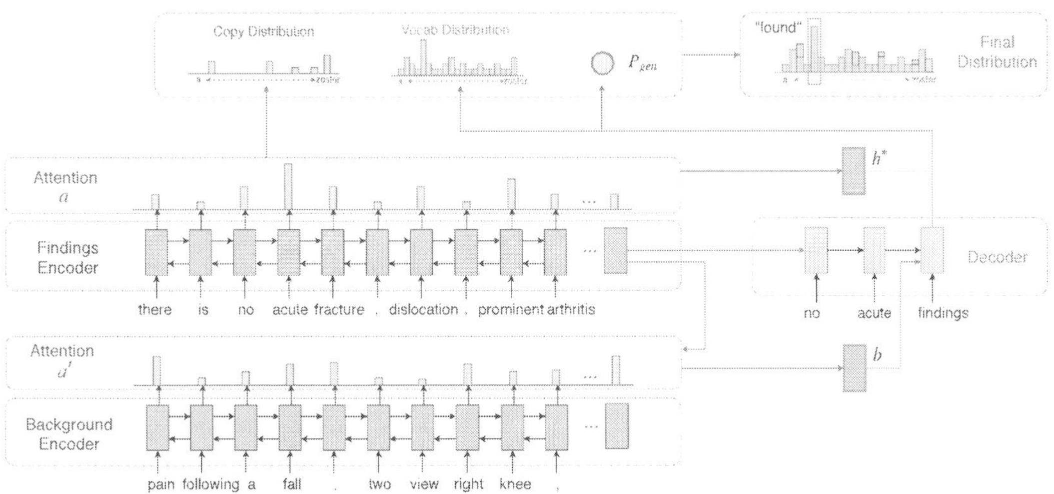

Figure 2: Overall architecture of our summarization model.

where $P_{\text{vocab}}(y_t)$ is the same as the output distribution in (6).

4.3 Incorporating Study Background Information

The background part of a radiology report is also important, since crucial information such as the purpose of the study, the body part involved and the condition of the patient are often mentioned only in the background. A straightforward way of incorporating the background information is to prepend all the background text to the findings, and treat the entire sequence as input to the pointer-generator network. However, as we show in Section 6, this naive method in fact hurts the summarization quality, presumably because the model cannot sufficiently distinguish between the findings and the background information, which as a result leads to insufficient modeling of both the findings and the background. To solve this, we propose to encode the background text with a separate attentional encoder, and use the resulting background representation to guide the decoding process in the summarization model (Figure 2).

For clarity we now use $\mathbf{x}^b$ to denote the background token sequence, and $\mathbf{x}$ to denote the actual findings section. Our goal is then to find $\mathbf{y}$ that maximizes $P(\mathbf{y}|\mathbf{x}, \mathbf{x}^b)$. To do this, we again obtain the hidden state vectors $\mathbf{h}$ of the findings section as in (1). Similarly, we obtain the hidden state vectors of the background text with $\mathbf{x}^b$ as input using a separate Bi-LSTM encoder:

$$\mathbf{h}^b = \text{Bi-LSTM}^b(\mathbf{x}^b). \qquad (9)$$

Next, we calculate a distribution over $\mathbf{h}^b$ as:

$$e'_i = v'^\top \tanh(W_b h^b_i + W_h h_N), \qquad (10)$$
$$a' = \text{softmax}(e'), \qquad (11)$$

where W_b and W_h are learnable parameters and h_N the last hidden state of the findings encoder. The distribution a' models the importance of tokens in the background section. We then obtain a weighted representation of the background text as:

$$b = \sum_i a'_i h^b_i, \qquad (12)$$

where vector b has the same size as h^b, and encodes the salient background information.

Lastly, we use the background vector b to guide the decoding process, by modifying the recurrent kernel of the decoder LSTM in (2) to be:

$$\begin{bmatrix} i_t \\ f_t \\ o_t \\ u_t \end{bmatrix} = \begin{bmatrix} \sigma \\ \sigma \\ \sigma \\ \tanh \end{bmatrix} W \cdot \begin{bmatrix} s_{t-1} \\ y_{t-1} \\ b \end{bmatrix}, \qquad (13)$$
$$c_t = f_t \cdot c_{t-1} + i_t \cdot u_t, \qquad (14)$$
$$s_t = o_t \cdot \tanh(c_t), \qquad (15)$$

where i_t, f_t, o_t denotes the input, forget, and output gates, W the weight matrix and c_t the internal cell of the LSTM repectively, and $\cdot$ represents an element-wise multiplication. Again for clarity we leave out the bias terms in (13). As a result, every state in the decoding process is directly influenced by the information encoded by the background vector b. The rest of the model, including

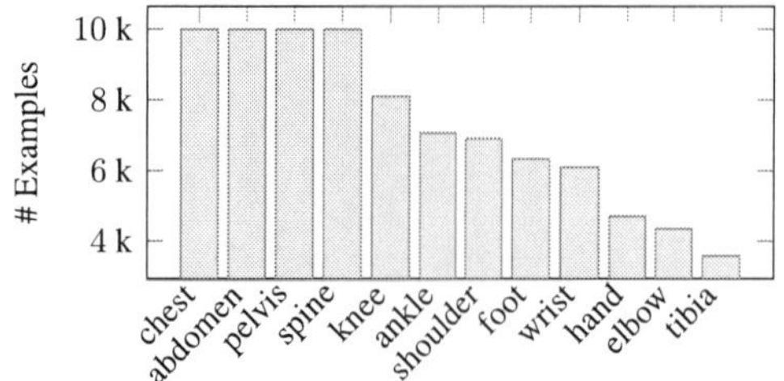

Figure 3: Number of examples split by body part in the collected Stanford Hospital dataset.

the calculation of the vocabulary distribution and the copy distribution, remains the same.

5 Experiments

To test the effectiveness of our summarization model, we collected reports of radiographic studies from the picture archiving and communication system (PACS) at the Stanford Hospital. We describe our data collection process, baseline models and experimental setup in this section, and present the results and discussions in Section 6.

5.1 Data Collection

Reports of all radiographic studies from 2000 to 2014 were collected. We first tokenized all reports with Stanford CoreNLP (Manning et al., 2014), and filtered the dataset by excluding reports where (1) no findings or impression section can be found; (2) multiple findings or impression sections can be found but cannot be aligned; or (3) the findings have fewer than 10 words or the impression has fewer than 2 words.

We removed body parts where only a small number of cases are available, and included reports of the top 12 body parts in the PACS system to maintain generalizability. For common body parts with more than 10k reports (e.g., chest), we subsampled 10k reports from them.

This results in a dataset with a total of 87,127 reports. We further randomly split the dataset into a 70% training (60,990), a 10% development (8,712) and a 20% test set (17,425). We show the dataset statistics split by body part in Figure 3.

5.2 Baseline Models

For our main experiments, we compare our model against several competitive non-neural and neural systems on the collected dataset. Unless otherwise stated, the baseline models take only the findings

section as input.[3]

S&J-LSA. This is an extractive approach described by Steinberger and Jezek (2004), which applies Latent Semantic Analysis (LSA) to summarization. It first scores "concept" clusters by applying singular value decomposition to the term-by-sentence co-occurence matrix derived from the passage. Sentences with the top scored concepts are then kept as the summaries.

LexRank. LexRank is another popular extractive model introduced by Erkan and Radev (2004). In LexRank, a passage is represented as a graph of sentences, and a connectivity matrix based on intra-sentence cosine similarity is used as the adjacency matrix of the graph. Sentences are scored by the eigenvector centrality in the graph, and the highest scored sentences are kept.

Pointer-Generator. We also run the baseline pointer-generator model introduced by See et al. (2017). We find the "coverage" mechanism described in the paper did not improve summary quality in our task and therefore did not use it for simplicity. We compare our model with two versions of the pointer-generator model: one with only the findings section as input and another one with the background sections prepended to the findings section as input.

5.3 Experimental Setup

Evaluation Metrics. In our main experiments we evaluate the models with the widely-used ROUGE metric (Lin, 2004). We report the F_1 scores for ROUGE-1, ROUGE-2 and ROUGE-L, which measure the word-level unigram-overlap, bigram-overlap and the longest common sequence between the reference summary and the system predicted summary respectively.

Word Vectors. To enable knowledge transfer from a larger corpus, we applied the GloVe algorithm (Pennington et al., 2014) to a corpus of 4.5 million radiology reports of all modalities (e.g., X-ray, CT) and body parts. We used the resulting 100-dimensional word vectors to initialize all word embedding layers in our neural models, and empirically found this to improve the performance of our neural models by about 1 ROUGE-L score.

---

[3]We find that when the background section is prepended to the input, the extractive baseline models may select sentences in the background part as the summary, resulting in deteriorated performance.

System	ROUGE-1	ROUGE-2	ROUGE-L
Extractive Baseline: S&J-LSA	29.39	16.27	27.38
Extractive Baseline: LexRank	30.48	17.09	28.49
Pointer-Generator	46.51	33.39	45.07
Pointer-Generator ($\oplus$ Background)	45.39	32.60	44.05
Our model	**48.56**	**35.25**	**47.06**

Table 1: Main results on the test set of the Stanford reports. "$\oplus$ Background" represents prepending the background section to the findings section to form the input to the model. All the ROUGE scores have a 95% confidence interval of at most ± 0.50 as calculated by the official ROUGE script.

Implementations & Model Details. For the two non-neural extractive baselines, we use their open implementations.[4] For both of them, we select the top N scored sentences to form the summary and treat N as a hyperparameter. We use $N = 3$ in our experiments as it yields best scores on the dev set. We implemented all neural models with Py-Torch.[5] To train the neural models we append a special <EOS> token to the end of every reference summary. We then employ the standard teacher-forcing with the reference summaries and optimize the negative log-likelihood loss using the Adam optimizer (Kingma and Ba, 2015). We tune all hyperparameters on the dev set. We use 2-layer Bi-LSTM for all encoders, and set the hidden size to be 100 for each direction; 1-layer LSTM for the decoder and set the hidden size to be 200. During inference, we employ the standard beam search with a beam size of 5. We stop decoding whenever a <EOS> token is predicted, and otherwise use a maximum output sequence length of 100.

6 Results & Analysis

6.1 Main Results

We present results of our main experiments in Table 1. We find that the two non-neural extractive models perform comparably, and both are able to obtain non-trivial subsequence overlap with the reference summaries as measured by ROUGE scores. However, a baseline neural pointer-generator that combines the sequence generation and the copy mechanism beats the non-neural baselines substantially on all metrics. We confirm that naively incorporating the study background information by prepending the background section directly to the input findings in the pointer-generator model in fact hurts the performance

(noted by $\oplus$ Background). In comparison, our model benefits from using the separately encoded background vector to guide the decoding process, and achieves best scores on all ROUGE metrics.

We also present sampled test examples and system output in Figure 4. We find that compared to the non-neural extractive baselines, the neural models are not limited by sentences in the findings section and therefore generate summaries of better quality. For example, the neural models learn to compose the summary by combining observation phrases from different sentences, or by generating new conclusive phrases such as "negative study". Compared to the pointer-generator model, our model learns to correctly utilize relevant background information (e.g., previous study or exam information) to improve the summary.

6.2 Clinical Validity with Radiologist Evaluation

One potential shortcoming of the ROUGE metrics is that they only measure the similarity between the predicted summary and the reference summary, but do not sufficiently reflect the overall grammaticality or utility of the predictions. Therefore, we also conducted evaluations with a board-certified radiologist to understand the clinical validity of our system generated summaries.

In this evaluation, we randomly sampled 100 examples from our test set. We ran our best model over these 100 examples, and presented each example along with the corresponding system predicted summary and reference human-written summary to the radiologist. We randomly ordered the predicted and reference summary such that the correspondence cannot be guessed from the order. The radiologist was asked to select which of the two summaries was better, or that they have roughly equal quality.

Table 2 presents the result. For 51 examples, the

[4]https://github.com/miso-belica/sumy
[5]https://pytorch.org/

Background: radiographic examination of the abdomen. clinical history: xx years of age, male, please obtain upright and lateral decub. comparison: abdominal x-ray <date>. procedure comments: two views of the abdomen. **Findings:** median sternotomy wires are seen in the anterior chest wall in addition to several mediastinal clips and an aicd. trace bilateral pleural effusions are noted. interval increase in small bowel dilatation compared to previous study with multiple air-fluid levels, consistent with small bowel obstruction. there is a paucity of colonic gas. no pneumoperitoneum.	**Background:** three views of the right shoulder and three views of the left shoulder: <date>. clinical history: an xx-year-old female with bilateral shoulder pain. **Findings:** three views of the right shoulder consisting of external rotation, axillary, and scapular views demonstrate no evidence of fracture or dislocation. the joint spaces are well-maintained without evidence of degenerative change. there is normal mineralization throughout. three views of the left shoulder . . . are well-maintained without evidence of degenerative change. mineralization is normal throughout.	**Background:** three views of the abdomen: <date>. comparison: <date>. clinical history: a xx-year-old male status post hirschsprung's disease repair. **Findings:** the supine, left-sided decubitus and erect two views of the abdomen show increased dilatation of the small bowel since the prior exam on <date>. there are multiple air-fluid levels, suggesting bowel obstruction. no free intraperitoneal gas is present.
Human: small bowel dilatation with multiple air-fluid levels and colonic decompression consistent with small bowel obstruction.	**Human:** unremarkable radiographs of bilateral shoulders.	**Human:** increased dilatation of the small bowel with multiple air-fluid levels, suggesting bowel obstruction. no free intraperitoneal gas.
Extractive Baseline: median sternotomy wires are seen in the anterior chest wall in addition to several mediastinal clips and an aicd.	**Extractive Baseline:** three views of the right shoulder consisting of external rotation, axillary, and scapular views demonstrate no evidence of fracture or dislocation.	**Extractive Baseline:** the supine, left sided decubitus and erect two views of the abdomen show increased dilatation of the small bowel since the prior exam on <data>.
Pointer-Generator: interval increase in bowel dilatation, consistent with bowel obstruction.	**Pointer-Generator:** no evidence of fracture or dislocation of the right shoulder.	**Pointer-Generator:** increased dilatation of small bowel, suggesting small bowel obstruction.
Our model: interval increase in small bowel dilatation compared to abdominal x-ray dated <date> with multiple air-fluid levels, consistent with small bowel obstruction.	**Our model:** unremarkable bilateral shoulders.	**Our model:** increased dilatation of small bowel, suggesting bowel obstruction. no free intraperitoneal gas.

Figure 4: Sampled test examples and system predictions from the Stanford dataset. First example: our model learns to relate the summary with a previous study mentioned only in the background section. Second: our model correctly summarizes the body part involved in the study. Third: our model correctly includes more crucial information as found in the human summary.

Category	Percentage
Human Summary Wins	33
System Prediction Wins	16
Roughly Equal Quality	51

Table 2: Radiologist evaluation result on 100 sampled test examples. For a total of 67 examples, the radiologist indicated that the system summary is at least as good as the human-written summary.

System	ROUGE-1	ROUGE-2	ROUGE-L
LexRank	15.42	5.65	14.60
Our model	35.02	20.79	34.56

Table 3: Cross-organization evaluation results on the Indiana University chest x-ray dataset. All the ROUGE scores have a 95% confidence interval of at most ± 1.10 as calculated by the official ROUGE script.

radiologist indicated that the human-written and system-generated summaries are equivalent. For 16 examples, the radiologist preferred the system summary, and for the remaining 33 examples, the radiologist preferred the human-written summary. Note that under our setting, a randomly generated sequence would have almost zero chance to be indicated as good as the human-written summary. We therefore believe the result suggests significant clinical validity of our system.

6.3 Does the model transfer to reports from another organization?

Deploying a clinical NLP system at an organization different from the one where the training data comes from is a common need. However, this is challenging in that medical practitioners including radiologists from different organizations tend to go through different training and follow different templates or styles when writing medical text. Here we aim to understand the cross-organization transferability of our summarization model.

We use the publicly available Indiana University Chest X-ray Dataset (Demner-Fushman et al., 2015), which consists of chest X-ray images paired with the corresponding radiology reports. We filtered the reports with the same set of rules and arrived at a collection of 2,691 unique reports. We used this dataset as the test set, and ran our best model trained on our own dataset directly on it. The results are shown in Table 3 and sampled examples are shown in the first two columns of Figure 5. We find that our model again outperforms the baseline extractive model substantially in this transfer setting, and the generated summaries are both grammatical and clinically meaningful.

Cross-organization	Cross-organization	Cross-body part: Knee
Background: indication: xxxx year old male with end-stage renal disease on hemodialysis **Findings:** the heart size is mildly enlarged. there is tortuosity of the thoracic aorta. no focal airspace consolidation, pleural effusions or pneumothorax. no acute bony abnormalities.	**Background:** indication: xxxx year old female, hypoxia. comparison: pa lateral views of the chest dated xxxx. **Findings:** bilateral emphysematous again noted and lower lobe fibrotic changes. postsurgical changes of the chest including cabg procedure, stable. stable valve artifact. there are no focal areas of consolidation. no large pleural effusions. no evidence of pneumothorax. . . . contour abnormality of the posterior aspect of the right 7th rib again noted, stable.	**Background:** radiographic examination of the knee: <date> <time>. clinical history: xx-year-old man with right knee pain. comparison: none. procedure comments: 2 views of the right knee were performed. **Findings:** there is no visible fracture or malalignment. likely small joint effusion. mild fullness in the popliteal region of the right knee may represent a baker 's cyst. mild soft tissue swelling along the medial aspect of the knee is present.
Human: cardiomegaly without acute pulmonary findings.	**Human:** no acute cardiopulmonary abnormality. stable bilateral emphysematous and lower lobe fibrotic changes.	**Human:** no acute bony abnormality. likely joint effusion and soft tissue swelling along the medial aspect of the knee.
Our model: mild cardiomegaly. no radiographic evidence of acute cardiopulmonary process.	**Our model:** stable postsurgical changes of the chest as described above. no evidence of pneumothorax.	**Our model:** mild soft tissue swelling along the medial aspect of the knee. no fracture or malalignment.

Figure 5: First two columns: sampled examples from the Indiana University dataset and system output in the cross-organization evaluation. Last column: sampled test example of a "knee" study in our cross-body part evaluation.

Body Part	ROUGE-1	ROUGE-2	ROUGE-L
Chest	31.24	17.99	30.38
Abdomen	28.90	17.23	27.83
Knee	48.78	35.07	47.49

Table 4: Cross-body part evaluation results of our neural model on the Stanford dataset. All the ROUGE scores have a 95% confidence interval of at most ± 0.75 as calculated by the official ROUGE script.

Category	Percentage
Good Summary	63
Missing Critical Info.	24
Inaccurate/Spurious Info.	8
Redundant	4
Ungrammatical	6

Table 5: Error analysis on 100 sampled dev examples from the Stanford dataset.

6.4 Does the model transfer to body parts unseen during training?

Radiology studies conducted on different body parts often include vastly different observations and diagnosis. For example, while "lung base opacity" is a common observation in chest radiographic studies, it does not exist in musculoskeletal studies. In practice, an organization may not have adequate report data that covers some rare body parts. It is therefore interesting to test to what extent our summarization model can generalize to reports for body parts unseen during training.

We study this by simulating the condition where a specific body part is not present in the training data. Given the entire dataset $\mathcal{D}$, and a subset of the dataset $\mathcal{D}_B$ that corresponds to a body part B, we reserved the entire subset $\mathcal{D}_B$ as test data, and used $\mathcal{D} - \mathcal{D}_B$ for training (90%) and validation (10%). Table 4 presents the evaluation results for body part "chest", "abdomen" and "knee". We find that for "chest" and "abdomen", the system summaries degrade substantially when the corresponding data were not seen during training. However, the predicted summaries degrade less for "knee" when reports of it were not seen during training, presumably because the model can learn to summarize reasonably well from reports of other close musculoskeletal studies such as "ankle" or "elbow" studies. We confirm this by examining the model predictions: in the example shown in the last column of Figure 5, the model learns to compose the summary with salient observations such as "tissue swelling" and "fracture", while being able to copy the anatomy "knee" (unseen during training) from the findings section.

6.5 What is the model missing on?

Lastly, we run a detailed error analysis on 100 sampled dev examples. We focus on four types of errors: (1) missing critical information, if the predicted summary fails to include some clinically important information; (2) inaccuate/spurious information, if the predicted summary contains observations or conclusions that are inaccurate, or that do not exist in the findings; (3) redundant summary, if the predicted summary is repetitive or over-verbose; and (4) ungrammatical summary, if the predicted summary contains significant gram-

Error type: missing critical information	Error type: redundant summary	Error type: ungrammatical summary
Background: radiographic examination of the lumbar spine: <time>. clinical history: <age>, lower back pain. comparison: none. procedure comments: 4 views of the lumbar spine. **Findings:** five non-rib bearing lumbar type vertebral bodies are present. there is trace retrolisthesis of l5 on s1. there is no evidence of instability on flexion and extension views. the spinal alignment is otherwise normal. the disc spaces and vertebral body heights are preserved. there is no visible fracture. no visible facet joint arthropathy or pars defects.	**Background:** radiographic examination of the shoulder: <time>. clinical history: <age> years of age, pain in joint involving shoulder region. comparison: outside study dated <date>. procedure comments: single axillary view of the left shoulder. **Findings:** single axillary view of the shoulder again demonstrates a highly comminuted fracture of the humeral head and likely fracture of the scapular body. the humeral head appears located on the glenoid.	**Background:** radiographic examination of the shoulder: <time>. clinical history: <age> years of age, xray exam of lower spine 2 or 3 views. x-ray exam of right shoulder complete. comparison: none. procedure comments: three views of the right shoulder. **Findings:** a calcification of the rotator cuff is seen above the greater tuberosity. there is no fracture or malalignment. the soft tissues and visualized lung are unremarkable.
Human: trace retrolisthesis of l5 on s1 with no evidence of instability with motion. otherwise normal lumbar spine.	**Human:** redemonstration of a highly comminuted fracture of the humeral head and likely fracture of the scapular body . the humeral head appears to be located on the glenoid .	**Human:** no acute bony or joint abnormality, but there is calcification of the rotator cuff that may be due to calcific tendinitis.
Our model: no acute bony or articular abnormality.	**Our model:** highly comminuted fracture of the scapular body and likely fracture of the scapular body.	**Our model:** calcification acute bony or joint abnormality.

Figure 6: Examples of different types of errors that our system makes on the Standord dataset. Words that are missing from or are erroneously included in the model predictions are highlighted in red.

matical errors. For each example, we examine whether it contains any of the errors by comparing it with the reference summary; otherwise we classify it as a good summary. Note that an example can be assigned to more than one error categories.

We include examples of different error types in Figure 6, and present the result of error analysis in Table 5. We find that 63% examples are qualitatively close to the reference summary, which aligns well with the radiologist evaluation result. Among the four error categories, missing critical information is the most common error with 24% examples, suggesting that the summaries may be improved with explicit modeling of the importance of different radiology findings. We also find through qualitative analysis that the model tends to miss on followup procedures recommended by the human radiologist, since these procedures are often not included in the findings section and generating them needs significant understanding of the study and domain knowledge.

7 Conclusion

In this paper we proposed to generate radiology impressions from findings via neural sequence-to-sequence learning. We proposed a customized neural model for this task which uses encoded background information to guide the decoding process. We collected a dataset from actual hospital studies and showed that our model not only outperforms non-neural and neural baselines, but also generates summaries with significant clinical validity and cross-organization transferability.

Acknowledgments

We thank Peng Qi and the anonymous reviewers for their helpful suggestions.

References

Dzmitry Bahdanau, Kyunghyun Cho, and Yoshua Bengio. 2015. Neural machine translation by jointly learning to align and translate. *The 2015 International Conference on Learning Representations*.

Jan ML Bosmans, Joost J Weyler, Arthur M De Schepper, and Paul M Parizel. 2011. The radiology report as seen by radiologists and referring clinicians: results of the COVER and ROVER surveys. *Radiology*, 259(1):184–195.

Yen-Chun Chen and Mohit Bansal. 2018. Fast abstractive summarization with reinforce-selected sentence rewriting. *The 2018 Annual Meeting of the Association of Computational Linguistics (ACL 2018)*.

Savelie Cornegruta, Robert Bakewell, Samuel Withey, and Giovanni Montana. 2016. Modelling radiological language with bidirectional long short-term memory networks. *Proceedings of the Seventh International Workshop on Health Text Mining and Information Analysis (LOUHI)*.

Dina Demner-Fushman, Marc D Kohli, Marc B Rosenman, Sonya E Shooshan, Laritza Rodriguez, Sameer Antani, George R Thoma, and Clement J McDonald. 2015. Preparing a collection of radiology examinations for distribution and retrieval. *Journal of the American Medical Informatics Association*, 23(2):304–310.

Jacob S Elkins, Carol Friedman, Bernadette Boden-Albala, Ralph L Sacco, and George Hripcsak. 2000.

Coding neuroradiology reports for the northern manhattan stroke study: a comparison of natural language processing and manual review. *Computers and Biomedical Research*, 33(1):1–10.

Günes Erkan and Dragomir R Radev. 2004. LexRank: Graph-based lexical centrality as salience in text summarization. *Journal of Artificial Intelligence Research*, 22:457–479.

Carol Friedman, George Hripcsak, William Du-Mouchel, Stephen B Johnson, and Paul D Clayton. 1995. Natural language processing in an operational clinical information system. *Natural Language Engineering*, 1(1):83–108.

Esteban F Gershanik, Ronilda Lacson, and Ramin Khorasani. 2011. Critical finding capture in the impression section of radiology reports. In *AMIA Annual Symposium Proceedings*. American Medical Informatics Association.

Daniel J Goff and Thomas W Loehfelm. 2018. Automated radiology report summarization using an open-source natural language processing pipeline. *Journal of Digital Imaging*, 31(2):185–192.

Saeed Hassanpour and Curtis P Langlotz. 2016. Information extraction from multi-institutional radiology reports. *Artificial Intelligence in Medicine*, 66:29–39.

George Hripcsak, John HM Austin, Philip O Alderson, and Carol Friedman. 2002. Use of natural language processing to translate clinical information from a database of 889,921 chest radiographic reports. *Radiology*, 224(1):157–163.

George Hripcsak, Gilad J Kuperman, and Carol Friedman. 1998. Extracting findings from narrative reports: software transferability and sources of physician disagreement. *Methods of Information in Medicine*, 37(01):01–07.

Charles E Kahn Jr, Curtis P Langlotz, Elizabeth S Burnside, John A Carrino, David S Channin, David M Hovsepian, and Daniel L Rubin. 2009. Toward best practices in radiology reporting. *Radiology*, 252(3):852–856.

Diederik P Kingma and Jimmy Ba. 2015. Adam: A method for stochastic optimization. In *The 2015 International Conference for Learning Representations*.

Julian Kupiec, Jan Pedersen, and Francine Chen. 1995. A trainable document summarizer. In *Proceedings of the 18th annual international ACM SIGIR conference on Research and development in information retrieval*.

M Lafortune, G Breton, and JL Baudouin. 1988. The radiological report: What is useful for the referring physician? *Canadian Association of Radiologists*, 39(2):140–143.

Chin-Yew Lin. 2004. ROUGE: A package for automatic evaluation of summaries. *Text Summarization Branches Out: ACL Workshop*.

Hans Peter Luhn. 1958. The automatic creation of literature abstracts. *IBM Journal of Research and Development*, 2(2):159–165.

Minh-Thang Luong, Hieu Pham, and Christopher D. Manning. 2015. Effective approaches to attention-based neural machine translation. In *Proceedings of the 2015 Conference on Empirical Methods in Natural Language Processing*.

Christopher D. Manning, Mihai Surdeanu, John Bauer, Jenny Finkel, Steven J. Bethard, and David McClosky. 2014. The Stanford CoreNLP natural language processing toolkit. In *Association for Computational Linguistics (ACL) System Demonstrations*.

Stephen Merity, Caiming Xiong, James Bradbury, and Richard Socher. 2017. Pointer sentinel mixture models. *The 2017 International Conference on Learning Representations*.

Rada Mihalcea and Paul Tarau. 2004. TextRank: Bringing order into text. In *Proceedings of the 2004 Conference on Empirical Methods in Natural Language Processing*.

Ramesh Nallapati, Bowen Zhou, Caglar Gulcehre, Bing Xiang, et al. 2016. Abstractive text summarization using sequence-to-sequence RNNs and beyond. *The SIGNLL Conference on Computational Natural Language Learning (CoNLL), 2016*.

Romain Paulus, Caiming Xiong, and Richard Socher. 2018. A deep reinforced model for abstractive summarization. *The 2018 International Conference on Learning Representations*.

Jeffrey Pennington, Richard Socher, and Christopher D Manning. 2014. GloVe: Global vectors for word representation. In *Proceedings of the 2014 Conference on Empirical Methods in Natural Language Processing*.

Alexander M Rush, Sumit Chopra, and Jason Weston. 2015. A neural attention model for abstractive sentence summarization. In *Proceedings of the 2015 Conference on Empirical Methods in Natural Language Processing*.

Abigail See, Peter J Liu, and Christopher D Manning. 2017. Get to the point: Summarization with pointer-generator networks. In *The 2017 Annual Meeting of the Association of Computational Linguistics (ACL 2017)*.

Josef Steinberger and Karel Jezek. 2004. Using latent semantic analysis in text summarization and summary evaluation. *Proceedings of the 2004 International Conference on Information System Implementation and Modeling*.

Author Index